8000008315
TREMOUGH CAMPUS LIBRARY

AF843643

University College
FALMOUTH
Incorporating Dartington College of Arts

UNIVERSITY OF EXETER

Tremough Campus Library
STANDARD LOAN
This book may be borrowed for 28 days, subject to recall

Please return or renew this book on or before the due date, after which daily fines will be added to your account. Please check your issue slip or log onto 'my account' at http//voyager.falmouth.ac.uk

Library & Information Services
Tremough Campus, Penryn, Cornwall TR10 9EZ
01326 370441

http//library.falmouth.ac.uk or email library@falmouth.ac.uk

FX PLUS LIBRARY
WITHDRAWN

Fracture and In-Situ Stress Characterization of Hydrocarbon Reservoirs

Geological Society Special Publications
Society Book Editors
R. J. PANKHURST (CHIEF EDITOR)
P. DOYLE
F. J. GREGORY
J. S. GRIFFITHS
A. J. HARTLEY
R. E. HOLDSWORTH
A. C. MORTON
N. S. ROBINS
M. S. STOKER
J. P. TURNER

Special Publication reviewing procedures

The Society makes every effort to ensure that the scientific and production quality of its books matches that of its journals. Since 1997, all book proposals have been refereed by specialist reviewers as well as by the Society's Books Editorial Committee. If the referees identify weaknesses in the proposal, these must be addressed before the proposal is accepted.

Once the book is accepted, the Society has a team of Book Editors (listed above) who ensure that the volume editors follow strict guidelines on refereeing and quality control. We insist that individual papers can only be accepted after satisfactory review by two independent referees. The questions on the review forms are similar to those for *Journal of the Geological Society*. The referees' forms and comments must be available to the Society's Book Editors on request.

Although many of the books result from meetings, the editors are expected to commission papers that were not presented at the meeting to ensure that the book provides a balanced coverage of the subject. Being accepted for presentation at the meeting does not guarantee inclusion in the book.

Geological Society Special Publications are included in the ISI Index of Scientific Book Contents, but they do not have an impact factor, the latter being applicable only to journals.

More information about submitting a proposal and producing a Special Publication can be found on the Society's web site: www.geolsoc.org.uk.

GEOLOGICAL SOCIETY SPECIAL PUBLICATION NO. 209

Fracture and In-Situ Stress Characterization of Hydrocarbon Reservoirs

EDITED BY

M. AMEEN

Saudi Aramco, Saudi Arabia

2003

Published by

The Geological Society

London

THE GEOLOGICAL SOCIETY

The Geological Society of London (GSL) was founded in 1807. It is the oldest national geological society in the world and the largest in Europe. It was incorporated under Royal Charter in 1825 and is Registered Charity 210161.

The Society is the UK national learned and professional society for geology with a worldwide Fellowship (FGS) of 9000. The Society has the power to confer Chartered status on suitably qualified Fellows, and about 2000 of the Fellowship carry the title (CGeol). Chartered Geologists may also obtain the equivalent European title, European Geologist (EurGeol). One fifth of the Society's fellowship resides outside the UK. To find out more about the Society, log on to www.geolsoc.org.uk.

The Geological Society Publishing House (Bath, UK) produces the Society's international journals and books, and acts as European distributor for selected publications of the American Association of Petroleum Geologists (AAPG), the American Geological Institute (AGI), the Indonesian Petroleum Association (IPA), the Geological Society of America (GSA), the Society for Sedimentary Geology (SEPM) and the Geologists' Association (GA). Joint marketing agreements ensure that GSL Fellows may purchase these societies' publications at a discount. The Society's online bookshop (accessible from www.geolsoc.org.uk) offers secure book purchasing with your credit or debit card.

To find out about joining the Society and benefiting from substantial discounts on publications of GSL and other societies worldwide, consult www.geolsoc.org.uk, or contact the Fellowship Department at: The Geological Society, Burlington House, Piccadilly, London W1J 0BG. Tel: +144 (0)20 7434 9944; Fax: +144 (0)20 7439 8975; E-mail: *enquiries@geolsoc.org.uk*.

For information about the Society's meetings, consult *Events* on *www.geolsoc.org.uk*. To find out more about the Society's Corporate Affiliates Scheme, write to *enquiries@geolsoc.org.uk*.

Published by The Geological Society from:
The Geological Society Publishing House
Unit 7, Brassmill Enterprise Centre
Brassmill Lane
Bath BA1 3JN, UK

(*Orders:* Tel: +144 (0)1225 445046
Fax: +144 (0)1225 442836)
Online bookshop: http//bookshop.geolsoc.org.uk

The publishers make no representation, express or implied, with regard to the accuracy of the information contained in this book and cannot accept any legal responsibility for any errors or omissions that may be made.
© The Geological Society of London 2002. All rights reserved. No reproduction , copy or transmission of this publication may be reproduced, copied or transmitted save with the provisions of the Copyright Licensing Agency, 90 Tottenham Court Road, London W1P 9HE. Users registered with the Copyright Clearance Center, 27 Congress Street, Salem, MA 01970, USA: the item-fee code for this publication is 0305-8719/03/$15.00.

British Library Cataloguing Publication Data
A catalogue record for this book is available from the British Library.

ISBN 1-86239-130-0

Typeset by Exe Valley Dataset Ltd, Exeter, UK
Printed by Alden Press, Oxford, UK
Project management by Swales & Willis, Exeter, UK

Distributors
USA
AAPG Bookstore
PO Box 979
Tulsa
OK 74101-0979
USA
Orders: Tel: +1 918 584-2555
Fax: +1 918 560-2652
E-mail: bookstore@aapg.org

India
Affiliated East-West Press PVT Ltd
G-1/16 Ansari Road, Daryaganj,
New Delhi 110 002
India
Orders: Tel: +91 11 327-9113
Fax: +91 11 326-0538
E-mail: affiliat@nda.vsnl.net.in
Japan
Kanda Book Trading Company
Cityhouse Tama 204
Tsurumaki 1-3-10
Tama-shi
Tokyo 206-0034
Japan
Orders: Tel: +81 (0)423 57-7650
Fax: +81 (0)423 57-7651
E-mail: geokanda@ma.kcom.ne.jp

Contents

It is recommended that reference to all or part of this book should be made in one of the following ways:

AMEEN, M. (ed.) 2003. *Fracture and In-Situ Stress Characterization of Hydrocarbon Reservoirs.* Geological Society, London, Special Publications, **209**.

CHANCHANI, S.K., ZOBACK, M.D. & BARTON, C. 2003. A case study of hydrocarbon transport along active faults and production-related stress changes in the Monterey Formation, California. *In* AMEEN, M. (ed.) *Fracture and In-Situ Stress Characterization of Hydrocarbon Reservoirs.* Geological Society, London, Special Publications, **209**, 17–26.

Fracture and in-situ stress characterization of hydrocarbon reservoirs: definitions and introduction

MOHAMMED S. AMEEN

Saudi Aramco, PO Box 2817, Dhahran 31311, Saudi Arabia (e-mail: mohammed.ameen@aramco.com)

Fracture and in-situ stress characterization is fast-evolving as an essential part of characterizing hydrocarbon reserviors. In this book, the Geological Society presents a selection of sixteen chapters that demonstrate the tools, methods, analysis, interpretation and application of this subject. These are of great interest to researchers and scientists in the industry and academia. This chapter includes definitions pertinent to the subject of the book and gives an overview of the papers, classified into seven major themes according to the nature of the studies.

Definitions

Fractures are defined here as all discontinuities that occur in rocks due to brittle/semi-brittle deformation.

Natural fractures are those related to natural deformation of the rock. They include faults, cracks, joints, veins and stylolites.

Induced fractures are those induced artificially, e.g. by core handling, coring, drilling, fluid injection etc.

In-situ *stresses* are the current-day natural stresses present in the earth's crust. They are the result of a few components :

- *I* *Gravitational stresses* due to the weight of the overburden.
- *II* *Current tectonic stresses* related to present-day tectonic forces such as those resulting from the active collision of the Arabian and the Eurasian continents in the Arabian Gulf region.
- *III* *Remnant/residual stresses* locked in the rock during past episodes of tectonic and gravitational stresses.

Fracture characterization is the science that deals with the detection (whether fractures exist?), diagnosis (identification of natural vs induced and type of each), and quantification of fractures (single fracture properties such as orientation, aperture, fault offset and length, and fracture population properties such as number of sets, orientation of each set, density etc.). Fracture characterization is a multi-disciplinary subject that integrates multi-tools and multi-scale observations like microscopic, borehole images, cores and 3D seismic data.

In-situ stress characterization is the science of determining the orientation and estimating the relative and absolute magnitudes of the three principal *in-situ* stresses (maximum, intermediate and mimimum, in-situ stresses referred to as σ_1, σ_2 and σ_3 respectively). This is done using an integrated suite of data like borehole images, borehole logs, extended leakoff tests, hydrofracturing tests and active seismicity analysis. In addition, using a cartesian system relative to earth surface, the stresses are also referred to as: maximum horizontal stress (σ_H) , minimum horizontal stress (σ_h) and vertical stress (σ_V).

Fractures play an important role in hydrocarbon reservoirs. They can impact permeability and porosity and thence reservoir performance. This is becoming more apparent with the advancement of technology and the shifting of frontiers to deeper and tighter reservoirs, in increasingly high temperature/high pressure environments. Therefore the need for fracture characterization is no longer limited to classical fractured reservoirs (in which fractures are the main source of permeability). Deep, tighter reservoirs rely on fractures as a source of porosity too. In addition, reservoirs with high matrix porosity and permeability can show negative symptoms related to un-wanted high permeability in fracture zones.

In both fractured and non-fractured reservoirs in-situ stresses can impact reservoir properties (porosity and permeability) and reservoir performance. In addition these stresses play an important role in issues of borehole stability, productivity and injectivity. Therefore, there is a need to characterize these stresses and model their

From: AMEEN, M. (ed.) *Fracture and In-Situ Stress Characterization of Hydrocarbon Reservoirs.* Geological Society, London, Special Publications, **209**, 1–6. 0305–8719/03/$15.00
© The Geological Society of London 2003.

impact on reservoir properties and performance. This has led to the emergence of a new concept to reservoir characterization: *geomechanical characterization*. This accounts for the stress-sensitivity of reservoirs (i.e. the changes in reservoir petrophysics and performance as a function of in-situ stresses).

The current volume

This volume stems from the Petroleum Group meeting on Fracture and *In-Situ* Stress Characterisation of Hydrocarbon Reservoirs, convened by the author at the Geological Society, Burlington House, London, 21–29 June, 1999. The current volume includes sixteen chapters that fall broadly into seven themes according to scale of observation, tools, and subject.

Borehole-scale characterization

In this theme three papers are presented with diverse objectives and level of data integration. Until recently it was always assumed that in a fractured rock mass, the most effective fluid conduits are fractures that parallel the maximum in-situ stress and are perpendicular to the minimum in-situ stress. The hypothesis behind this is that these fractures have the lowest normal stresses across them and therefore attain the maximum hydraulic apertures. However more recent work led to a major revision in this concept. In addition to fractures that are subjected to current-day tensile regime, fractures that are orientated such that they are subjected to both shear and tensile stresses can be critical fluid coduits, depending on the ratio of shear to normal stress and associated dilation. This dilation results in a significant increase in fracture hydraulic aperture and permeability. It is now accepted that critically sheared fractures (i.e. those close to frictional failure in the current-day stress field) serve as conduits for fluid flow.

The first chapter by S. F. Rogers and C. J. Evans demonstrates, through a case study from Sellafield in the UK, how the critical stress technique was successfully applied for predicting zones of enhanced fracture-permeability at the well scale. Difficulties for implementing the technique on a wider, field-scale are discussed. Although the study was not done for hydrocarbon reservoirs, it is of direct application to reservoir characterization.

In the second chapter Sneha K. Chanchani, Mark D. Zoback and Colleen Barton present a case study of hydrocarbon transport along active faults and production-related stress changes in the Monterey Formation, California. They studied the Antelope Shale, a low permeability siliceous shale hydrocarbon reservoir in the Buena Vista Hills field in the southern San Joaquin Valley to determine the influence of the stress state on the relative hydraulic conductivity of the fractures and faults. Because production has both lowered reservoir pressure and the horizontal stresses, it was necessary to 'restore' the reservoir stress state to initial conditions in order to identify correctly the most highly productive intervals. This analysis demonstrates that prior to production, faults in the reservoir were active in a transitional reverse/strike-slip faulting stress state, consistent with regional tectonics. Initial production rates in the field were 2000 bopd, principally from intervals where critically stressed faults were encountered.

In addition to predicting fracture permeability and its stress sensitivity, discussed above, fracture porosity assessment is another challenging subject. This issue is gaining more significance as advances in technology have extended the threshold in terms of the economical feasibility of tighter and deeper reservoirs. In the third chapter Aliverti *et al.* discuss methods used in an integrated borehole-scale characterization aimed at predicting fracture network and assessing fracture porosity in a tight reservoir.

The work demonstrates the incorporation of oriented core and wellbore images to calculate the geometrical parameters of each feature (dip direction, dip, size, terminations) and to classify them according to their filling (e.g. oil, water, shale, and calcite). In addition the total fracture population was studied to derive the number of fracture sets, their orientation, spatial distribution along the cored/logged interval, distribution laws of the fracture length and relevant minimum radii and fracture aperture estimate. Based on the results a stochastic simulation of fracture network at wellbore scale was built. Fracture porosity evaluation, matrix block size, and fracture network connectivity at wellbore scale constitute the outputs of such simulations. The application of the method is demonstrated through a study on a tight, platform carbonate reservoir, Southern Apennines, Italy.

Integrated borehole to reservoir scale characterization

This theme demonstrates the essential need to integrate multi-source, multi-scale data in fracture characterization. In this context it links borehole-scale observations, like those covered in the previous section, to reservoir-scale/interwell-scale observations (seismic data). This section includes two chapters.

The first paper by Hesthammer and Fossen illustrates an example from the North Sea Gullfaks Field, using integrated structural interpretation to optimize fracture characterization utilizing all available seismic surveys, well log correlation, dipmeter and core data. Interpretation of seismic data helps understand large-scale structural and stratigraphic geometries. Time-lapse seismic (4D-seismic) helps to identify changes in reservoir properties caused by operational activities like injection and production. Well log correlation is used to document variations in sequence thickness caused by sedimentological or structural changes whereas dipmeter data tie observations of bedding orientation from subseismic scale (borehole scale) to seismic scale. Core data represents the most detailed (millimeter to meter scale) data available and can yield information on rock properties as well as sedimentological and structural features. Small-scale deformation structures such as deformation bands and fractures can be identified and characterized effectively. In addition, it is possible from unoriented cores to find the orientation of bedding and deformation structures. This information is compared to observations from dipmeter data, well log correlation data and seismic data to improve the interpretation.

The second chapter in this section, by Yale, deals with fault and stress magnitude controls on variations in the orientation of in situ stress. Over the last decade in-situ stress characterization has been gaining impetus, due to the realization of the impact of current-day *in-situ* stresses on the performance, of both fractured and non-fractured reservoirs. Fracture orientation (for both natural and induced fractures, e.g. hydrofracturing stimulation of reservoirs), well stability, well placement and design, and permeability anisotropy in reservoirs are all strongly affected by variations in the current-day stress field.

Using stress orientation data from a number of fields in different tectonic environments, Yale tried to determine some of the tectonic and geologic controls on variations in *in-situ* stress orientation. The findings show that distance to large faults, fault structure, and magnitude of current-day tectonic stress play primary roles in determining whether the regional stress field will be perturbed in a given reservoir, and whether small-scale variations in the stress field can be expected.

Highly stressed terraines (with large horizontal differential stress) are characterized by a more consistent stress field than tectonically quiescent areas. Faults can play a significant role in rotating the local stress field. The smaller the difference between the maximum and minimum horizontal stress magnitude (i.e. lower tectonic stress), the larger a fault's zone of influence is. Large-scale faults, which compartmentalize the reservoir into discrete fault blocks, can lead to significant stress orientation variations across the reservoir, even in areas of large differential stress.

Reservoir-scale characterization using seismic data

In this theme two chapters illustrates the application of two 'unconventional' seismic data analyses and interpretation of reservoir-scale fracture characterization.

The first work by Trappe and Hellmich addresses the utilization of seismic volume attributes for fracture analysis. Seismic volume attributes are useful in the characterisation of subtle faults and fractures that cannot be recognized on seismic sections (their vertical offset is below seismic resolution), incorrectly referred to as 'subseismic' faults. The seismic attributes are based on neighbourhood analysis of the seismic signal and uses 2D or 3D data. To detect fractures, attributes such as coherency, that quantify change in the seismic signal, are used. Case studies are presented to show the impact of the attribute analysis on fracture characterization of reservoirs. Seismic multi-trace filtering to compute 2D and 3D seismic volume attributes are used for this purpose. This includes advanced seismic processing algorithms with extensive use of higher order statistics and feature extraction methods. These methods are as applicable in fields such as medical image processing or material science, where direct measurements of desired properties are either costly, or impossible.

In the second chapter of this section, Jupe *et al.* discuss case studies in which microseismic (passive seismic) monitoring contributed to the development and management of hydrocarbon and geothermal reservoirs in which fractures play an essential role. Microearthquakes (microseismic events) are induced during hydrocarbon and geothermal fluid production operations in naturally fractured reservoirs. The technology is based on the monitoring and analysis of microearthquakes. They typically result from shear-stress release on pre-existing faults and fractures due to production/injection induced perturbations to the effective stress conditions. These stress changes may be due to reservoir depletion, flooding or stimulation operations. Over a number of years it has been shown that microseismic monitoring has the potential to provide valuable time-lapse 3D information on the geomechanical processes taking place within a reservoir. These processes include the distribution of fluid flow and pressure fronts within naturally fractured systems, production-related compaction and the re-activation of faults. With the advent of permanent reservoir monitoring systems (e.g. intelligent wells), microseismic monitoring has the potential to become a practicable means of time-lapse (pseudo-real time) imaging of hydrocarbon reservoirs.

Outcrop-based characterization and modelling, on plate tectonic scale to predict in-situ stress regime

In this section, Claudio Lima discusses present-day compression across the South American plate. He presents observations, numerical modelling and some implications for petroleum geology. Stress and tectonic data (space-based geodetic results, seismicity, fission track analysis, and analysis of leakoff and hydraulic fracturing data) and intraplate stress field numerical models indicate that the maximum principal stress is horizontal for most Brazilian basins. The compression/shortening is most probably due to the convergence of the South American and the Nazca plates and the divergence of the South American and the African plates.

A conceptual model for the plate-wide deformation is presented and numerically tested using elasto-plastic rheologies. The model states that in response to the compression, the lithosphere as a whole (or only the crust if thermal gradients are high enough) tends to fold and fracture. This tendency is stronger during peaks of the Andean orogeneses. The forming antiforms are responsible for uplift along the erosional basin borders, whereas the forming synforms are sites of continental sedimentation, at basin centers. The denudation of sedimentary covers promotes the exhumation of deeper and deeper rocks, outcropping at the foot of retreating scarps. Consequently, the erosional borders of the basins form local topography highs with respect to the adjacent basement.

In exploration, neotectonics, including in-situ stresses, are usually disregarded, especially in 'passive' margin basins. However, neotectonics impact the distribution and preservation of petroleum accumulations, since (i) accumulations are ephemeral in a geologic time scale, being strongly dependent on seals fine geometry and biodegradation (ii) a strong positive correlation exists between permeability anisotropy and SHmax (iii) the source rocks of the most important Subandean and intraplate petroleum systems are still in the oil generation window and (iv) the disruption of 'kitchens' of generation is a positive factor to primary migration. This inference seems to be confirmed, since the most important South American petroleum accumulations are found along the actively deforming border between South American and Caribbean plates. Indeed, the most important accumulations in marginal basins are found in the Southeastern Brazilian margin (the Campos basin), which has been deformed the most with respect to other margins during the Cenozoic, and continues to be the most seismically active. Fracture networks on the Belluno Syncline, a fault-propagation fold in the footwall of the Belluno Thrust, Venetian Alps, northeast Italy.

Outcrop-based fracture characterization

Outcrop studies remain the most important source of our conceptual structural models of subsurface structural traps and associated fracture patterns. The distribution of fractures in space and their evolution relative to the host structures are vital for exploration and development (e.g. borehole design). In this theme, Zampieri and Grandesso studied the Belluno Syncline, in the Venetian fold and thrust belt in Italian Southern Alps. This SSE-vergent chain mainly developed during the Neogene as a back-thrust belt of the Alps deforming the northern margin of the Adria plate. The syncline is asymmetric, with a steep backlimb, as the forelimb of the paired Monte Coppolo Anticline, resulted from fault-propagation folding during the development of the underlying Belluno thrust. In plan view the Belluno thrust shows a prominent curvature interpreted as reactivation of inherited Mesozoic extensional structures. The hinge of the Belluno syncline presents a similar curvature. Rocks outcropping in the syncline core are Upper Chattian to Langhian fine-grained molassic sediments, mainly siltstones and marls.

Most fractures on the Belluno Syncline have formed at high angles to bedding. The dominant fracture trends are sub-parallel and orthogonal to the fold hinge. Sometimes these fractures show plumose structures characteristic of extension (mode I) fractures and are referred to as longitudinal (bc) joints and cross (ac) joints respectively. Other fracture trends oblique to the hinge line are found on limbs distant from the fold hinge. They are referred to as oblique (shear) fractures. Throughout the western part of the Belluno Syncline the fracture network maintains a roughly symmetrical distribution with respect to the fold hinge, suggesting a development in association with folding.

Geomechanical and geodynamic modelling on reservoir scale

This is discussed through four key chapters. The first presents a method for stress modelling from local data. The study by Muller *et al.* presents a new method for smoothing orientation data in general and demonstrates its application to in-situ stress modelling.

Smoothing algorithms provide a means of identifying significant patterns in sets of oriented data, eliminating local perturbations within the observations and predicting patterns of oriented data in places which lack observations. Muller *et al.* smooth orientation data with a distance-related method of data weighting as an alternative to previous weighting algorithms. This method is developed on the basis of a statistical smoothing algorithm. They can be applied to orientation data of 180° periodicity such as maximum horizontal tectonic stresses (S_H) as compiled in the World Stress Map database. This method enables the user to discriminate between local (<250 km of lateral extent) and regional stress fields (approximately 250–5000 km of lateral extent), and to compare S_H with other directional data such as fault trends or strain data. The technique is demonstrated by producing stress models for NE America, the Himalayas and Western Europe. The accuracy and smoothness sensitivity is tested by varying the scale and smoothing. They also give recommendations for appropriate choice of these parameters.

The second chapter in this section, by Wynn and Stewart, deals with the role of spectral curvature mapping in characterizing subsurface strain distributions. The curvature of key geological surfaces can be used to assess the strain they have undergone. In many hydrocarbon reservoirs, this strain is expressed as brittle fracturing that may significantly impact reservoir performance. Wynn and Stewart describe the development of an algorithm for measuring the curvature of gridded surfaces derived from seismic data. For any grid node, the algorithm calculates the magnitude and orientations of the two principal curvatures, K_1 and K_2, from which other curvature measurements can be derived, such as Gaussian curvature and summed absolute curvature (K_1+K_2). The algorithm has also been used to generate plots of summed absolute curvature as a function of grid node separation (k versus λ). These 'spectral' or kλ plots can be generated for each grid node and allow the definition of short wavelength, high amplitude noise cut-off lengths. They also deliver intermediate wavelength features such as fault drag or buckle folding and the identification of long wavelength (basin scale) curvatures. Portions of this data can be collapsed into single values by calculating the integral of the kλ curve. Further filters designed to screen the effects of background tectonic, or non-tectonic curvatures can be applied to the kλ integral.

This algorithm has been verified using data from several North Sea chalk fields. A range of alternative types of curvature and curvature spectra are compared with other approaches to curvature calculation and other factors relevant to the calibration of such techniques in terms of the distribution of brittle fractures in sedimentary rocks.

The kλ integral provides a relatively simple approach to calculating the degree of multi-wavelength strain present at a particular grid node. Freeing algorithms from the restriction of the 'arbitrarily' selected minimum grid node spacing is a key step towards calibrating measured curvature against strain mechanisms. However, care must be taken to separate intrinsic and tectonic curvatures when generating and interpreting kλ plots and their integrals.

The third chapter within this topic, by Smart *et al.*, deals with reservoir characterization for the management of stress-sensitivity. In spite of the well-established effort to effectively characterize and model borehole-scale sensitivity, reservoir-scale studies are lagging behind and have only recently gained some impetus. In this chapter the authors argue that fractured reservoirs must be considered stress-sensitive, with this possibly being masked in the past by high productivities and abundant reserves. A basic conceptual model for all reservoirs is proposed, consisting of intact blocks of rocks bounded by discontinuities (fractures and bedding-parallel), loaded by an anisotropic stress state. It is argued that such a model should be used to drive the reservoir characterization, with the premise that the new generation of coupled simulators can usefully accommodate this data set.

Tuncay *et al.* discuss a method to simulate 3D fracture network dynamics in reservoirs, faults and salt tectonic systems. The method is based on the numerical solution of rock deformation processes coupled to the myriad of other basin reaction, transport and mechanical (RTM) processes. Seismic, well log and surface geological data are used to quantitatively assess distribution of fractures, stress, petroleum and porosity, grain size and other textural information.

The model uses an incremental stress rheology that accounts for poroelasticity, nonlinear viscosity with yield/faulting, pressure solution and fracturing. It couples mechanics to multiphase flow and diagenesis (through their influence on effective stress and rock rheological properties, respectively). The resulting model is 3D in terms of the full range of fracture orientations and the tensorial nature of stress, deformation and permeability. All rock properties (rheologic, multi-phase fluid transport, grain shape, etc.) are coevolved with the other variables. Examples are used to illustrate the relative importance of various overpressuring mechanisms, lithology and flexure on the location and characteristics of a fracture network.

Prediction and modelling of fluid flow in fractured reservoirs

The main objective of fracture characterization is to constrain their impact on reservoir properties such as permeability. Jolly and Cosgrove use geological evidence of fluid flow in an attempt to understand fluid flow patterns in fracture networks. Because many fluids (e.g. gas, water etc.) leave little or no evidence of their passage through the rock others such as magmas and fluidized sediments preserve the pathways they follow by forming dykes and sills. In their study Jolly and Cosgrove found that the pathways preserved by the two types of fluids are different (i.e. the spacing of the clastic dykes follows a power-law distribution and that of the igneous dykes a lognormal distribution). It is suggested that this in part might reflect the different properties of the dyke material (specifically its permeability) which determines whether or not the fracture containing the dyke can continue to act as a channel of easy fluid migration once the dyke has been emplaced.

In the second chapter in this theme Ewing and Spagnuolo discuss the difficulties and uncertainties in mathematical/numerical modelling of fluid flow in fractured media. The chapter especially discusses uncertainties arising in reservoirs, which have large, dominant pathways for high velocity flow. In such environments, conventional single and dual-porosity/dual-permeability models for simulation cannot adequately represent the flow processes. Instead, a major paradigm shift in both mathematical modelling and numerical methods to discretize these models are required. The author argues that new simulators must include non-Darcy flow models that allow channels and non-traditional multiphase flow models. In order to accurately describe complex localized phenomena that may dominate the flow process, the simulators must be able to accurately describe complex localized geological structures in the same content as enormous, field-scale reservoirs. This requires the ability to have complex, non-orthogonal flexible localized grid capabilities tied to larger, more-uniform grids via domain decomposition and various mortar finite element, finite volume element formulations. Furthermore, localized time-stepping procedures coupled with adaptive gridding may be needed to deal with such reservoirs. Ewing and Spagnuolo presents aspects of mathematical modelling and numerical methods that address each of these difficult areas of simulation. These techniques will be essential in trying to simulate the complex systems of large vertical faults and fracture systems that dominate the flow in many of the heavily fractured reservoirs worldwide.

Simulation of water-gas flow in fractured porous media by AL-Khlaifat and Arastoopour is the subject of the third chapter in this group. The chapter tackles porous gas reservoirs characterized by a low-permeability matrix, and a high-permeability fracture system. The transient, three-dimensional, two-phase numerical model is presented for simulating the simultaneous flow of gas and water through porous and fractured media. It assumes a single horizontal fracture perpendicular to the flow direction and considers both fluids as incompressible and immiscible. The study takes into consideration the effect of capillary pressure and relative permeability. Technically, the finite volume approach is used to discretize the equations. The set of discretized and linearized equations are solved using the IPSA (Inter-Phase Slip Algorithm) method. The results of this study indicate that the multi-layer reservoir provides better estimates of post-fracture performance compared to a more conventional, single-layer reservoir description.

Critical stress-related permeability in fractured rocks

STEPHEN F. ROGERS

British Geological Survey, Keyworth, Nottingham, NG12 5GG, UK
Current address: Golder Associates, Landmere Lane, Edwalton, Nottingham, NG12 4DG, UK
(e-mail: srogers@golder.com)

Abstract: Conventional wisdom has always assumed that fluids prefer to flow along fractures that are orientated parallel to the maximum *in-situ* horizontal stress direction. These fractures have the lowest normal stresses across them and therefore provide the least resistance to flow movement. However more recent work (Barton *et al.* 1995) has suggested that those fractures that are orientated such that they experience a high ratio of shear to normal stress are most likely to flow. These critically stressed fractures are in a state of stress that is close to failure, allowing them to undergo a degree of shear. Even small shear displacements can cause significant dilation along a fracture surface as the two surfaces un-mate during sliding. This dilation results in a significant increase in fracture permeability, resulting in an increase in flow from these features. A review of past work reveals that this shear mechanism may be more important than previously thought. A case study from Sellafield in the UK is presented, where the critical stress technique has been successfully applied for predicting zones of enhanced permeability at the well scale. Difficulties for implementing the technique on a wider field wide scale are discussed.

Characterizing fluid flow pathways within fractured rocks, hinges upon trying to understand a complex flow system within a poorly defined volume. The use of borehole imaging tools reveals a myriad of potential fracture pathways but production logging indicates that a relatively small number of these features actually contribute to flow. Obviously fracture properties such as aperture, extent and connectivity play a major role in determining the permeability of a pathway. However, dynamic inputs such as stress also strongly influence this heterogeneous system.

Heffer & Lean (1993) demonstrated a strong empirical relationship between the direction of fluid flow and the direction of the maximum, horizontal, *in-situ* stress component (σ_{hmax}) within strike slip and normal faulting stress regimes. This is consistent with the notion that fractures orientated parallel to σ_{hmax} will experience the least degree of closure (i.e. the lowest normal stress) and would thus be the most permeable. The corollary is that fractures experiencing the highest normal stress would offer the greatest resistance to fluid flow and therefore have the lowest permeability. However Barton *et al.* (1995) suggest that this is not always correct and state that it is fracture surfaces experiencing localized failure as a result of a large shear component that are the most conductive. Their work is supported by considerable field evidence from a number of boreholes where those fractures whose state of stress was considered to have reached a failure criterion, were almost universally found to be flowing (see also Hickman *et al.* 1997; Finkbeiner *et al.* 1997; Barton & Moos, 1997; Trice 1999).

The majority of fracture surfaces are not planar, but rough or undulating. In their natural undisturbed state, fracture surfaces mate to produce the smallest aperture. However, when the state of stress is such that the fracture undergoes a degree of shearing, the asperites will ride up over each other resulting in dilation and an increase in permeability (Heffer & Koutsabeloulis 1995). Yeo *et al.* (1998) demonstrated this in laboratory tests, showing that the resultant permeability increase is distinctly anisotropic, with the maximum permeability direction perpendicular to the direction of shear, in the plane of the fracture. This case obviously assumes that the fracture surface is strong enough to undergo shear without failure. Clearly the situation can arise where this shearing motion results in failure of the surface and the production of gouge material which will act to reduce the permeability. This paper considers the situation of non- or pre-gouge forming microshear failure only.

From: AMEEN, M. (ed.) *Fracture and In-Situ Stress Characterization of Hydrocarbon Reservoirs*. Geological Society, London, Special Publications, **209**, 7–16. 0305-8719/03/$15.00
© The Geological Society of London 2003.

Determination of stress magnitude and orientation

If the stress dependency of fracture permeability is a result of fracture surfaces experiencing incipient shear failure, then no longer will the stress orientation alone be sufficient to define the most conductive pathways. In order to calculate the state of stress on a surface, the full stress tensor needs to be defined. The determination of reliable measurements of crustal stress generally comes from one of two methods of measurement: directly via hydraulic fracturing and indirectly using borehole breakouts. The former technique can be used to deduce the magnitude and direction of stress within a borehole. The breakout method is possible in any borehole with a standard dipmeter (or borehole image log) and many workers have used this technique to develop regional, reservoir or even local-scale stress maps (Plumb & Hickman 1985; Evans & Brereton 1990). However, the breakout method can generally only be used for determining the stress orientation and not magnitude.

The determination of the magnitude and direction of the stress field using hydro-fracturing works by overpressuring a section of borehole in order to induce a vertical tensile crack in the plane of the maximum horizontal stress direction. Obviously in discontinuous media like rock, it is important to induce tensile failure fractures rather than simply opening existing fractures. This makes the interpretation and analysis of the results less than trivial if an accurate determination of the stress tensor is to be made. The breakout method involves measuring borehole ellipticity with either the calliper or travel-time waveform of an acoustic televiewer. It is generally considered that stress-related breakouts are formed by the concentration of compressive stresses at the borehole wall in the direction of the minimum horizontal stress. This results in a shear failure zone developing, resulting in an elliptical or 'dog ear' shaped borehole being formed (Dart & Zoback 1989). This shear mechanism can result in the production of a bimodal distribution of breakouts, formed symmetrically about σ_{hmax}. Thus the interpretation of breakouts as well as hydrofractures requires some care.

It is generally believed that relative plate motions determine the predominant regional horizontal stress components of a rock mass at the largest scale. Within the UK, the maximum horizontal stress is widely reported as 145/325° or approximately NW–SE. Research by Muller *et al.* (1992) found this direction to be regionally extensive throughout Western Europe and deduced that the stress originates from the Mid-Atlantic ridge push and the relative motion of Europe and Africa.

Some of the most recent stress measurements undertaken within the UK were carried out as part of the UK's radioactive waste disposal programme near Sellafield in NW England. A total of 28 hydraulic fracture tests, five overcoring measurements and breakout data from ten boreholes were analysed to define the stress tensor in the Sellafield area (Nirex 1997). For the most part the results demonstrated the stress field at Sellafield to be largely in accord with the rest of England with a mean direction of 340°. It is these stress results, which have been used with the critical stress theory described below to examine the nature of stress-related flow in fractured rock.

Critical stress theory

The critical stress theory for fracture flow (Barton *et al.* 1995) involves calculating the shear and normal stresses experienced by all planar fracture and fault surfaces as a result of the *in situ* stress field acting upon it.

Assuming that both the magnitude and orientation of the stress field has been defined, then the resultant shear (τ) and normal stress (σ_n) experienced on a fracture surface can be calculated from:

$$\tau=\beta_{11}\,\beta_{21}\,\sigma_1+\beta_{12}\,\beta_{22}\,\sigma_2+\beta_{13}\,\beta_{23}\,\sigma_3 \quad (1)$$

and

$$\sigma_n=\beta_{11}^2\,\sigma_1+\beta_{12}^2\,\sigma_2+\beta_{13}^2\,\sigma_3 \quad (2)$$

(Jaeger & Cook 1984), where β_{ij} are the directional cosines between the fracture surfaces and the stress tensor, and σ_1, σ_2 and σ_3 are the magnitudes of the maximum, intermediate and minimum principal stresses.

In order to display this stress and orientation data, a three-dimensional Mohr diagram is used. When considering borehole data, the magnitudes of the stresses increase with depth. To allow all of the data from the same borehole to be plotted within the same data space of the Mohr circle, all of the data are normalized with respect to the vertical stress component (Fig. 1). To test whether a fracture surface is likely to shear, a failure criterion can be added to the graph. Barton *et al.* (1995) use the Coulomb failure criterion with a coefficient of friction of 0.6:

$$\tau=\mu\,(\sigma_n-P_p) \quad (3)$$

where P_p is the pore pressure and μ is the coefficient of friction. The coefficient of friction

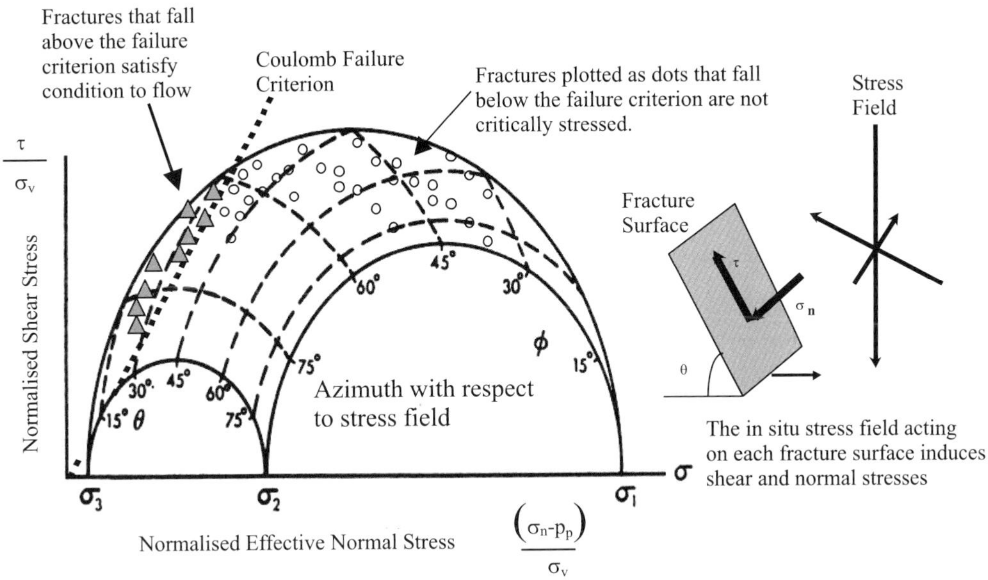

Fig. 1. Three-dimensional Mohr diagram showing shear and normal stress on a fracture surface as a function of the *in situ* stress field. Adapted from Jaeger & Cook (1984). The dark triangles represent fractures that are experiencing a state of stress sufficient to induce shear failure whereas the dots represent fractures in a more stable state of stress.

of 0.6 is used after the work of Byerlee (1978), who compiled a wealth of shear test data and found a remarkably consistent pattern. Fractures with an intermediate and above confining stress (greater than 200 MPa) are best described by a coefficient of friction of 0.6. However, the applicability of this value to most fractured reservoirs is limited given that the 200 MPa threshold is in the range of 5–6 km depth. Byerlee showed that for fractures with a normal stress less than 200 MPa, a value of m=0.85 better defined the failure state although there was a much greater scatter in the data. Intuitively the higher coefficient should be used as this better fits the typical stress state for most reservoirs. However, there is considerable empirical evidence to indicate that the lower value should be used. It should be noted that the failure criterion lacks a value for cohesion or intrinsic shear strength. This is justified in that the magnitude of this value is very small compared to the stresses experienced by fracture surfaces.

Barton *et al.* (1995) used the term 'critically stressed' fractures to refer to those fractures whose shear and normal stresses fall in excess of the 0.6 failure criterion and therefore exist in states likely to result in enhanced permeability. However, geology is seldom binary with the onset of a phenomenon happening when a threshold is crossed. It is considered that the coefficient of friction calculated for each fracture should be used as an indicator of the likelihood of failure and hence flow rather than simply applying a test for failure or non-failure.

For a typical UK onshore stress field (strike slip stress regime), the ratio of shear to normal stress can be computed for all orientations. These data can be plotted stereographically as a polar plot, with the data contoured with respect to the stress ratio (see Fig. 2). This will then indicate those orientations most likely to experience enhanced permeability. Given a maximum horizontal stress direction of approximately 340°, it is clear from Figure 2 that there are two main strike orientations that cause the highest ratio of shear to normal stress. These correspond to fracture and fault surfaces striking in two distinct directions: NW–SW and also almost north–south. So in contrast to the empirical evidence of Heffer & Lean (1993), which suggests that flow is greatest in the direction of σ_{hmax}, given this particular stress tensor the optimum fracture orientations for flow would be approximately $\sigma_{hmax} \pm 30°$.

Case study: Sellafield RCF3 borehole

To test the hypothesis of shear-stress-dependent flow being a significant controlling mechanism

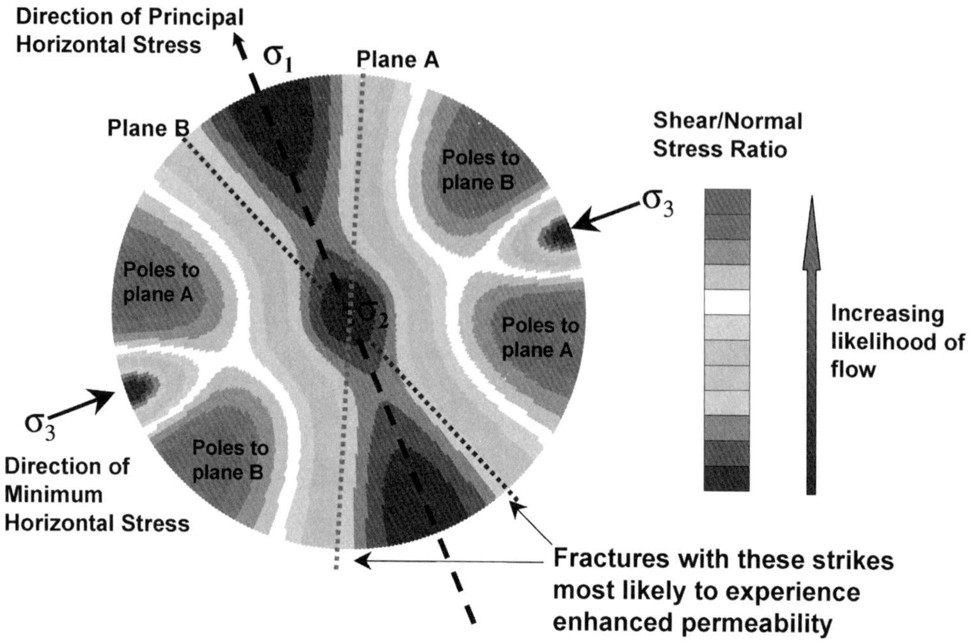

Fig. 2. Coefficient of friction as a function of orientation, displayed stereographically as a lower hemisphere polar projection.

for flow through fractured reservoirs, detailed fracture, stress and flow data are required. Borehole RCF3 was drilled as part of the UK radioactive waste disposal programme through 525 m of Permo-Triassic sandstone and breccia and approximately 450 m of Borrowdale Volcanic Group basement rocks. These rocks are a thick pile of mixed intrusive and extrusive volcanics and volcaniclastics with generally little primary porosity. The vast majority of flow in these rocks is through the complex fracture and vuggy vein system. In addition to the almost complete coring and full wireline logging of this borehole, 100 short interval tests (SITs) were carried out over zones 1.56 m long, using Schlumberger's wireline conveyed testing string, the Modular Dynamics Tester™ (Gutmanis *et al.* 1998). This provides a detailed continuous record of borehole transmissivity over a total distance of approximately 150m of fractured basement rocks. This hydraulic testing identified seven zones where the transmissivity was several orders of magnitude higher than the background level. If the critical stress theory is to account for the locations of flow within the borehole, then there needs to be a state of stress within these seven zones that is different from the borehole as a whole.

Orientated borehole discontinuity information was available from the interpretation of the Fullbore MicroImager FMI™ and Ultrasonic Borehole Imager UBI™. Careful picking of all discontinuities in the test zone from 640 to 790 mbRT provided a continuous record of fracturing in the borehole. All discontinuities were picked rather than just the larger ones, as it was the importance of orientation, rather than apparent aperture that was being tested.

Having produced a data set of all discontinuity information from the test interval, the shear and normal stresses on every discontinuity were calculated. These results are shown in Figure 3a, with a Mohr diagram showing the ratio of shear to normal stress calculated for all discontinuities from this section. The distribution of these values is also displayed as a histogram alongside. To test whether there was a stress dependency to the origin of the zones where the transmissivity was significantly higher than the background, discontinuity data from just these zones were also analysed. Figure 3b shows a Mohr diagram with the shear and normal stresses for all discontinuities from the seven transmissive zones. Again, a histogram of the distribution of coefficients of friction for all of these fractures is displayed alongside. Comparison of the data for

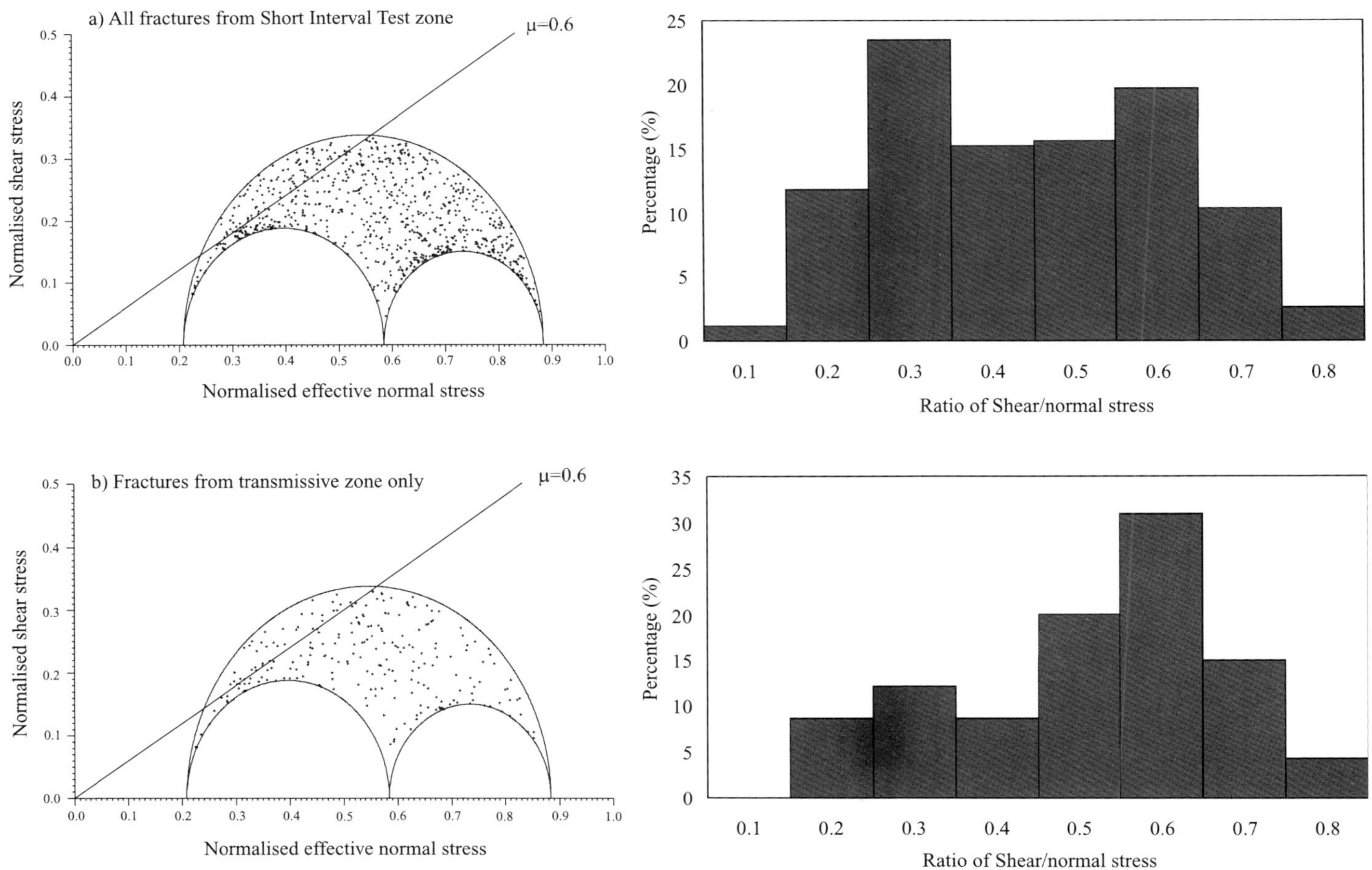

Fig. 3. (**a**) Mohr diagram of shear and normal stresses for all fracture data from the short interval test zone with histogram showing distribution of ratio of shear to normal stress for a random sample of fractures from the short interval test zone. (**b**) Mohr diagram of shear and normal stresses for all fracture data from the seven transmissive zones with histogram showing distribution of ratio of shear to normal stress for these fractures.

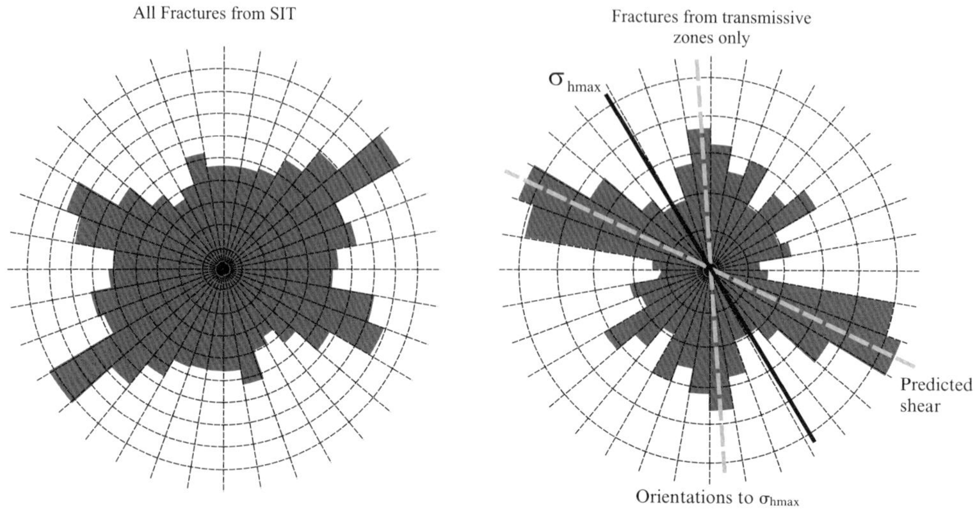

Fig. 4. Rose diagrams showing the relationship between non-flowing and flowing fracture orientations from borehole RCF3.

the whole test interval and the seven transmissive zones reveals a definite shift in the state of stress on the fracture surfaces, with a far higher percentage of the fractures falling on or above the 0.6 failure criterion. Note how the histogram for the transmissive zone data is skewed towards higher ratios when compared to the whole test interval. Statistical testing of the data reveals that at the 95% level, the mean stress ratio for the flow zone intervals is significantly different than the background rock mass showing lower transmissivities.

When the orientations of discontinuities from the flow zones are plotted on rose diagrams and the inferred shear orientations added, a strong relationship can be seen between the predicted and observed data (see Fig. 4). This is contrasted with the orientations of all discontinuities from the whole of the short interval test zone, which shows little relationship.

In order to use these data in a predictive way, a log of coefficient of friction was generated by taking a moving average over ten values of the μ values for all of the fractures within the test interval. This provides an indicative vertical log of the state of stress on the fractures with the most likely zones being identified by the presence of a strong kick towards a higher coefficient of friction. However, the log indicates more zones than the seven identified by the short interval testing.

The Stoneley wave reflection coefficient (REFC) is generally considered to be a good indicator of zones with the potential for flow. In the Sellafield boreholes the Stoneley wave data have shown a strong response from faults and fault zones intersecting the borehole. All of the seven flow zones have a strong reflection coefficient and a high stress index value (see Fig. 5). There is strong evidence that small faults and their associated fault-damaged zone are the primary flow path for contemporary flow. This is supported by the presence of a late-forming diagnostic carbonate mineralization known as a potential flowing feature, indicating the location of present-day flow (Milodowski *et al.* 1998). The mineralization has a tendency to cluster around faults within the borehole although there are many other features seemingly unrelated to major faults. Whilst Gutmanis *et al.* (1998) observed that faults themselves were generally not associated with flow, adding a stress component to the borehole observations helps provide a mechanism to account for the distribution of flow in the boreholes. If the *in situ* stress acts upon a feature, such as a fault or fracture zone that is orientated in such a manner to induce a high shear to normal stress ratio, this fault can become critically stressed. Any small-scale reactivation or movement on this surface can generate dilation, which can be experienced within the fault or damaged zones of the fault system. This dilation results in preferential pathways for fluid flow and the precipitation of mineralization seen in the drill core as the potential flowing features. As the fractures seen within the damage zones represent a synthetic (fault-parallel) set overprinted upon an existing scattered set within the host rocks, the orientation of the potential flowing features would be expected to show little directional consistency.

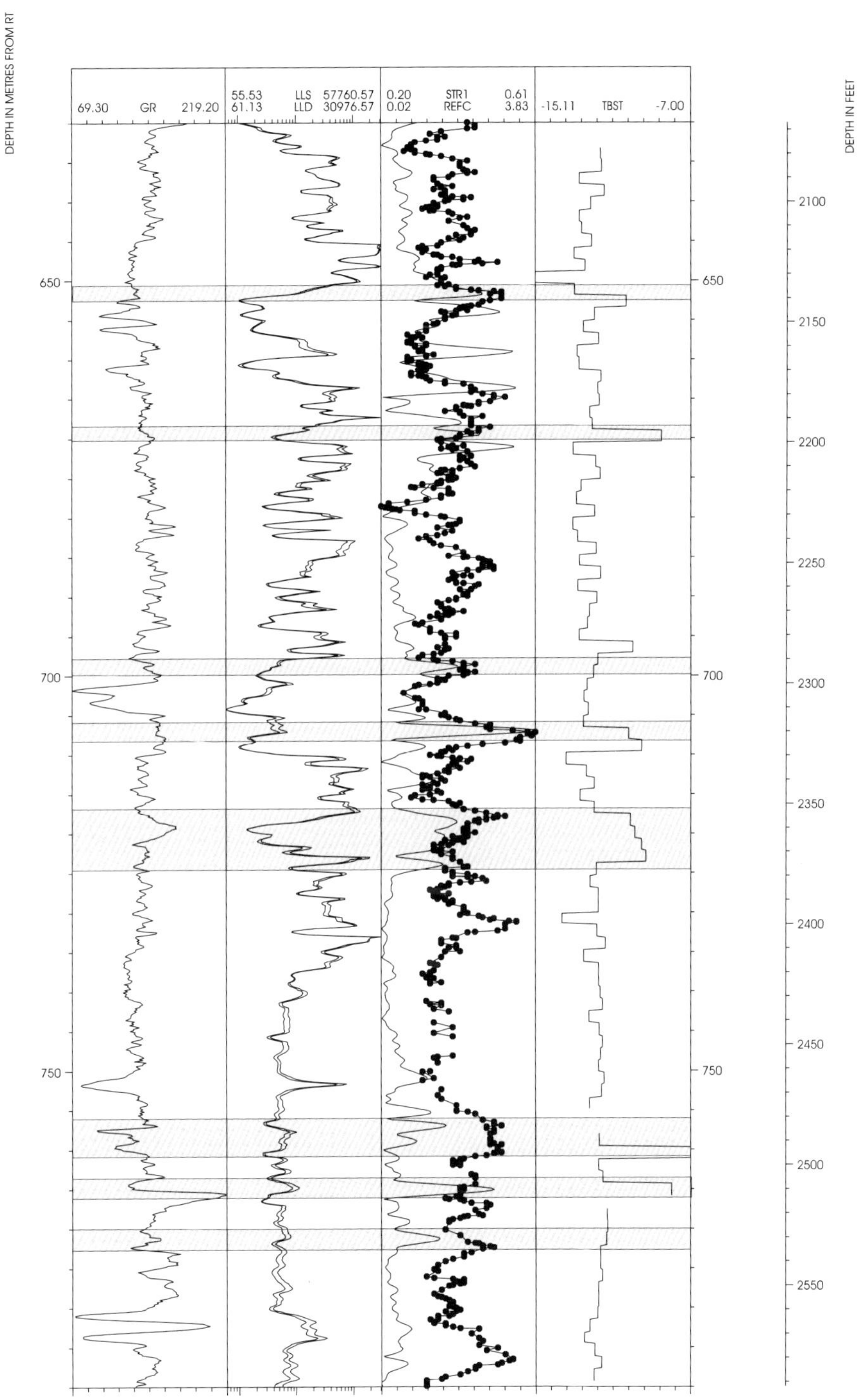

Fig. 5. Composite log of the short interval test zone of RCF3. From left: track 1, gamma; track 2, deep Laterlog; track 3, Stoneley wave reflection coefficient (light line) and stress index (dark line); track 4, interval transmissivity (m s^{-2}). Shaded intervals indicate zones where there is a high Stoneley wave reflection coefficient and also a high stress index value. Note the close agreement between these zones and the transmissivity as shown in track 4.

Discussion

Determining the orientation of preferential flow pathways in a fractured rock mass experiencing an anisotropic stress field is a critical element of the development and management of reservoirs. The critical stress theory indicates that the shear orientations can be highly significant in controlling fluid flow. Whilst the original empirical data on water flood direction implied a flow direction parallel to the principal horizontal stress component (Heffer & Lean, 1993), re-examination of the data shows two clear flow directions inclined either side of σ_{hmax}. This is also supported by numerical modelling (Heffer & Koutsabeloulis, 1995) which points towards a shear-related permeability pathway controlling flood migration. These data sets indicate that the flow pathways are obliquely orientated at approximately 20–30° to σ_{hmax}. This has been shown from a number of critical stress focused studies but is also seen in data from the Clare field off the west of Shetland where two fracture orientations approximately 30° either side of σ_{hmax} are seen to control the flow in this fractured reservoir (Coney *et al.* 1986).

One of the implications that arises from having stress-induced permeability anisotropy is how the fracture connectivity of the rock mass is significantly reduced. Jolly *et al.* (2000) examined the implications of critical-stress-controlled permeability on simulated fracture networks by removing all fractures with a shear to normal stress ratio of less than 0.6 from the network. Simulated well tests through a network of fractures that satisfied the critical stress conditions for flow revealed how a geologically well-connected network becomes hydraulically poorly connected. The associated well test pressure derivatives showed how the formation transmissivity falls and the flow through the fracture network becomes increasingly linear. The actual reduction in hydraulic connection of a rock mass obviously depends on whether the permeability anisotropy is controlled by a shear-related process with its twin preferred orientations or a normal stress mechanism with its single preferred orientation. Clearly the twin orientations of the shear model provide a greater degree of connection.

The choice of an appropriate failure criterion to describe the onset of critical stress behaviour is not clear. As discussed in an earlier section, the failure criterion used by Barton *et al.* (1995) was established by Byerlee (1978). This coefficient of friction of 0.6 was derived from tests where the confining stress was considerably higher than those typically encountered in most reservoir situations. Additionally, examination of Byerlee's data reveals that there is considerable lithological control upon the coefficient of friction under lower confining stresses. This suggests that the application of a universal threshold of 0.6 needs to be applied with some caution and that where possible it should be conditioned by local shear measurements.

Implementation of this technique in realistic flow simulations is particularly difficult given that the *in situ* stress field can vary considerably over the scale of the reservoir. Many of both the seismic and sub-seismic-scale faults within a reservoir act to perturb the stress field by significant amounts (Barton & Zoback, 1994). This results in a stress field that varies according to the scale of the measurement. Generally there are relatively few stress orientation measurements available across a reservoir to accurately define a field-scale stress pattern. As both the stress magnitudes and orientations are needed to compute those fault surfaces that are critically stressed, upscaling the stress measurements from local well observations to field scale in order to undertake permeability modification calculations is highly problematic. The best way to undertake this stress upscaling exercise is to undertake numerical stress analysis upon the reservoir model, calibrating the results against the stress orientation measurements available. Once this is completed, a stress field map is available for wider reservoir-scale use.

One of the problems in resolving the difference between shear-related phenomena and normal-stress-related phenomena is that the orientation of these oblique planes to σ_{hmax} is relatively small with the result that the direction of any flow may be inappropriately assigned along the stress axis rather than inclined to it. A more accurate reappraisal of flood data may reveal more incidences where the direction of flow is not parallel to σ_{hmax}. A review of both breakout studies and hydrofracture results provides many cases where the workers report a discrepancy of 20–30° between their results and what was expected (e.g. Hillis & Williams 1992; Brereton 1992; Batchelor & Pine 1986). There is some evidence that rather than determining the orientation of σ_{hmax}, the orientation of the induced shear planes has been measured resulting in these 30° rotations.

Conclusions

The critical stress theory has been shown to predict the occurrence of flow on the well scale in a number of studies. Previously this has been achieved by integrating fluid temperature and conductivity logs with discontinuities as seen on borehole image logs. This study related the transmissivity of the formation as determined by short interval hydraulic tests to the discontinuity population as measured from borehole image logs. Highly transmissivity zones were shown to have a mean shear to normal stress ratio significantly higher than that seen in the background rock mass.

It has been considered that fluid moving through a fractured rock mass experiencing strong stress anisotropy would flow preferentially in the direction of the maximum horizontal stress direction. However, the critical stress theory shows that fracture orientations inclined to the maximum horizontal stress direction can show the maximum permeability orientation as a result of shear-related processes.

Upscaling this technique from the well bore to a reservoir-wide model has some difficulties. Perturbation of the stress field on the scale of the reservoir by faults, fractures and varying physical properties, makes it hard to define a reliable field wide stress field as input to the critical stress calculation. Therefore some caution is advised in applying this technique widely.

This paper is based upon a short study carried out subsequent to the main Sellafield investigations and sought to investigate whether the *in situ* stress helped to explain observed flow patterns in one of the boreholes. However, the work of Gutmanis *et al.* (1998) and Milowdowski *et al.* (1998) remains the more complete interpretation of contemporary flow patterns within the Sellafield area. This paper is published with the permission of UK Nirex Ltd.

References

BARTON, C. A., MOOS, D. & ZOBACK, M. D. 1997. *In situ* stress measurements can help define local variations in fracture hydraulic conductivity at shallow depth. *The Leading Edge*, **16**, 1653–1656.

BARTON, C. A, AND ZOBACK, M. D. 1994. Stress perturbations associated with active faults penetrated by boreholes: possible evidence for near complete stress drop and a new technique for stress magnitude measurement. *Journal of Geophysical Research*, **99** (B5), 9373–9390.

BARTON, C. A., ZOBACK, M. D. & MOOS, D. 1995. Fluid flow along potentially active faults in crystalline rock. *Geology*, **23**, 683–686.

BATCHELOR, A. S. & PINE, R. J. 1986. The results of *in situ* stress determinations by seven methods to depths of 2500m in the Carnmenellis Granite. *Proceedings of the International Symposium on Rock Stress and Rock Stress Measurements*, Stockholm, Sept. 1986, 467–478.

BRERETON, N. R. 1992. *Rock stress orientation measurements in Sellafield boreholes 2,3 & 4*. BGS Technical Report **WK/92/13C**.

BYERLEE, J. 1978. Friction on rocks. *Pure & Applied Geophysics*, **116**, 615–626.

CONEY , D., FYFE, T. B., RETAIL P. & SMITH, P. J. 1986. Claire appraisal: the benefits of a co-operative approach. *In*: PARKER, J. R. (ed.) *Petroleum Geology of Northwest Europe: Proceedings of the 4th Conference.* Geological Society, London, 1409–1420.

DART, R. L. & ZOBACK, M. L. 1989. Wellbore breakout stress analysis within the central and eastern continental United States. *Log Analyst*, **30**, 12–25.

EVANS, C. J. & BRERETON, N. R. 1990. *In situ* crustal stress in the United Kingdom from borehole breakouts. *In*: HURST, A., LOVELL, M. A. & MORTON, A. C. (eds) *Geological Applications of Wireline Logs.* Geological Society, London, Special Publications, **48**, 327–338.

FINKBEINER, T, BARTON, C. A. & ZOBACK, M. 1997. Relationships among *in situ* stress, fractures and faults, and fluid flow: Monterey Formation, Santa Maria Basin, California. *AAPG Bulletin*, **81**, 1975–1999.

GUTMANIS, J., LANYON, G. W., WYNN, T. J. & WATSON, C. R. 1998. Fluid flow in faults: a study of fault hydrogeology in Triassic sandstone and Ordovician volcaniclastic rocks at Sellafield, north-west England. *Proceedings of the Yorkshire Geological Society*, **52**, 159–175.

HEFFER, K. J. & KOUTSABELOULIS, N. C. 1995. Stress effects on reservoir flow: numerical modelling used to reproduce field data. *In*: DE HAAN, H. J. (ed.) *New Developments in Improved Oil Recovery*. Geological Society, London, Special Publications, **84**, 81–88.

HEFFER, K. J. & LEAN, J. 1993. Earth stress orientation – a control on, and a guide to, flooding directionality in a majority of reservoirs. *In*: LINVILLE, W. (ed.) *Reservoir Characterisation III*. PennWell Books, Tulsa, 799–822.

HICKMAN, S. H., BARTON, C. A., ZOBACK, M. D., MORIN, R., SASS, J. & BENOIT, R. 1997. *In situ* stress and fracture permeability along the Stillwater fault zone, Dixie Valley, Nevada. *International Journal of Rock Mechanics and Mining Science*, **34**, 414.

HILLIS, R. R. & WILLIAMS, A. F. 1992. Borehole breakouts and stress analysis in the Timor Sea. *In*: HURST, A., GRIFFITHS, C. M. & WORTHINGTON, P. F. (eds) *Geological Applications of Wireline Logs II*. Geological Society, London, Special Publications **65**, 157–168.

JAEGER, J. C. & COOK, N. G. W. 1984. *Fundamentals of Rock Mechanics* (3rd edition). Chapman & Hall, London.

JOLLY, R. J. H, WEI, L & PINE, R. J. 2000. *Stress-sensitive fracture-flow modelling in fractured reservoirs*. SPE Paper 59042.

MILODOWSKI, A. E., GILLESPIE, M. R., NADEN, J., FORTEY, N. J., SHEPHERD, T. J., PEARCE, J. M. & METCALFE, R. 1998. The petrology and paragenesis of fracture mineralisation in the Sellafield area, west Cumbria. *Proceedings of the Yorkshire Geological Society*, **52**, 215–241.

MULLER B., ZOBACK, M. L. *et al.* 1992. Regional patterns of Tectonic Stress in Europe. *Journal of Geophysical Research*, **97** (B8), 11783–11803.

NIREX. 1997. *Sellafield Geological and Hydro-geological Investigations*. Assessment of the *in situ* Stress Field at Sellafield. Nirex Report **S/97/003**.

PLUMB, R. A. & HICKMAN S. H. 1985. A comparison between the 4 arm Dipmeter and the Borehole Televiewer in Auburn Geothermal Well. *Journal of Geophysical Research*, **90**, 5513–5521.

TRICE, R. A. 1999. Application of borehole image logs in constructing 3D static models of productive fracture networks in the Apulian Platform, Southern Apennines. *In*: LOVELL, M. A., WILLIAMSON, G. & HARVEY, P. K. J. (eds) *Borehole Imaging; Applications and Case Histories*. Geological Society, London, Special Publications, **159**, 77–90.

YEO, I. W., DE FREITAS, M. H. & ZIMMERMAN, R. W. 1998. Effect of shear displacement on the aperture and permeability of a rock fracture. *International Journal of Rock Mechanics and Mining Sciences & Geomechanical Abstracts*, **35**, 1051–1070.

A case study of hydrocarbon transport along active faults and production-related stress changes in the Monterey Formation, California

SNEHA K. CHANCHANI[1], MARK D. ZOBACK & COLLEEN BARTON[2]

Department of Geophysics, Stanford University, Stanford, CA 94305, USA
[1]*Current address: BP-Amoco Exploration, 501 Westlake Park Boulevard, Houston, TX 77479, USA*
[2]*Current address: GeoMechanics International, Inc., 250 Cambridge Avenue, Palo Alto, CA 94306, USA*

Abstract: Recent field studies show that critically stressed faults, that is, faults that are close to frictional failure in the current stress field, serve as conduits for fluid flow. Similarly, geological field studies of the low permeability siliceous shales in the Monterey Formation, California, clearly indicate that faults influence hydrocarbon transport. We report here a study of the Antelope Shale, a low permeability siliceous shale hydrocarbon reservoir in the Buena Vista Hills field in the southern San Joaquin Valley to determine the influence of the stress state on the relative hydraulic conductivity of the fractures and faults present in the subsurface. Because production has both lowered reservoir pressure and the horizontal stresses, it was necessary to "restore" the reservoir stress state to initial conditions in order to identify correctly the most highly productive intervals. This analysis demonstrates that prior to production, faults in the reservoir were active in a transitional reverse/strike-slip faulting stress state, consistent with regional tectonics. Initial production rates in the field were 2000 barrels of oil per day, principally from intervals where critically stressed faults were encountered.

The relationship between the *in situ* state of stress and the orientation of hydraulically conductive fractures is conventionally viewed in the context of mode I extension fractures orientated perpendicular to the least principal stress (Secor 1965; du Rouchet 1981; Nur & Walder 1990). Such a relationship has straightforward implications for utilizing geophysical techniques such as seismic velocity anisotropy, shear-wave splitting, and amplitude versus offset (AVO) to identify *in situ* directions of permeability anisotropy in regions where fracture permeability is important (e.g. Crampin 1993, Winterstein & Meadows 1991). In this paper, we investigate an alternative hypothesis in which critically stressed faults, that is faults close to frictional failure in the *in situ* stress state, may be responsible for hydrocarbon transport in reservoir rocks with low primary permeability.

Barton *et al.* (1995) demonstrated that in crystalline rock, fractures and faults that were critically stressed for frictional failure in the current stress field, were also hydraulically conductive. Finkbeiner *et al.* (1997) documented a similar increase in formation permeability in four wells in the Santa Maria basin, California, at depth intervals of brittle lithologies and critically stressed fracture and bedding planes. In a field study of sedimentary rocks, Dholakia *et al.* (1998) showed the predominant association between faulting, fault-related fractures and the occurrence of petroleum in siliceous shale deposits of the Monterey Formation. Petroleum-staining in surface exposures of the Monterey Formation was observed primarily within fault zones and fault-related fractures. This field study identified fault development in surface exposures of low permeability siliceous shale along the central California coast and in the San Joaquin Valley. As faulting develops, the low permeability rocks systematically fragment and subsequently brecciate to increase permeability locally, and facilitate hydrocarbon transport. These studies support the hypothesis that fluid flow is not simply controlled by mode I extension fractures orientated perpendicular to the minimum horizontal stress, but rather that shear faults are important for hydrocarbon transport.

Efforts to understand fracture controls on hydrocarbon transport and accumulation in tight,

From: AMEEN, M. (ed.) *Fracture and In-Situ Stress Characterization of Hydrocarbon Reservoirs*. Geological Society, London, Special Publications, **209**, 17–26. 0305–8719/03/$15.00
© The Geological Society of London 2003.

low permeability siliceous shales are being pursued as part of a multidisciplinary project to develop enhanced oil recovery methods for further extraction of trapped hydrocarbons within the siliceous shale. As part of this study, we have investigated the Buena Vista Hills field to determine the potential for hydrocarbon production from fracture permeability and the influence of reservoir depletion on the current stress field.

The Buena Vista Hills field, discovered in 1952, is located in the southern San Joaquin Valley, approximately 40 km (25 miles) SW of Bakersfield, California (Fig. 1). The field was drilled initially in the 1950s and 1960s, and is situated on a NW–SE doubly plunging anticline paralleling the San Andreas Fault. Initial production rates from individual wells peaked at 2000 barrels of oil per day (bopd) with daily production rates dropping to 6 bopd in the 1970s. Recent well stimulation efforts have brought the rate to 20 bopd. Total production to date from the field is 9 million barrels of oil, only 6.5% of the estimated oil in place in the Antelope Shale in the Buena Vista Hills field. The Antelope Shale pay thickness is about 244 m (800 ft) at a depth ranging from 1189 to 1524 m (3900–5000 ft). Porosity ranges from 20 to 40% and matrix permeability is 0.5–1 mD. Initial pressure in the reservoir was 13.8 MPa (2000 psi), and has since dropped to 5.5 MPa (800 psi) leading to further production difficulties (Chevron 1994). In a recent paper, Montgomery & Morea (2001) describe the Antelope Shale reservoir in some detail.

Stress orientations and active tectonics

Wellbore breakouts, reverse fault earthquake focal mechanisms, and hydraulic fracture orientations along the length of the Coast Ranges in central California indicate a NE–SW direction of maximum horizontal compression (S_{Hmax}) nearly perpendicular to the San Andreas Fault system (Zoback *et al.* 1987; Mount & Suppe 1987; Wentworth & Zoback 1989). Axes of young, active folds along the Coast Ranges lie nearly perpendicular to the NE–SW directed S_{Hmax} suggesting anticlinal development driven by seismogenic blind thrusts is an active process in the current stress field. Figure 2 is a regional stress map showing the S_{Hmax} direction determined from various stress indicators (e.g. wellbore breakouts, earthquake focal mechanisms, and hydraulic fractures) and geological structures in the San Joaquin Valley. The light symbols reflect earlier studies on the state of stress in California (Mount & Suppe 1987; Zoback *et al.* 1987; Castillo & Zoback 1995). The dark symbols represent new data collected for this study from tiltmeter measurements of hydraulic fracture tests, which are described in more detail below. The lengths of each stress indicator are weighted according to a quality-ranking scheme (Zoback & Zoback 1991) based on the standard deviation, depth of observation, and number of observations.

In this study, *in situ* stress orientations were considered in the area shown by the box in of Figure 1A using new data from tiltmeter surveys

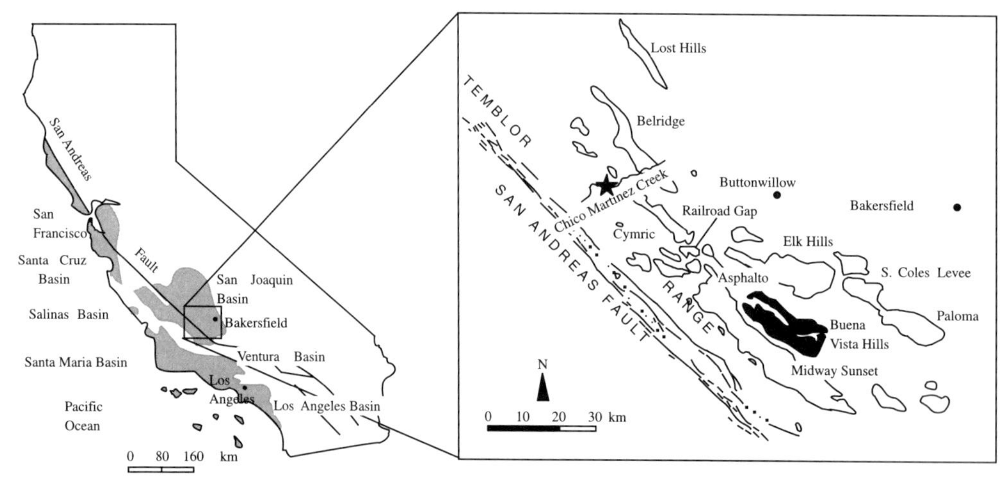

Fig. 1. (**a**) Location of major Neogene basins in California (from Graham & Williams 1983). (**b**) Location of the Buena Vista Hills field and other major oil fields in the southern San Joaquin basin.

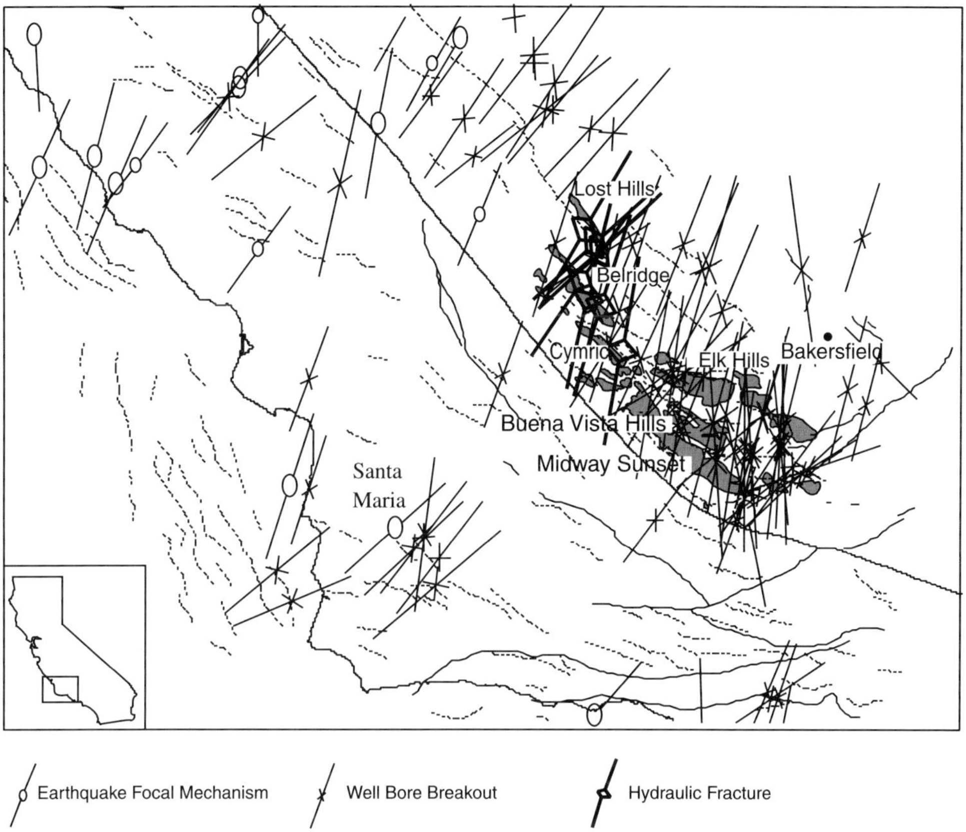

Fig. 2. Regional stress map in southern San Joaquin and Santa Maria basins, California. The symbols in black represent new data collected for this study from tiltmeter measurements of hydraulic fracture tests.

of hydraulic fractures in 56 wells from Belridge, Lost Hills, and Cymric fields by Pinnacle Technologies (written comm., C. Wright). Hydraulic fractures propagate in a direction perpendicular to the minimum principal compressive stress in the crust (Hubbert & Willis 1957). The deformation, or local displacement gradients, at the surface induced by the hydraulic fracture is a function of the fracture orientation. Utilization of tiltmeters for determining the azimuth of large-scale, near-vertical, hydraulic fractures at relatively shallow depth is described by Wright (1998). Because of the high density of data available, the hydraulic fracture orientations were averaged with depth within an individual well and with other wells located within a one mile radius in a given field.

The seven hydraulic fracture orientations indicated in black with the open diamond symbol in Figure 2 represent fracture orientations from wells within the Cymric field and the northern, central and southern areas of both Lost Hills and Belridge fields. The average hydraulic fracture orientations were computed for vertical or near-vertical fractures, dipping more than 70° from horizontal, and occurring below 300 m depth. The statistical mean fracture orientation and standard deviations were computed using the method described by Mardia (1972) for azimuthal data and are shown in Table 1.

Wellbore breakout analyses of FMS (Formation Micro Scanner) data were also conducted within the Buena Vista Hills field. Wellbore breakouts refer to spalled regions of the borehole in which the circumferential stress around the borehole exceeds the compressive strength of the rock. The borehole is elongated in the direction of the minimum horizontal stress (Bell & Gough 1979; Zoback *et al.* 1985) and is considered a reliable indicator of horizontal stress directions (e.g. see discussion is Zoback & Zoback 1991). The orientation of S_{Hmax} determined from the breakout analyses of FMS data of two wells in

Table 1. *Summary of SHmax orientations determined from tiltmeter measurements of deformation caused by hydraulic fractures. Each orientation reflects the average of several fractures wells within a one-mile radius.*

Field name	S_{Hmax} direction	Number of hydraulic fractures
Cymric	N12.5°E±3.2°	6
North Belridge	N40.3°E±3.6°	21
Central Belridge	N15.7°E±3.7°	15
South Belridge	N20.5°E±2.9°	59
North Lost Hills	N35.5°E±1.2°	5
Central Lost Hills	N53.7°E±4.3°	34
South Lost Hills	N52.7°E±3.2°	28

the Buena Vista Hills field are shown on the stress map in black by the two inward pointed arrows (Fig. 2). The breakout information was obtained from the caliper surveys of the FMS log. The azimuthal mean orientation and standard deviation of the breakouts are listed in Table 2 and were computed using the method described by Mardia (1972). Like the tiltmeter surveys of the hydraulic fracture, the points on the map are length-weighted according to data quality. Data quality factors for wellbore breakout data include the number of breakout intervals, the total length of breakouts within the wellbore, the standard deviation, and the depth of observation (Zoback and Zoback, 1991).

The breakout analyses and tiltmeter surveys indicate a NE–SW directed S_{Hmax}, orientated nearly perpendicular to the San Andreas Fault and the axes of young, active folds along the Coast Ranges. This near-perpendicular relationship suggests active structural development and deformation in the current stress field (Zoback *et al.* 1987; Mount & Suppe 1987, 1992; Wentworth & Zoback 1989).

Constraining stress magnitudes

Knowledge of the orientation and magnitudes of *in situ* stress is required to determine which subsurface fractures and faults are most important for hydrocarbon transport. In this section we first constrain the current horizontal principal stress magnitudes and then attempt to extrapolate back in time to assess the stress state prior to reservoir pressure depletion. In Buena Vista Hills field, we determine the current stress orientation and magnitude to evaluate the influence of the *in situ* stress state on fractures in the reservoir identified using borehole image data.

Stress magnitudes in 1989

As supported by tiltmeter data (Castillo *et al.* 1997; Wright 1998), we assume that the principal stresses act in three orthogonal directions, one near-vertical, normal to the surface of the Earth, and in two orthogonal horizontal directions (S_{hmin} and S_{Hmax}). The vertical stress (S_v) is assumed to be equal to the weight of the overlying rock. Density logs were integrated with depth to determine S_v. The overlying lithologies consist primarily of low density diatomite and siliceous shale, and the densities reported in the density log are consistent with laboratory measurements of siliceous rocks (2.0–2.3 g cm^{-3}) (Isaacs 1981).

Hydraulic fracturing of the wells to enhance production provides direct measures of the minimum principal stress magnitude. The fluid pressure required to propagate a hydraulic fracture is a measure of the minimum horizontal stress magnitude (S_{hmin}) when $S_{hmin} \leq S_v$ (Hubbert & Willis 1957; Kehle 1964; Haimson & Fairhurst 1970). S_{hmin} values corresponding to three hydraulic fracturing tests at different depths conducted in 1989 are provided in Table 3.

In Buena Vista Hills field we constrain the magnitude of the maximum compressive stress (S_{Hmax}) by employing simple elastic failure criterion and wellbore failure analysis following Moos & Zoback (1990). Breakouts in this study were detected using four-arm dipmeter data and analysed in accordance with criteria proposed by

Table 2. *Summary of* S_{Hmax} *directions*

Well name	S_{Hmax} direction	Breakout length
621–25B	N8°E±6°	65 m
723–9D	N21°E±7°	229 m

Data were determined from wellbore breakout analysis of two wells in Buena Vista Hills field.

Table 3. *Summary of* S_{hmin} *with depth.*

Depth (m)	S_{hmin} (MPa)
1204	18
1289	20
1341	22
	Average S_{hmin}=20

Data were determined from hydraulic fracture tests. The average Shmin is used in this study.

Plumb & Hickman (1985). A wellbore breakout will form when the stress concentration around the wellbore exceeds the compressive strength of the rock, C_o (Zoback *et al.* 1985; Moos & Zoback 1990) and is described by the following equation:

$$S_{Hmax}+S_{hmin}-2(S_{Hmax}-S_{hmin})\cos 2\theta-2P_o\geq C_o \quad (1)$$

Consideration of the breakout's width ($2\theta_B$) can be used to better constraint S_{Hmax}, as for a given rock strength S_{Hmax} must be sufficiently high to have caused the breakout to have a given width (Barton *et al.* 1988). In order for breakouts to be detected in the FMS data, the breakout must be at least as wide as the FMS pad or the pad would not have fallen into the breakout. We assume that the minimum breakout width is greater than or equal to 54°, the width of the FMS pad, and is centred at the azimuth of S_{hmin}. Equation 1 leads to the following:

$$2.18S_{Hmax}-0.18S_{hmin}-2P_o\geq C_o \quad (2)$$

which leads to

$$S_{Hmax}\geq 0.46\,(C+0.18S_{hmin}+2P_o) \quad (3)$$

where P_o is the pore pressure in the formation at the time the well is drilled and the mud weight is essentially equal to the pore pressure.

Uniaxial compressive rock strength values, C_o, for the Antelope Shale were measured directly in triaxial compression tests at the depths listed in Table 3. Rock strength values were found to be quite high and range from 92 to 126 MPa (Chevron 1994). The shear modulus computed from the shear wave and compressional wave velocities determined from the sonic log at the depth from which the laboratory samples were tested are much higher (4.2–4.9 GPa) than the values from the intervals at which breakouts occur (3.4 GPa). This difference in shear moduli suggests that the compressive rock strength value used to constrain S_{Hmax} is likely to be lower than the average laboratory measured value of 110 MPa. Compressive rock strength tests on siliceous sediments from an offshore well in the Monterey Formation yield much lower values between 25 MPa and 117 MPa (C. Bovberg, pers. comm.). Thus, there is considerable uncertainty in the appropriate rock strength in this formation. Using a compressive rock strength value of 90 MPa establishes a lower bound for S_{Hmax} of about 48 MPa.

Considering the state of stress in the earth's crust to be limited by its frictional strength, the allowable stress states are given by

$$\frac{S_1-P_o}{S_3-P_o}\leq[(1+\mu^2)^{1/2}+\mu]^2 \quad (4)$$

where P_o is the pore pressure and μ is the coefficient of sliding friction (e.g. Jaeger & Cook 1979). Frictional equilibrium of the Earth's crust where $\mu=0.6$ suggests an upper limit of S_{Hmax} of 50 MPa. Thus, in 1989, the average state of stress appears to be a strike-slip regime in which the stresses are as follows: $S_{hmin}=20$ MPa; $S_v=25$ MPa; 48 MPa$\leq S_{Hmax}\leq$50 MPa.

Stress magnitudes in 1952

Drilling in the Buena Vista Hills field in the Antelope Shale began in 1952 and continued to 1989. Reservoir depletion and drilling through time lowered the pore pressure and thus lowered the horizontal stresses poroelastically. During this time the reservoir pressure has declined significantly from 13.8 MPa (2000 psi) in 1952 to approximately 5.5 MPa (800 psi) in 1989 (Fig. 3). From the hydraulic fracture tests in 1989, we know the average S_{hmin} is 20 MPa (Table 3). Because of the large pressure decline in the reservoir, the state of stress in 1989 reflects lower magnitudes of stress than that which existed prior to reservoir depletion.

The magnitude of S_{hmin} measured in 1989 reflects the minimum horizontal stress at the time of hydraulic fracture testing and not the stress state in the undisturbed reservoir. In order to extrapolate the minimum horizontal stress to initial conditions in the Antelope Shale, we employ the theory of poroelasticity which describes the change in the horizontal stresses as a function of pore pressure, P_o (Biot 1941; Segall 1992). Approximating the reservoir shape as flat and infinite in lateral extent, the change in the horizontal stresses as a function of the change in P_o may be expressed as follows:

$$\Delta Sh=\Delta P_o\alpha\,\frac{1-2v}{1-v} \quad (5)$$

where α is the Biot coefficient of effective stress and v is the undrained Poisson's ratio (Segall 1992). The Biot coefficient of effective stress, a, is defined as follows:

$$\alpha=\left(1-\frac{k_b}{k_g}\right)$$

where k_b is bulk modulus of the dry porous frame and k_g is the bulk modulus of the grains. When k_b is small, and α approaches 1, the largest

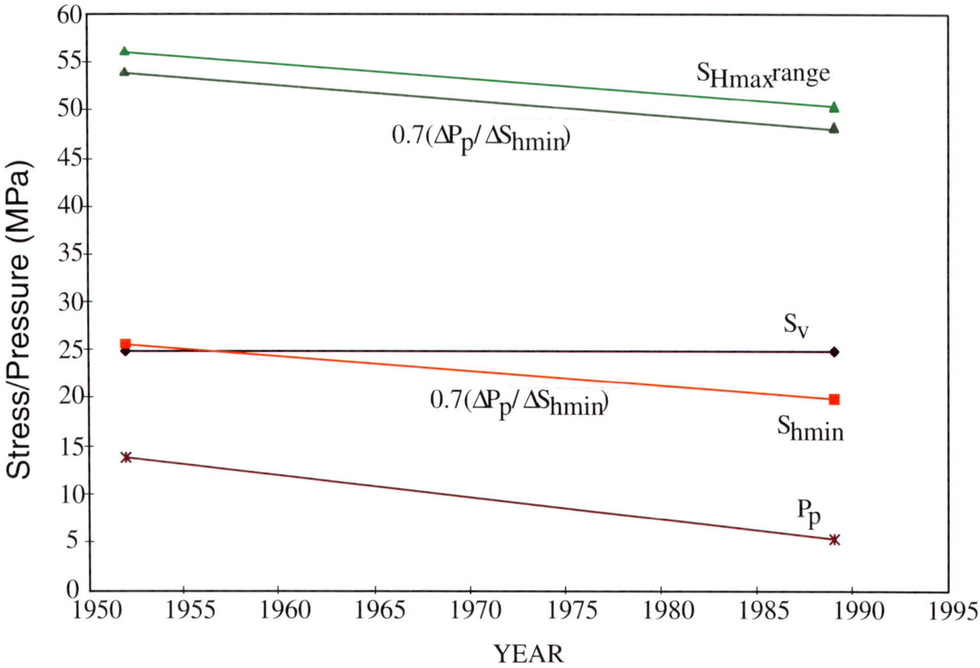

Fig. 3. General pressure decline curve for the Buena Vista Hills field that illustrates the change in horizontal stresses and pore pressure through time.

fluid effects are observed. A high porosity, soft sediment rock similar to the siliceous shale of the Monterey Formation will have $\alpha \sim 1$ (e.g. Krief *et al.* 1990; Segall *et al.* 1994). We use $\alpha \sim 1$, and from laboratory triaxial compression tests, ν is approximately 0.23 (Table 4). Substituting these values in equation 4, the ratio of the change in horizontal stress to pore pressure in the reservoir is $\Delta S\mathrm{h}/\Delta P_\mathrm{o} \sim 0.70$, similar to values reported by Engelder & Fischer (1994) in other areas. Figure 3 illustrates the decline in the horizontal stresses through time as a function of reservoir depletion. The horizontal stresses decreased by about 5.8 MPa. Note that while the state of stress in 1989 appears to be a strike-slip stress state ($S_\mathrm{Hmax}>S_\mathrm{v}>S_\mathrm{hmin}$), the original stress state in the reservoir appears to have been a reverse faulting regime ($S_\mathrm{Hmax}>S_\mathrm{hmin}>S_\mathrm{v}$). Because S_hmin is only slightly greater than S_v, it is indicative of a transitional reverse/strike-slip faulting regime. Thus, in 1952, the state of stress appears to be as follows: S_v =25 MPa; S_hmin=26 MPa; $S_\mathrm{Hmax}\geq 54$ MPa

Table 4. *Compressive rock strengths measured in triaxial compression.*

Sample depth (m)	Poisson's ratio, ν	Strength, C (MPa)
1172	0.23	92
1306	0.22	113
1395	0.25	126

Fractures, faults and production

Microresistivity borehole image (FMS) data from the Buena Vista Hills field provide an opportunity to identify fractures and fracture zones and to determine the effect of the stress state prior to reservoir production in 1952 on fluid flow. The FMS tool consists of four pads, two orthogonal pads with microresistivity imaging electrodes and two pads with standard dipmeter electrodes (Ekstrom *et al.* 1987). The image data are typically displayed as if the walls of the borehole were unwrapped and laid flat. Fractures intersecting the borehole would thus appear as a sinusoid cutting across the individual image pad displays.

Fractures, bedding, and other zones of disrupted appearance are identified in the FMS images from the Chevron well 621–25B. Only a partial section of approximately 343 m (1125 ft)

of the Antelope Shale section, below the opal A/opal CT diagenetic contact, was logged. Over 300 fractures were identified in the logged interval of image data of the Antelope Shale. The relative fracture frequency with depth in the 621–25B well (Fig. 4) is related to lithology differences. It is known from the core that the Antelope Shale beds dip at a low angle, therefore low angle (<10°) planar features in the image data are interpreted to be bedding. The Antelope Shale contains more fractures than the Brown Shale. The Brown Shale is more clay-rich, whereas the Antelope Shale consists of thin graded siliceous shale beds, porcelanite and chert beds. The Antelope shale is more brittle and contains more fractures (Chevron 1994).

In this study we are concerned with the influence of the present-day stress state and the impact of reservoir depletion on production. Measured matrix permeabilities of core plugs from the Antelope Shale are very low, 0.1–0.5 mD. The low permeability, tight nature of the Antelope Shale suggests production is primarily from fracture permeability. The stress analysis shows the environment to be in a transitional strike-slip to reverse faulting regime in which the intermediate and minimum, S_2 and S_3, principal stresses are nearly equal. Transitional strike-slip/reverse stress state is consistent with the stress inversion of the 1952 Kern County magnitude 7.8 earthquake and contemporary seismicity occurring approximately 30 km SW of Buena Vista Hills field (Castillo & Zoback 1995).

The poles to the fracture planes identified in the image data and directions of S_{Hmax} determined from the wellbore breakout analysis are plotted in the stereonet in Figure 5. The fractures observed concentrate in two broad regions as conjugate reverse faults. Colours on the stereonet illustrate the normalized Coulomb failure function (*CFF*) and critical P_p (that required to induce slip) as a function of fracture orientation, and poles in white indicated the critically stressed faults (fracture poles falling in the red-shaded region). The Coulomb failure function is defined in the following equation:

$$CFF = \tau - \mu(S_n - P_p) \qquad (6)$$

where τ is the shear stress acting on a fault plane, S_n is the normal stress and P_p is the ambient pore pressure. μ is the coefficient of friction and is

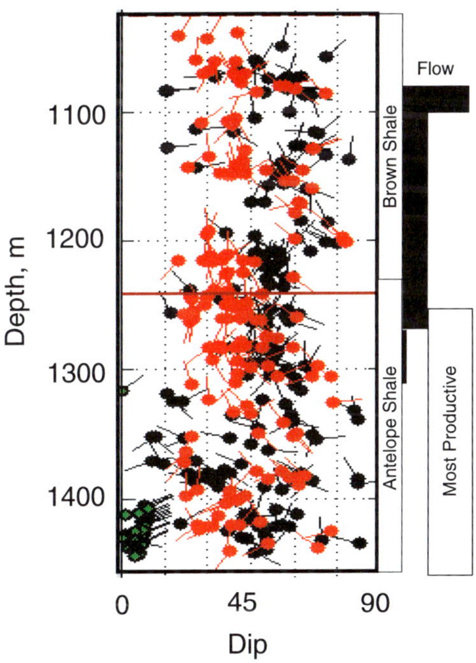

Fig. 4. Tadpole plot of the fractures with depth. The red tadpoles are the ones that are critically stressed. The green ones are mode I cracks, normal to the least principal stress. These are plotted in white in Figure 5.

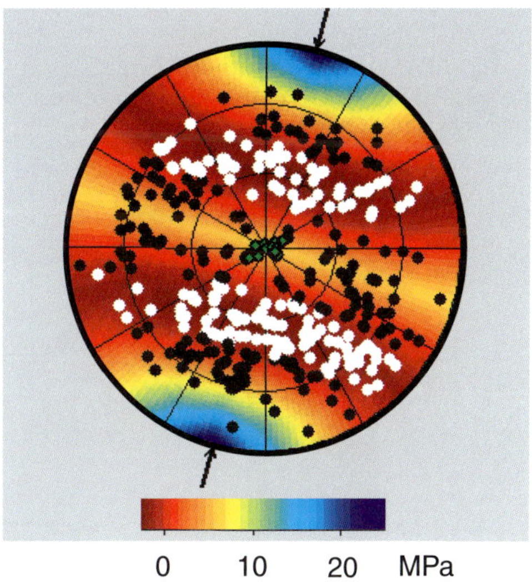

Fig. 5. The stereonet illustrates the degree to which fractures and faults are critically stressed as the excess pore pressure required to cause fault. Critically stressed fractures are highlighted in white (require no excess pore pressure to be active) and correspond to the red tadpoles in Figure 4. The critically-stressed fractures observed in the 621–25B well concentrate in two broad regions as conjugate reverse/strike-slip faults.

taken as 0.6. Thus, when $CFF \geq 0$ faults are critically-stressed and when $CFF \leq 0$ they are not. The figure indicates that conjugate reverse faults or fractures are more likely to slip in the transitional strike-slip/reverse stress regime. It is clear to see that there are many critically stressed faults in the reservoir.

Comparison of the critically stressed fracture occurrence with production data suggests a qualitative relationship between critically stressed fractures and fluid flow (Fig. 4). According to Chevron (1994), the most productive interval in the 621–25B is in the Upper and Lower Antelope Shale between the Pa and P2a markers. The Lower Antelope Shale from P1b to P2a accounts for two-thirds of the fluid produced from the well. Historically, in the Buena Vista Hills field, the P-P1b is the most productive interval. Spinner flowmeter production logging in the perforated intervals in Chevron well 621–25B (Fig. 4) shows the distribution of flow rates in the wellbore. Spinner flowmeter tools measure the flow rate (revolutions per second) below the tool, and the black bars in Figure 4 show the relative magnitudes of flow rate as a function of depth. Overall, the productive interval in the Antelope Shale interval contains many critically stressed faults. There are also critically stressed faults in the Brown Shale where the flowmeter survey indicated significant flow. However, as the flowmeter survey did not cover both formations, it is not clear what the response would have been in the most productive section of the Antelope Shale and the respective abundance of critically stressed faults.

Discussion

The Antelope Shale at Chico Martinez Creek (Dholakia *et al.* 1998) serves as a surface analogue to the subsurface reservoir at Buena Vista Hills field. Not only are both areas structurally similar anticlines associated with the San Andreas Fault, but also the section studied at Chico Martinez Creek correlates to portions of the producing reservoir in the BVH field. The Chico Martinez Creek outcrop, 56 km NW of BVH, exposes the Antelope Shale Member of the Monterey Formation, the reservoir unit at BVH, in the northeastern limb of the anticline. In addition, the diagenetic grade of the Antelope Shale at both localities consists of siliceous shale and porcelanite. The Antelope Shale outcrop study by Dholakia *et al.* (1998) at Chico Martinez Creek in the San Joaquin Valley documents a predominant association between hydrocarbon transport and brecciated fault zones. Petroleum staining occurs in bed-parallel breccia zones. Observations of hydrocarbon-stained breccia and rubble zones in core from Buena Vista Hills and neighbouring fields, along with evidence from surface exposures, directly support the relationship between shearing deformation and hydrocarbon flow pathways and storage in the subsurface. Figure 6 (modified from Dholakia *et al.* 1998) shows an oil-saturated breccia zone

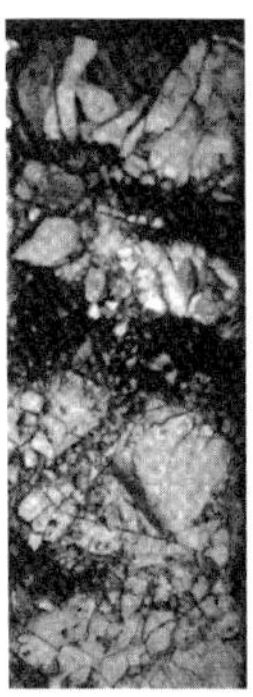

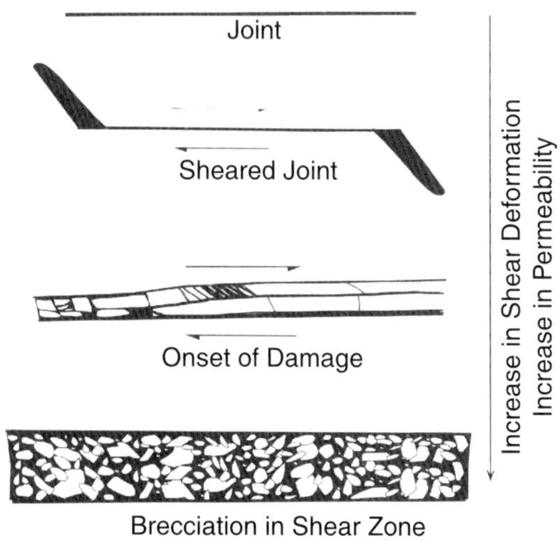

Fig. 6. Photograph of oil-stained breccia zone in Antelope Shale core from the Buena Cymric field, left, and a schematic diagram, right, illustrating the way in which progressive shear deformation leads to development of breccia zones and enhanced permeability (modified after Dholakia *et al.* 1998).

in core from a well in the Antelope Shale in the Cymric field as well as a diagrammatic illustration of the way in which shear deformation induces brecciation and enhanced permeability.

A major shortcoming of this work is our inability to compare, in detail, the affect of individual, or sets, of critically stressed faults on productivity. This would also make it possible to further investigate other factors such as lithology, fault orientation, apparent aperture, etc.

Conclusions

This case study of the Antelope Shale, Monterey Formation, demonstrates that faults active in the current stress state correlate with intervals of higher productivity. Stress analysis indicates that prior to production, faults in the reservoir were active in a transitional reverse/strike-slip faulting regime, consistent with regional tectonics. Because production has both lowered reservoir pressure and the horizontal stresses, the initial reservoir stress state was 'restored' to identify correctly the most highly productive intervals. Comparison of the critically stressed fracture occurrence with production data suggests a qualitative relationship between critically stressed fractures and fluid flow. This subsurface geophysical study at Buena Vista Hills field supports the hypothesis that fractures close to frictional failure, *active faults*, are important for hydrocarbon transport.

This work was supported by Chevron USA, the Department of Energy contract DE-PS22-94BC 1114973, the Stanford Borehole Geophysics Laboratory, the Stanford Rock Fracture Project and the Phillips Fellowship. We thank C. Wright of Pinnacle Technologies for additional stress data in the San Joaquin basin from tiltmeter surveys of hydraulic fracture tests. We thank B. Bilodeau of Chevron Production Co. for subsurface data and discussion critical for this study

References

BARTON, C. A., ZOBACK, M. D. & BURNS K. L. 1988. In situ stress orientation and magnitude at the Fenton Geothermal site, New Mexico, determined from wellbore breakouts. Geophysical Research Letters, **15**, 467–470.

BARTON, C. A., ZOBACK, M. D. & MOOS, D. 1995. Fluid flow along potentially active faults in crystalline rock. Geology, **23**, 683–686.

BELL, J. S. & GOUGH, D. I. 1979. Northeast-southwest compressive stress in Alberta: evidence from oil wells. Earth and Planetary Science Letters, **45**, 475–482.

BIOT, M. A. 1941. General theory of three-dimensional consolidation. Journal of Applied Physics, **12**, 155–164.

CASTILLO, D. A. & ZOBACK, M.D. 1995. Systematic stress variations in the southern San Joaquin Valley and along the White Wolf fault: Implications for the rupture mechanics of the 1952 Ms 7.8 Kern County earthquake and contemporary seismicity. Journal of Geophysical Research, **100**, 6249–6264.

CASTILLO, D. A. *et al.* 1997. Deep hydraulic fracture imaging: recent advances in tiltmeter technologies. *International Journal of Rock Mechanics & Mining Services*, **34**.

CHEVRON PRODUCTION COMPANY 1994. *Advanced reservoir characterization in the Antelope Shale to establish viability of CO2 enhanced oil recovery in California's Monterey siliceous shales*, US Department of Energy Technical Proposal. Vol. IIA.

CRAMPIN, S. 1993. A review of the effects of crack geometry on wave propagation through aligned cracks. *Canadian Journal of Exploration Geophysics*, **29**, 3–17.

DHOLAKIA, S. K., AYDIN, A., POLLARD, D. D. & ZOBACK, M. D. 1998. Fault-controlled hydrocarbon pathways in the Monterey Formation, California. *AAPG Bulletin*, **82**, 1551–1574.

DU ROUCHET, J. 1981,. Stress fields, a key to oil migration. *AAPG Bulletin*, **65**, 74–85.

EKSTROM, M.P. *et al.* 1987. Formation Imaging with microelectrical scanning arrays. *The Log Analyst*, 294–306.

ENGELDER, T. & FISCHER, M. P. 1994. Influence of poroelastic behavior on the magnitude of minimum horizontal stress, S_h, in overpressured parts of sedimentary basins. *Geology*, **22**, 949–952.

FINKBEINER, T., BARTON, C. A. & ZOBACK, M. D. 1997, Relationship among in-situ stress, fractures and faults, and fluid flow: Monterey Formation, Santa Maria Basin, California, *AAPG Bulletin*, **12**, 1975–1999.

GRAHAM, S. A. & WILLIAMS, L. A. 1983. The Monterey Formation tapped for big pays. *AAPG Explorer*, **5**, 1–5.

HAIMSON, B. AND FAIRHURST, C. 1970. In situ stress determination at great depth by means of hydraulic fracturing. *In: Rock Mechanics—Theory and Practice*. Society of Mining Engineers, 559–584.

HUBBERT, M. K. & WILLIS, D. G. 1957. Mechanics of hydraulic fracturing, *AIME Petroleum Transactions*, **210**, 153–166.

ISAACS, C. M. 1981. Porosity reduction during diagenesis of the Monterey Formation, Santa Barbara coastal area, California. *In* Isaacs, C. M. & Garrison, R. E. (eds) *Petroleum Generation and Occurrence in the Miocene Monterey Formation, California*. Pacific Section SEPM, 257–271.

JAEGER, J. C. & COOK, N. G. W. 1979. *Fundamentals of Rock Mechanics* (2nd edn) Methuen and Co., London.

KEHLE, R. O. 1964. The determination of tectonic stresses through analysis of hydraulic well fracturing. *Journal of Geophysical Research*, **69**, 259–273.

KRIEF, M., GARAT, J., STELLINGWERFF, J. & VENTRE, J. 1990. A petrophysical interpretation using the velocities of P and S waves (full-waveform sonic). *The Log Analyst*, 355–369.

MARDIA, R. V. 1972. Academic Press, *Statistics of Directional Data*. London.

MONTGOMERY, S. L. & MOREA, M. F. 2001. Antelope shale (Moneterey Formation), Buena Vista Hills field: Advanced reservoir characterization to evaluate CO2 injection for enhanced oil recovery, *AAPG Bulletin*, **85**, 561–585.

MOOS, D. & ZOBACK, M. D. 1990. Utilization of wellbore failure to constrain the orientation and magnitude of crustal stresses: Application to Continental, Deep Sea Drilling Project, and Ocean Drilling Program boreholes. *Journal of Geophysical Research*, **95**, 9305–9325.

MOUNT, V. S. & SUPPE, J. 1987. State of stress near the San Andreas fault: implications for wrench tectonics. *Geology*, **15**, 1143–1146.

MOUNT, V. S. & SUPPE, J. 1992. Present-day stress orientations adjacent to active strike-slip faults: California and Sumatra, *Journal of Geophysical Research*, **97**, 11,955–12,013.

NUR, A. & WALDER, J. 1990. Time-dependent hydraulics of the earth's crust. *In*: *The Role of Fluids in Crustal Processes*. Geophysics Study Committee, National Academy of Sciences.

PLUMB, R. A. & HICKMAN, S. H. 1985. Stress-induced borehole elongation: A comparison between the four-arm dipmeter and the borehole televiewer in the Auburn geothermal well. *Journal of Geophysical Research*, **90**, 5513–5521.

SECOR, D. T. 1965. Role of fluid pressure in jointing. *American Journal Science*, **263**, 633–646.

SEGALL, P. 1992. Induced stresses due to fluid extraction form axisymmetric reservoirs. *Pure and Applied Geophysics* **139**, 535–560.

SEGALL, P., GRASSO, J. R. & MOSSOP, A. 1994. Poroelastic stressing and induced seismicity near the Lacq gas field, southwestern France. *Journal of Geophysical Research*, **99**, 15,423–15,438.

WENTWORTH, C. M. & ZOBACK, M. D. 1989. The style of Late Cenozoic deformation at the eastern front of the California Coast Ranges. *Tectonics*, **8**, 237–246.

WINTERSTEIN, D. F. & MEADOWS, M. A. 1991. Shear-wave polarizations and subsurface stress directions at Lost Hills field. *Geophysics*, **56**, 1331–1348.

WRIGHT, C. 1998, Tiltmeter fracture mapping: from surface and now downhole. *Petroleum Engineer International*, **71**, 50–63.

ZOBACK, M. D. & ZOBACK, M. L. 1991. Tectonic stress field of North America and relative plate motions. *In* SIEMMONS, D. B., ENGDAHL, E. R., ZOBACK, M.D. & BLACKWELL, D. D. (eds), *Neotectonics of North America*. Geological Society of America, Boulder, Colorado.

ZOBACK, M. D., MOOS, D., MASTIN, L. & ANDERSON, R. N. 1985. Wellbore breakouts and *in situ* stress. *Journal of Geophysical Research* **90**, 5523–5530.

ZOBACK, M. D. *et al.* 1987. New evidence on the state of stress of the San Andreas fault system. *Science*, **238**, 1105–1111.

Data analysis, processing and 3D fracture network simulation at wellbore scale for fractured reservoir description

E. ALIVERTI, M. BIRON, A. FRANCESCONI, D. MATTIELLO, S. NARDON & C. PEDUZZI

ENI-Agip E&P Division, via Emilia 1, 20097 San Donato Milanese (MI), Italy (e-mail: sergio.nardon@agip.it)

Abstract: Three-dimensional fracture networks, combined from **seismic scale** to **wellbore scale**, can greatly enhance the knowledge of fracture contribution to hydrocarbon storage and flow inside the reservoir. This paper presents some techniques used at ENI-Agip for fracture network simulation at wellbore scale. First, new imaging techniques adopted for enhancing and facilitating the data acquisition from orientated cores will be shown. We will discuss how the fracture data, acquired from the combined analysis of orientated core and wellbore image logs, have to be processed in order to calculate the geometrical parameters of each feature (dip direction, dip, size, terminations) and to classify them according to their filling (oil, water, shale, calcite, etc.). Second, we will illustrate how these kind of data are processed in order to extract the fracture representative parameters needed for stochastic simulation of fracture network at the wellbore scale: spatial distribution along the cored/logged interval, number of fracture sets, representative orientation and statistical distribution of each set, distribution laws of the fracture length and relevant minimum radii and fracture aperture estimate. Fracture porosity evaluation, matrix block size, fracture network connectivity at wellbore scale constitute the outputs of such simulations: they are used to better characterize a fractured reservoir and to describe its behaviour. The synthetic results of the application of such a methodology on a real case (tight packstone/wackestone of carbonate platform from a southern Apennine italian oil reservoir) complete the paper.

The main target of the reservoir geologist is the petrophysical characterization of the reservoir in order to estimate:

- the original hydrocarbons in place;
- the fluid flow path; and thus
- the reserves.

The building of a geological model, which reasonably describes the lateral and vertical variations of the petrophysical characteristics, i.e. porosity, permeability and hydrocarbon saturation, stands at the root of these estimates. The hydrocarbons in place are, in fact, estimated on the basis of this geological model, and very often a three-dimensional (3D) numerical model, which schematically represents the geological reality, is built in order to estimate the reserves.

The petrophysical characterization of fractured reservoirs is more complicated than conventional non-fractured reservoirs because of the presence of two different pore systems: the matrix pore system and the fault-fracture network pore system

The recognition of the presence of a fracture network is, therefore, crucial because it has a significant effect on reservoir fluid flow. Considering values of fracture porosity that become progressively more significant when compared to the matrix porosity, there are four ways in which fractures can affect a reservoir. Schematically, four types of reservoir can be recognized according to the different contribution of fractures:

type 1: fractures impart no positive reservoir quality but create reservoir anisotropy and heterogeneity;
type 2: fracture porosity provides a permeability assist to an already producible reservoir;
type 3: fracture porosity provides the essential permeability;
type 4: fracture porosity represents the entire storage porosity and permeability of the reservoir.

For these different types of reservoirs different parameters are critical in order to evaluate both the amount of hydrocarbons in place and the potential production problems. This paper addresses type 4 fractured carbonate reservoirs where fractures are considered to represent the

From: AMEEN, M. (ed.) *Fracture and In-Situ Stress Characterization of Hydrocarbon Reservoirs*. Geological Society, London, Special Publications, **209**, 27–37. 0305-8719/03/$15.00
© The Geological Society of London 2003.

entire storage porosity and permeability of the field. In this kind of reservoir a precise description of the structural geometry of the fault and fracture network is therefore fundamental. The paper describes how fractures and microfractures recognized on logs and cores, have been modelled to define the fluid flow path and estimate the fracture pore volume. ENI has set a complete workflow for an integrated analysis of both core and log image to optimize the overall management, integration and interpretation of core and log data in fractured reservoirs in order to enhance reservoir characterization and fracture system modelling.

This paper has a methodological approach, even if the results of a case study are presented at the end of the work. A complete presentation of a real case application and relevant calibration with dynamic data can be found in Verga *et al.* (2001).

Problems for quantifying the fracture porosity

The reconstruction of the fracture network is strongly hampered by the limited points of view from which we can observe the network in wells, because the limited window of observation hardly encompasses the representative elementary volume.

Another problem springs from the difficulty of directly measuring the fracture porosity. The estimate of this parameter becomes really crucial in the case of reservoirs where fractures and microfractures represent the entire storage porosity and permeability. In this case the total porosity of the system, i.e. matrix+microfractures, can be very small and very often below the porosity log resolution, ranging around 1 or 2%. In this case matrix porosity can sometimes be neglected, as it is generally 100% water saturated or tight, whereas fracture porosity is fundamental. In some cases these samples maintain their consistence and shape, though microfractured, because the microfractures are relatively small and partly cemented. In these cases we can try to use the capillary pressure curves of these full-size samples in order to estimate the pore volume characterized by very low capillary forces, which can reasonably be attributed to the fracture network if the presence of vugs can be ruled out. On selected full-size samples, the fracture porosity is measured by the proprietary MAGRITTE software, which is built upon an automatic sequence of image processing steps for the magnetic resonance images (MRI).

MRI is a tomographic technique which produces 2D images of the 3D rock fluids distribution. Quantitative MRI produces both magnetization images and T2 relaxation time images. Since the MRI magnetization signal is proportional to the number of 1H nuclei present in the analysed volume, the magnetization image values can be considered as direct measurements of the quantity of fluid in the region imaged by each pixel, i.e. porosity, because the solid matrix does not interfere with measurements of fluids. The T2 images give indirect clues about the geometry of the pore system, since T2 values can be related to the size and geometry of the pore system imaged by each pixel. Moreover, other parameters computed by the MRI technique, such as the diffusion coefficient and chemical composition, can be related to the rock properties, giving further information about the pore system.

To properly measure porosity values, a known standard porosity reference is coupled to the sample (not shown).

Fracture contribution to total porosity can be computed by isolating features related to fracturing in the MRI images. The fracture porosity can then be computed by applying the following formula:

$$\Phi fr = \Phi tot * \frac{\bullet_{i,j \in fractures} I(i,j)}{\bullet_{k,h} I(k,h)} \quad (1)$$

where $I(i,j)$ is the magnetization value of the (i,j) pixel, and the sum on the numerator is computed on the pixels belonging to fractures, while the sum on the denominator is over all the image pixels. To find which pixels belong to fractures, we processed the MRI images to enhance the image features that can be related to fracturing, which mainly are intensity value and T2 relaxation time, then picked these features and used them in the numerator of Equation 1. The processing method is based on the assumption that there exists an underlying statistical model that describes the spatial and physical interaction between adjacent pixels. So fractures can be signalled by the combination of a high magnetization value and high T2 relaxation time for the same (I,j) index on the two images, and adjacent pixels belong to the same fracture if spatial continuity of the features is statistically significant. These, and other more sophisticated processing methods are currently implemented by MAGRITTE, a proprietary ENI software (Mattiello *et al.* 1977). Figure 1 shows an example of the magnetization image of a 2 mm slice of a fractured sample, with the corresponding fracture porosity image computed by MAGRITTE.

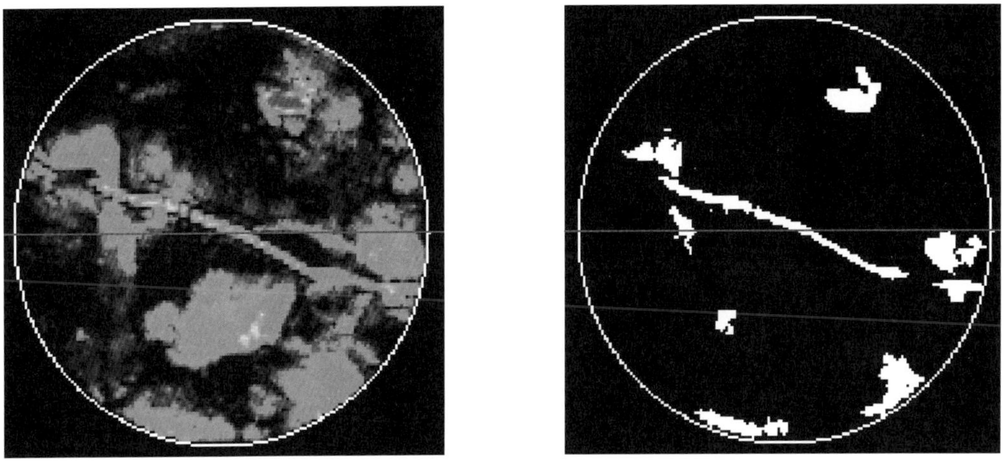

Fig. 1. The magnetic resonance image of a 2 mm slice of a vuggy, fractured reservoir sample. Colormap indicates: red to yellow→high magnitization value→higher porosity; green to blue→low magnetization values→lower porosity. The white zones are automatically found on the previous image by a correlation between continuty of magnetization values and the corresponding T2 relaxation time values (image not shown). This intermediate result image is the region used to compute fracture porosity.

Therefore measured total porosity, matrix+fracture, and some local values of the volume percentage related to fractures are the only information directly available.

Stochastic modelling at small scale

Because of this lack of reliable direct data an important contribution to porosity and permeability evaluation can come from the stochastic modelling of the fracture network at well scale. This modelling is based on the structural data that are collected on cores and logs, on which the presence of different types of fractures (hydrocarbon filled, water filled or cemented) is defined in terms of trace length and position. The general approach consists of building at this scale a stochastic model of the fracture network that is statistically consistent with the fracture data recognized on cores. The fracture porosity estimate is thus based on the calculation of the stochastic model fracture network surface and on the average fracture aperture, estimated on cemented fractures observed on thin sections.

Obviously, cemented fractures are not part of the porosity network; however they could represent the starting point for quantitative sensitivity on the aperture of the producing fracture network. This issue is very critical, the aperture being controlled by many factors and in particular by the *in situ* stress orientation and intensity and by the reservoir fluid pressure. As 'true' data on fracture aperture are not available and collectable, a trial-and-error approach seems to be the most reasonable way to investigate the phenomenon.

The estimated microfracture porosity has to be reasonably validated by the comparison of this value with the total porosity measured in the laboratory and with the available local fracture porosity values. The matrix block size of these models can then be inferred on the basis of the geometrical characteristics of this stochastic model.

Input data for the stochastic model are the statistical descriptions of the fracture sets observed on cores and logs. Each set is characterized by its orientation, the joint Fisher distribution of dip and azimuth (dip direction), by its characteristic dimension, expressed as the distribution of equivalent fracture radius, and by the fracture set intensity distribution along the sampled interval.

Data acquisition

For modelling purposes, each fracture trace observed on the core surfaces or on image logs is modelled by a lens, whose diameter is inferred from the trace length and whose width is related to the average cemented fracture aperture observed on thin sections. The procedure to calculate the equivalent minimum radius of each fracture is extremely simple as each trace length is related to the smallest circle whose diameter and orientation fit the trace length when inter-

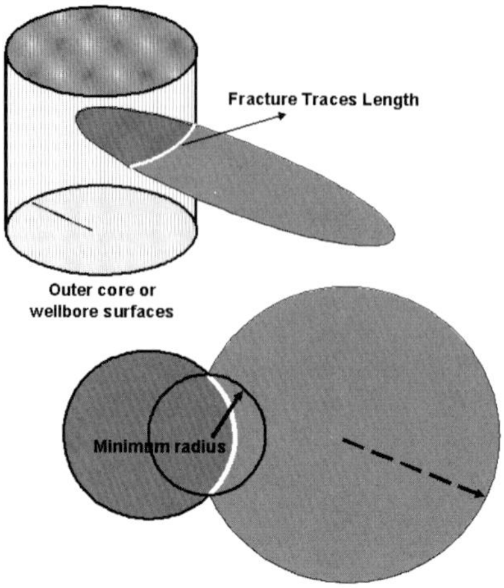

Fig. 2. Concept and calculation of the equivalent minimum radius.

secting a solid the external surface of which has the same shape and size of the core surface and/or the image log envelope (Fig. 2).

Fracture traces are identified and characterized on the wellbore image logs using the Z&S SW tools, which allow each trace on the image to be directly picked and labelled. For each picked trace, the tool calculates its geometric description, then outputs the calculated fracture intensity log. The equivalent minimum radii and fracture density along the sample are computed by a proprietary module embedded in the Z&S environment.

For cores, a proprietary hardware tool named AUTOCAR acquires a panoramic photographic image of the outer cylindrical surface, which can be processed with the same Z&S SW tool used for log image data. Figure 3 shows a panoramic core photograph in the Z&S environment.

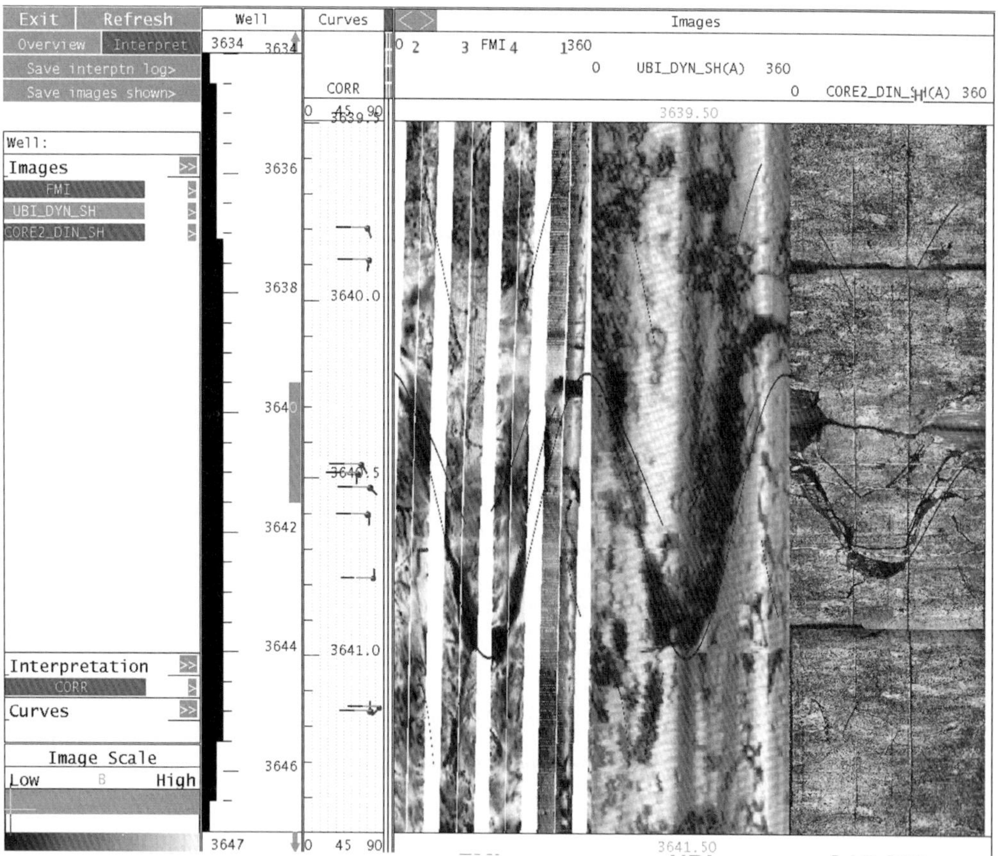

Fig. 3. Fracture characterization and azimuth computation from the contemporaneous display of the outer core surface image (right), UBI wellbore image (middle) and FMI image (left).

When cores are broken, or cannot be photographed, an alternative procedure based on the manual transcription of the traces on a transparency film can be applied. Traces are classified using colours as labels. The transparency is then scanned, and the resulting image is processed by the proprietary FREEFRACT software that automatically reconstructs the geometry of each trace, labelled by colour. This is accomplished by the automatic picking of the manual fracture transcription on the scanned transparency film, and by the automatic individuation and classification of different fracture termination types. This step avoids picking the traces again with the Z&S SW tool. The same geometrical descriptive parameters calculated with Z&S can be obtained with the proprietary tool, and the results can be merged.

The FREEFRACT software also allows automatic counting and classification of fracture intersections. Figure 4 shows fracture reconstruction, classification and intersection interpretation in FREEFRACT, while the conversion from image to geometries, both 2D and 3D, is shown in Figure 5.

The relevant parameters can be summarized as follows:

- fracture geometric description of each recognized feature (position, orientation, dimension measured as trace length and as equivalent minimum radius);
- characteristics of each recognized fracture (open, not fluorescent under UV light, and interpreted as water filled; open, fluorescent, interpreted as hydrocarbon filled; cemented);
- calculated true fracture intensity log.

Information concerning fracture width of cemented fractures can be collected by the visual analysis of thin sections performed with the microscope, giving some reasonable 'clues' about fracture aperture at reservoir conditions.

Data analysis

The subsequent data analysis can be subdivided into three steps.

The first step consists in the post-processing of geometrical fracture data as available from the core and log analysis. Post-processing consists in the statistical analysis of orientation, in order to define fracture sets, and dimensions (trace lengths and estimated radii) to provide a model for the distribution of fracture sizes.

The second step is the evaluation of fracture orientation along the whole interval in order to verify the homogeneity of fracture framework with depth. If fracture orientation varies along the wellbore, the interval must be subdivided into smaller homogeneous intervals.

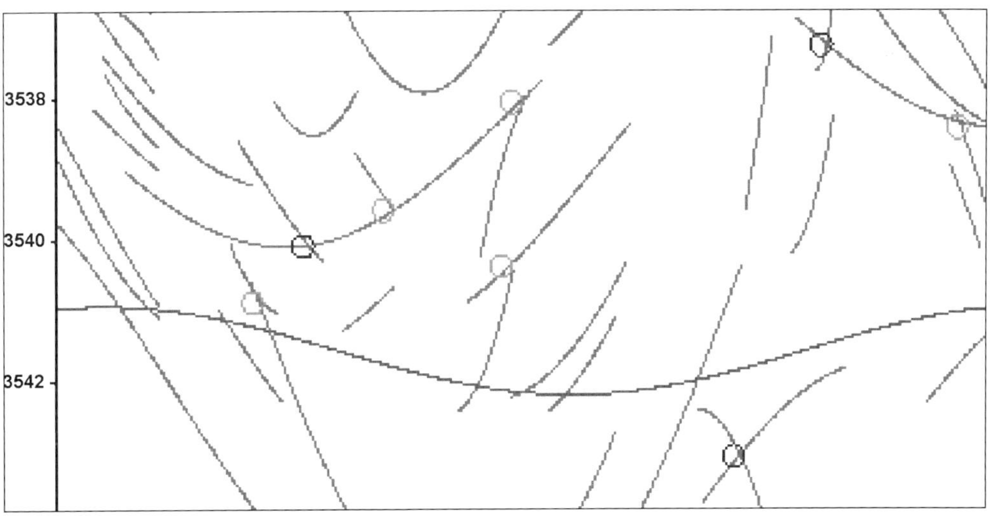

Fig. 4. Reconstruction of fracture traces and their intersections on the outer core surface and/or wellore image logs. Dark circles indicate reconstructed X-crossings, while the lighter circles are used for T-crossings. In such a case the intersection classification is performed only for the open fractures (magenta), while the lines identify bedding surfaces.

Fig. 5. Reconstruction of the fracture geometries (middle) from their traces on the outer core surface (left), using FREEFRACT software. The reconstructed geometries are plotted on a cylindrical surface, simulating a core sample, on the right.

The third step consists of a comparison of the geometric characteristics of the total fracture population with those of the open fracture population to check where open fractures can be trated as a representative subset of the overall fracture population. The similarity criterion is the first condition to be fulfilled to assume that the estimated cemented fracture width is representative of fracture aperture at reservoir conditions.

Fracture set definition

In order to simulate the fracture network in a stochastic way, the total fracture population has to be subdivided into different sets, which differ in mean orientation (dip and azimuth) and can be related to different deformation events. The geologist performs this operation on the open fracture population by plotting the fracture orientation parameters on a stereoplot and then selecting the more dense areas. The chronology of the deformation events can be inferred from the percentage of T-terminations of each selected set: fracture sets formed during earlier deformation phases will have a higher ratio of T to X crossings than those formed in later events. Therefore, each 'stratigraphical' horizon can be characterized by a number of different fracture sets, each of which is defined by a specific distribution of dip and azimuth and by the probability distribution of length and diameter. The characteristic parameters (i.e. average and standard deviation) of these distributions are computed for each fracture sets. This is illustrated on Figure 6.

Stochastic fracture network modelling

The fracture network of each stratigraphical horizon is modelled using a stochastic approach, whose input is the different fracture sets previously recognized. This process guarantees that the stochastic model of the fracture network is statistically consistent with the evidence observations on cores and logs. The stochastic modelling is considered the most appropriate approach to describe a complex phenomenon, such as a fracture network, that can be only partially directly observed. In fact, it should be noted that the fracture size data from core and wellbore image log will be biased, since the smallest and largest fractures will either not crop out at the outer core surface or will be underestimated, due to the limited sampling window. Tests on some studied core materials show that a deterministic modelling of the fracture network in this case could lead to a significant underestimation of fracture surface (about 30% less).

The model is built using a regular spatial distribution (stationary Poisson point process) for defining the discrete fracture network. ENI-Agip uses the proprietary Sibylfrac software package (Golder Associates 1997), assuming a model dimension of 0.14 m (such a size determines a modelization volume very similar to the dimension of both core samples and image logs). Some preliminary stochastic models must be run in order to match the fracture intensity measured on the core with the simulated fracture intensity. The same preliminary stochastic models are also used to validate the distribution of estimated fracture lengths. Validation is performed by comparing the statistics of the measured trace with the statistics of the simulated trace length calculated on a cylindrical sampling surface. Once the model of each homogeneous horizon has been validated

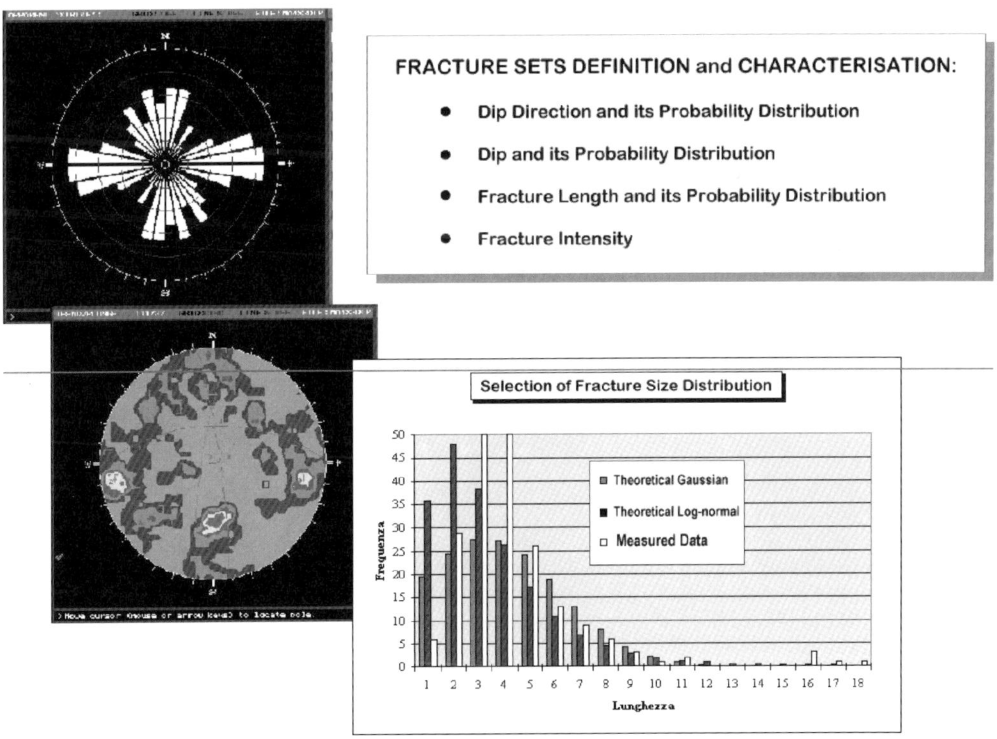

Fig. 6. Rosette diagrams and stereoplot contours of fracture poles are used for the statistical definition of the fracture orientation in order to identify the main fracture sets (left). On the right is an example of the histograms used for the statistical definition of fracture size distribuion.

and calibrated, about 500 multiple stochastic realizations are carried out. Figure 7 schematically shows the process of stochastic modelling.

Stochastic fracture network sampling

Each stochastic realization of the fracture network is then sampled to calculate the fracture density values, namely the P21 and total P32 density. These are defined as follows: P21 is the sum of trace lengths on a named surface divided by the area of the same surface (unit m^{-1}); P32 is the sum of the areas of the fractures contained in a named volume divided by the same volume (unit is m^{-1}).

The fracture spacing is sampled in three different directions, x-y-z, to calculate P10 and thus spacing. P10 is defined as follows: P10 is the number of fractures on a surface divided by the height of the surface (unit is m^{-1}). Spacing is calculated as 1/P10 unit m.

Plotting the P21 against the P32 values, a good correlation can usually be found using a regular spatial distribution. This correlation is specific to each homogeneous horizon and shows a direct linear trend with an increasing value of P32. Figure 8 shows the correlation between P21 and P32 for a fractured sample.

Another correlation, specific to each homogeneous horizon, can usually be defined among fracture density P21 and fracture spacing in the three different directions, x-y-z. This latter correlation generally shows an exponential increase in the average fracture spacing with reduction in fracture intensity. This correlation can, however, be different in the three different directions x-y-z, suggesting an irregular, elongated, shape of the 'surrounded blocks'.

Calibration of the fracture network modelling

Considering the relative reliability of some input data like the fracture aperture, all interpretations have to be calibrated with the available measurements performed on some selected full-size

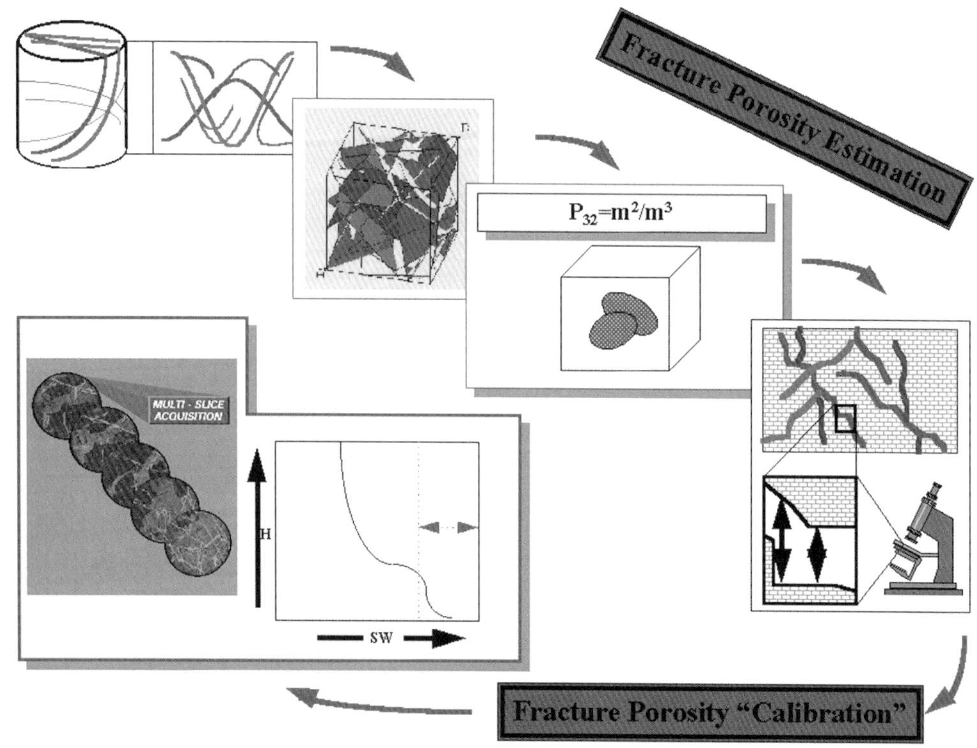

Fig. 7. The various steps performed during the stochastic modelling approach for fracture porosity estimation.

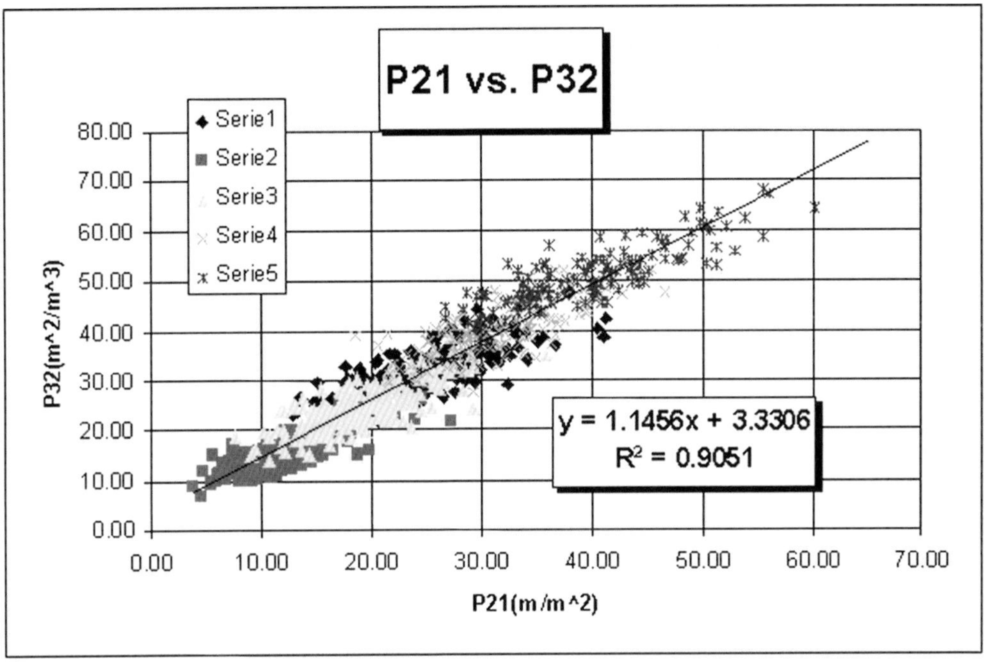

Fig. 8. Comparison between the calculated fracture intensit P21 and P23: a direct, strong correlation between the two density values is expected using the regular spatial distribution hypothesis.

samples. This kind of calibration has, however, some limitations, considering that the laboratory measurements are very often performed at room conditions, as in the cases analysed, while the estimated fracture aperture on the cemented fractures should represent the fracture width at reservoir conditions.

Keeping this problem in mind, the fracture porosity profile can be compared to the total fracture porosity measured on some selected samples and also with the pore volume affected by small capillary forces. The fracture connectivity/permeability can, on the other hand, be compared to the measured permeability of the available samples.

Results of well-scale modelling

Various applications of the previously described methodology have been applied to real cases. Here the results of this modelling used for the characterization of tight carbonates (Cretaceous packstone and wackestone with matrix porosity less than 2 m^3/m^3) of an Italian reservoir located in the southern Apennines (Van Dijk *et al.* 2000) are summarized. The above-mentioned elaborations have been performed as a starting point on 30 full-size samples and on the neighbouring cored intervals for which the P21 density data were available. In this way the information obtained corresponds to a wellbore interval 50 m long, calibrated with 28 m of cores.

Fracture porosity

Fracture porosity can be obtained using a constant fracture width, which is the average or median value of the population, observed on thin sections. Of course, the porosity value is generally very sensitive to the fracture aperture, which has to be carefully selected and calibrated against any available fracture porosity value.

In the above-mentioned example, the obtained fracture porosity ranges between 0 and 0.7%, with an average value of 0.14%. This value was obtained using a constant fracture width of 40 μm. Comparing the porosity results to the total fracture porosity, the pore volume is characterized by small capillary forces which are not always recognized. However, this average value of 40 μm for the fracture aperture might not always be suitable. Therefore, the porosity value can generally be considered useful although a more accurate analysis of the different cored wells which could lead to more differentiated values of fracture aperture ranging between 20 and 80 μm might alter this average estimate.

Fracture connectivity

The connectivity of the fracture network can be analysed in the three different directions, x-y-z, in relation to the average fracture size. When the values differ considerably in the three directions, a high anisotropy of the system can be inferred.

In the examined case study, the connectivity seems to be generally poor and sometimes effective only in the x and y direction and therefore suggests a high anisotropy of the system.

Matrix blocks size

As mentioned above, when the fracture intensity is locally high it suggests the development, at well scale, of a continuous network of isolated matrix blocks. The size and the shape of these blocks can be evaluated using the spacing along the x-y-z directions.

As mentioned above, the fracture intensity seems locally high enough to develop, at this scale, a continuous network of isolated matrix blocks. The size of these blocks, when defined, is generally small, ranging around 2–3 cm and sometimes elongated in the z direction.

Fracture permeability

The fracture permeability measured at room conditions on the selected samples can be reasonable matched with the simulated network using values of the throat apertures ranging between 6 and 31 μm. These values seem to be quite consistent with the fracture aperture estimated on the cemented events that show a mode of about 10 μm, and seems to support the reliability of these data. However, if one could measure the fracture permeability at reservoir conditions, the real throat aperture affecting the flow behaviour in the reservoir could be estimated and a better estimate made of the cemented fracture widths.

Conclusions

The present study shows that on the well scale, the combination of discrete stochastic and deterministic modelling of the fracture network can lead to a more reliable estimate of various basic parameters necessary for static and dynamic numerical modelling of a fractured reservoir.

Stochastic modelling of the well-scale fracture network provides reliable estimates of fundamental parameters such as the fracture porosity, connectivity and permeability and of the matrix block size and shape. Heterogeneity at this scale can be inferred from the variation of different parameters in various layers constituting the reservoir. The results of such models can be compared with production data, as successfully done by Verga *et al.* (2001).

We thank ENI-AGIP Division – SVAG Department for the highly appreciated continuing support to the project. The results of the present study were issued in a number of specific internal ENI reports related to the different items here discussed.

References

ALIVERTI, E., BERSANI A., CESARO M. & MALETTI, G. 1999. Wellbore and core image data integration using Z&S software. *In: Proceedings of the Recall User Meeting*, London, April 1999.

DERSHOWITZ, W. S. & HERDA, H. H. 1992. Interpretation of fracture spacing and intensity. *In*: Tillerson & Wawersik (eds) *Rock Mechanics*, Balkema, Rotterdam, 757–766.

DYKE, C. G., WU, B. & MILTON-TAYLER, D. 1995. *Advances in Characterising Natural-Fracture Permeability from Mud-Log Data*. Paper SPE 25022, **160**.

GILLESPIE, P. A., HOWARD, J. J. WALSH, J. J. & WATTERSON J. 1993. Measurement and characterisation of spatial distributions of fracture. *Tectonophysics*, **226**, 113–141.

GOLDER ASSOCIATES, 1995. *Fracman User Documentation*. Golder Associates Inc. Seattle, Washington.

GOLDER ASSOCIATES, 1997. *User Documentation SibylFrac v 3.2*. Golder Associates Inc. Seattle, Washington.

HORNBY, B. E., LUTHI, S. M. & PLUMB, R. A. 1990. Comparison of fracture apertures computed from electrical borehole scans and reflected Stoneley waves: an integrated interpretation. *The Log Analyst*, **33**, 55.

KAZEMI, H. 1969. Pressure transient analysis of naturally fractured reservoirs with uniform fracture distribution. *SPE Journal* (Dec.) 463–472.

KULANDER, B. R., DEAN, S. L. & WARD B. J. 1990. *Fractured core analysis. Interpretation, Logging and use of Natural and Induced Fractures in Core*. AAPG Methods in Exploration Series, no. **8**. Tulsa, Oklahoma, USA..

LIETARD, O., UNWIN, T., GUILLOT, D. & HODDER, M. 1996. Fracture width LWD and drilling mud/LCM selection guidelines in naturally fractured reservoirs. *European Petroleum Conference, Milan, Italy, Oct. 22–24*, Paper SPE 38832.

LUTHI, S. M. & SOUHAITÉ, P. 1990. Fracture apertures from electrical borehole scans. *Geophysics*, **55**, 821.

MATTIELLO, D., BALZARINI, M., FERRACCIOLI, L. & BRANCOLINI A. 1997. Calculation of constituent porosity in a dual porosity matrix. MRI and image analysis integration. *In: Proceedings of the 1997 International Symposium of the Society of Core Analysts, Calgary, 7–10 Sept. 1997*.

REISS, L. H. 1974. *The Reservoir Engineering Aspects of Fractured Formations*. Istitut francais du petrole, Editions Technip, Paris.

SANFILIPPO, F., BRIGNOLI M., SANTARELLI, F. J. & BEZZOLA, C. 1997. Characterization of conductive fractures while drilling, *SPE European Formation Damage Conference, The Hague, Netherlands, June 2–3*, Paper SPE 38177.

VAN DIJK, J. P. 1998. *Analysis and Modelling of Fractured Reservoirs*. Society of Petroleum Engineers, SPE 50750.

VAN DIJK, J. P., BELLO, M., TOSCANO, C., BERSANI, A. & NARDON, S. 2000. Tectonic model and three-dimensional fracture network analysis of Monte Alpi (southern Apennines). *Tectonophysics*, **324**, 203–237.

VERGA, F., CARUGO, C., CHELINI, V., MAGLIONE, R. & DE BACCO, G. 2000. Detection and characterization of fractures in naturally fractured reservoirs. *Proceedings 2000 SPE Annual Technical Conference and Exhibition, Dallas (TX), Oct.* Paper SPE 63266.

VERGA, F. M., GIGLIO, G., MASSERANO, F. & RUVO, L. 2001. Calibration of fractured reservoirs with dynamic data. *SPE Reservoir Simulation Symposium, Houston, Texas, 11–14 February 2001*. Paper SPE 66395.

From seismic data to core data: an integrated approach to enhance reservoir characterization

JONNY HESTHAMMER & HAAKON FOSSEN

Department of Earth Science, University of Bergen, Allegt. 41, N-5007 Bergen, Norway (e-mail: jonny.hesthammer@ifjf.uib.no)

Abstract: Integrated structural analyses of seismic and various well data are necessary to optimize hydrocarbon reservoir characterization. However, there are many published examples from the oil and gas industry where single data types are analysed but not integrated. This may lead to erroneous interpretations and drainage strategies. As illustrated by an example from the area around well 34/10–B-12 in the North Sea Gullfaks Field, integrated structural interpretation should typically utilize all available seismic surveys, well log correlation data, dipmeter data and core data. Interpretation of seismic data helps in the understanding of large-scale structural and stratigraphic geometries. Time-lapse (4D) seismic helps to identify changes in reservoir properties caused by injection and production. Well log correlation data are used to document variations in zonation thickness caused by sedimentological or structural changes. Dipmeter data tie observations of bedding orientation from seismic data to subseismic scale. Core data represent the most detailed (millimetre to metre scale) data available and can yield information on rock properties as well as sedimentological and (micro)structural features. Small-scale deformation structures such as deformation bands and fractures can typically be identified and characterized. In addition, it is possible from unorientated cores to find the orientation of bedding and deformation structures. This information is compared to observations from dipmeter data, well log correlation data and seismic data to improve the interpretation.

Well 34/10–B-12 is a hanging wall injector near one of the large-scale faults in the Gullfaks Field. Several 3D seismic surveys are available from the area, as are standard well log data, dipmeter information and cores. Together, the data range from millimetre to kilometre with some overlap between the data types. Through integrated analysis, pitfalls such as interpreting any linear feature on timedip attribute maps as faults has been avoided. Also, a geometric relation between core-scale and seismic-scale faults has been established, and it has been possible to relate small-scale and large-scale structures in a model which is consistent with all the available data.

Deformed reservoirs require integrated use of all available sources of information rather than 'stand-alone' interpretation of the individual data types. However, for various reasons, the simple principle of integration is commonly neglected in the interpretation of oil and gas fields, potentially leading to serious misinterpretations and erroneous drainage strategies (Hesthammer *et al.* 2001). This paper demonstrates how the analysis and integration of seismic data, stratigraphic log correlation data, dipmeter data, production (pressure) data and core data give a comprehensive understanding of the reservoir characteristics in the area around one of the wells drilled in the Gullfaks Field, northern North Sea (Fig. 1).

The Gullfaks Field covers an area of *c.* 55 km^2, and contains more than 200 wells with information from more than 125 km of drilled reservoir rocks. Among the available well data are 34 km of dipmeter data and 6 km of core data. The field is structurally complex, but large amounts of well data have helped to outline the structural geology of the field (Fossen & Hesthammer 1998a). The Gullfaks Field is divided into three structural domains: a western domino-style fault system with easterly dipping faults and westerly dipping bedding; an eastern horst complex of elevated subhorizontal layering and steep faults; and a transitional accommodation zone (graben system) which is in part identified as a modified anticlinal fold structure (Fig. 1b). Total reserves amount to 319×10^6 standard m^3 of oil and some 30×10^9 standard m^3 of gas in the Jurassic Brent Group, the Cook and Statfjord Formation (Fm.) reservoirs (Fig. 2).

Well log correlation has identified 290 faults with displacements ranging from 4 m to more than 500 m. Many of these faults are also identified from dipmeter analyses. If the faults are

From: AMEEN, M. (ed.) *Fracture and In-Situ Stress Characterization of Hydrocarbon Reservoirs*. Geological Society, London, Special Publications, **209**, 39–54. 0305-8719/03/$15.00
© The Geological Society of London 2003.

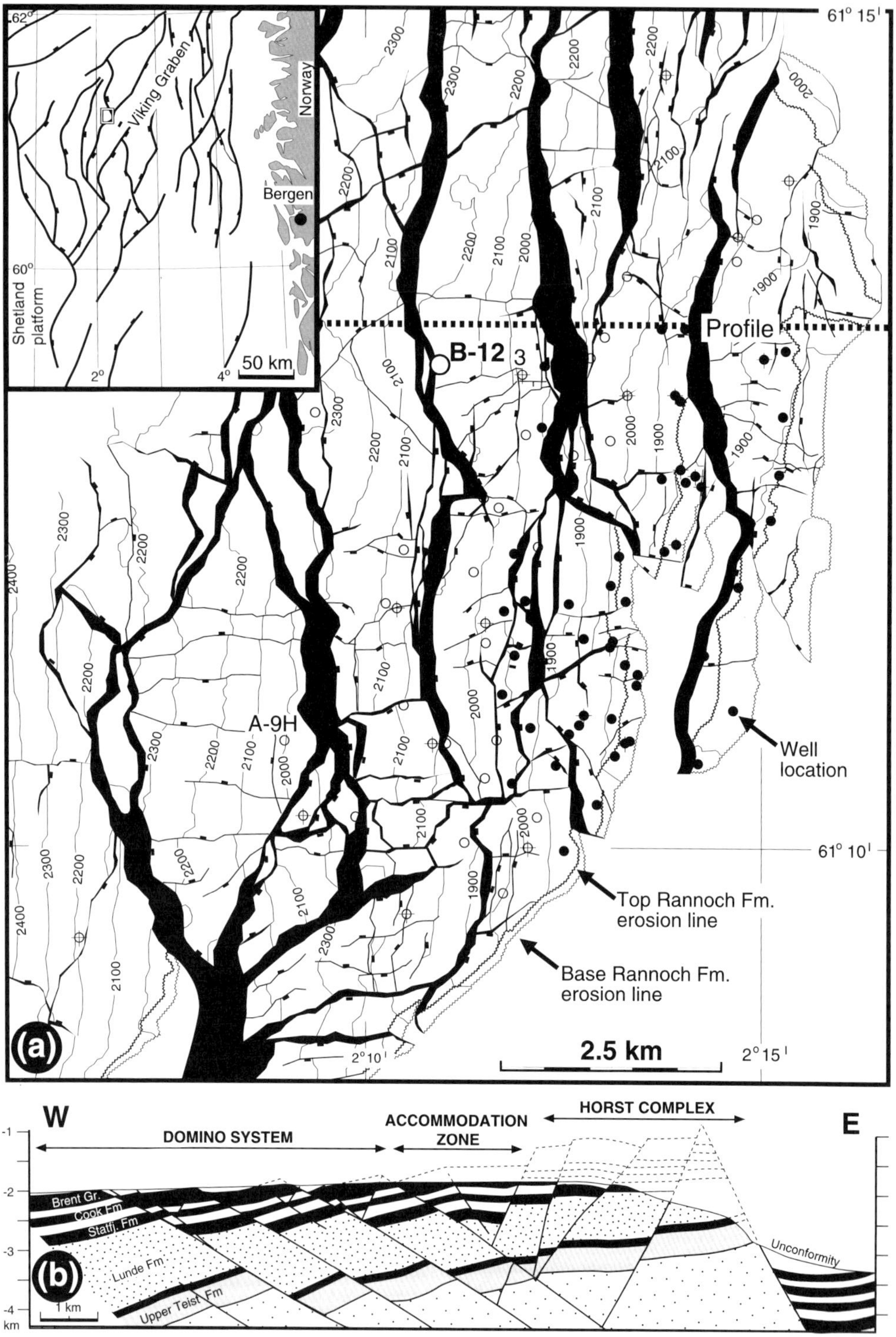

Fig. 1. (**a**) Structure map of the Rannoch Fm. in the Gullfaks Field; see inset map for location. (**b**) East–West profile through the three different structural domains; see (a) for location. Modified after Fossen & Hesthammer (1998a).

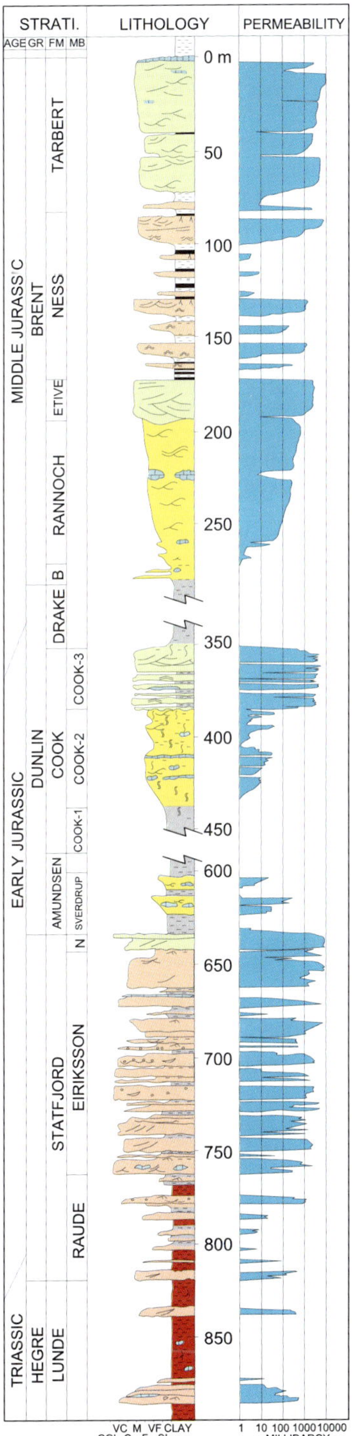

Fig. 2. Stratigraphic column for the Jurassic and Triassic reservoir units within the Gullfaks Field. Modified after Tollefsen *et al.* (1994).

associated with drag of bedding (which applies to more than 60% of all faults; Hesthammer & Fossen 1998), it is often possible to estimate the orientation of the faults. In addition, the width of the drag zones can be quantified. This can then be used for analyses of sealing potential of the faults. Comparison of faults identified from detailed well log correlation with seismic data demonstrates that only 25% of faults with displacement less than 30 m can be identified in the seismic data (Hesthammer & Fossen 2001). For faults with displacements that exceed 30 m, more than 67% of the faults can be observed on the seismic survey. This important observation can help the geologist to avoid overinterpreting seismic data as well as to identify the potential for structural interpretation.

Core analyses demonstrate that faults in the Gullfaks Field are associated with abundant deformation bands. Such zones of deformation bands are easily identified and quantified from fracture frequency diagrams. The deformation bands can reduce the permeability of the sandstones by up to three orders of magnitude (depending on the content of mica; Hesthammer 1999b; Hesthammer & Fossen 2001) and represent important barriers for fluid flow. Within the narrow fault zone, the deformation bands tend to become subparallel to the larger faults. Orientation analysis of deformation bands therefore also yields information on the orientation of the larger-scale faults. It is possible from unorientated cores to find the orientation of faults provided that the orientation of bedding is known.

Well 34/10–B-12

Well 34/10–B-12, which is an injector in the Middle Jurassic Tarbert Fm. (Fig. 2), is used in this example to demonstrate how seismic and well data can be integrated to enhance the structural understanding of an area. The well is located mainly in the hanging wall to two large-scale faults (Fig. 3) but penetrates the faults in the lower part of the well. In addition to seismic and standard well data, dipmeter and core data were collected for sedimentological and structural analyses.

Seismic data

Several seismic surveys have been collected over the Gullfaks Field. The reasons for collecting more than one survey is mainly related to improvements in data acquisition procedure. In addition, studies of time-lapse (4D) seismic lead to better oil recovery by identifying areas where

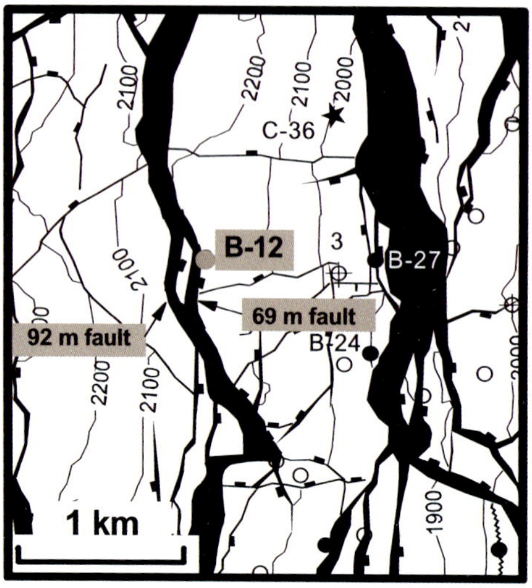

Fig. 3. Structure map of the Rannoch Fm. from the area around well 34/10–B-12. Two faults associated with 69 m and 92 m missing section, respectively, have been identified from well log correlation. Seismic data, dipmeter data and core data show that the faults dip 25–35° towards the ESE.

the oil has not been drained by existing producers. Once available, different seismic surveys may be combined to enhance data quality for structural interpretation (Hesthammer & Løkkebø 1997; Hesthammer 1999a).

As an example, Figure 4 shows two seismic profiles from where well 34/10–B-12 penetrates two main faults. Figure 4a is from the survey collected in 1985 (and reprocessed in 1992) whereas Figure 4b is from the 1996 survey. In general, the main faults and seismic reflections are better imaged in the oldest survey. This somewhat surprising observation is probably related to non-optimal acquisition geometries and processing procedures of the 1996 data.

Although two large-scale faults have been identified from well-log correlation, the faults are indistinguishable on the seismic data. This is due to the closeness of the two faults (located only 15 m apart). Within the hanging wall, bedding dip decreases somewhat towards the faults as a result of large-scale drag which is commonly observed within the domino system (Fossen & Hesthammer 1998*a*). The footwall does not display any gradual change in dip and thus

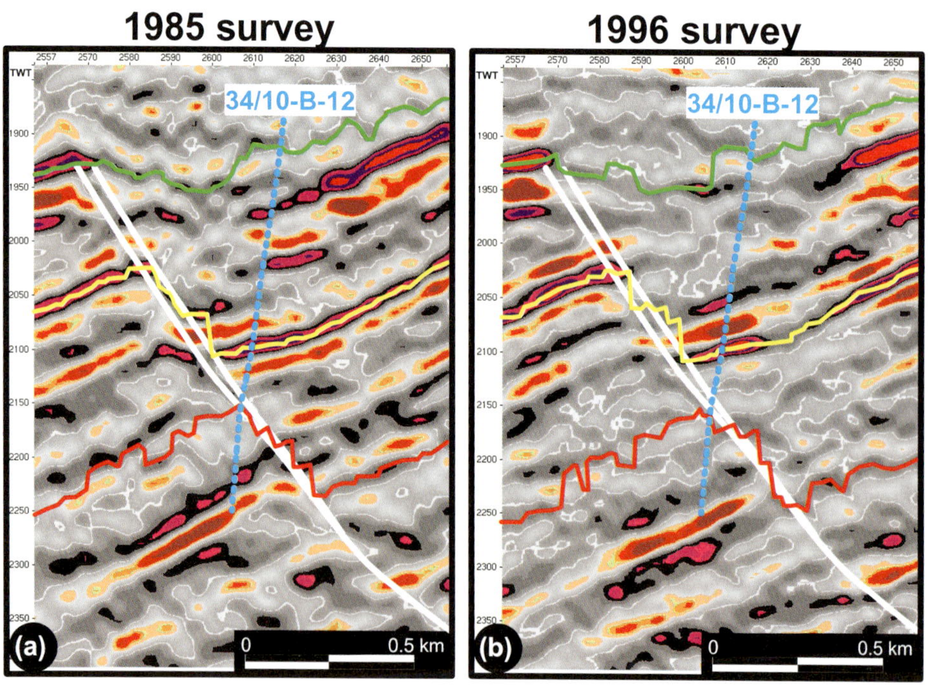

Fig. 4. (**a**) Seismic east–west profile from the area around well 34/10–B-12. The seismic data are from a 1985 survey that was reprocessed in 1992. A large-scale fault offsets the seismic reflections. Detailed well log correlation show that the fault actually consist of two faults associated with 69 m and 92 m missing section. The seismic resolution does not allow separation of the two faults. (**b**) Same as (a) but from a survey acquired in 1996. The fault is less obvious in this data set. In addition, the seismic reflections appear to be less continuous.

appears from seismic data to be unaffected by deformation.

Based on interpretation of the seismic reflections, several seismic attribute maps can be generated. For structural interpretation of the Gullfaks Field, the timedip map (Dalley *et al.* 1989; Hoetz & Watters 1992; Voggenreiter 1991) proved most useful. Figure 5a shows a timedip map of the intra-Ness Fm. (Fig. 2) reflection from the area around well 34/10–B-12. The map was created by snapping the interpretation to the nearest maximum and then calculating the derivative of the snapped horizon. Dark lines represent jumps in the interpretation that may be related to real features such as faults, or may be seismic artifacts caused by noise interference problems (Hesthammer & Fossen 1997*a*, *b*). The prominent dark centrelines are related to offset of the intra-Ness Fm. reflection caused by the main faults. The presence of two such distinct linear features may erroneously be interpreted to represent offset along the two main faults. To improve the visual impression of the attribute map along the fault, the interpreted horizon was interpolated across the faulted area prior to snapping. The linear features arise as the autotracked horizon attempts to follow the trace of the fault plane. Where the fault plane reflection is weak or absent, the autotracked horizon will tend to follow bedding plane reflections or multiples in a stepwise manner (Fig. 5c).

Within the footwall area to the main fault, several (curvi-)linear fault-like features are present. Whether these represent real faults or not cannot be determined from analyses of data from the 34/10–B-12 well alone. However, numerous other detailed studies of well and seismic data from the Gullfaks Field demonstrate that the vast majority of the (curvi-)linear features are caused by the interference of dipping coherent noise with real reflections (Hesthammer & Fossen 1997*a*, *b*, 1998; Hesthammer & Løkkebø 1997; Hesthammer 1998, 1999*a*). This interference causes the real reflections to break up and rotate in the direction of the dipping noise, thus giving the appearance of a faulted layer. The noise interference features are particularly abundant in areas of poor seismic data quality and along weak seismic reflections. The latter is especially obvious in Figure 5b, which is a timedip map of the Cook Fm. reflection from the same area as shown in Figure 5a. The reflection is associated with weaker amplitudes and, as a result, it is generally not possible to extract detailed structural information from the seismic attribute maps. The reason that the main fault can be observed (the through-going black centerlines) on the timedip map from the Cook Fm. reflection is that the snapping routine uses the subjective interpretation, which is in part based on observations from well log correlation, as a guide and is therefore forced to follow the fault plane. In this sense, the timedip map is only semi-objective. Efforts to image the faults by using fully objective mapping techniques, such as the coherence or correlation maps (Bahorich & Farmer 1995; Brenneke 1995; George 1996), fail when the data quality becomes this poor (Hesthammer 1998). By combining different seismic surveys and applying strong frequency and dip filters to the seismic data, much of the noise features can be removed or diminished (Hesthammer & Løkkebø 1997, Hesthammer 1999*a*).

One of the main motivations for collecting new seismic surveys over the Gullfaks structure in 1995 and 1996 was to explore the potential of time-lapse (4D) seismic. Figure 6 shows amplitude maps of the Tarbert Fm. from the 1985 (Fig. 6a) and 1996 (Fig. 6b) seismic surveys. In addition, Figure 6c displays the differences in amplitude values between the two surveys. There is a marked difference in the seismic amplitudes around injector 34/10–B-12 between the two surveys. In particular, the amplitude map from the 1996 seismic survey displays weaker values near the well. This is likely related to replacement of oil by water due to injection and production. This replacement changes the velocity properties of the rock and therefore the seismic reflectivity.

In 1997, a producer (34/10–C-36) was drilled into the footwall of a domino-style fault block that contains the B-12 injector (i.e. updip of well 34/10–B-12). The well experienced early water breakthrough, consistent with observations from the seismic attribute maps which indicate that the waterfront had almost reached 34/10–C-36 prior to drilling of the well. Unfortunately, processing and analysis of the 4D data were not finished when the C-36 well was planned. As a result, the well was planned and drilled without incorporating results from integrated studies of well data from the 34/10–B-12 well and the two seismic surveys. It is quite possible that the entire cost of the new seismic survey could have been covered by simply placing well 34/10–C-36 more optimally (farther north) based on results from 4D studies.

Well log correlation data

Well 34/10–B-12, which deviates 28° from vertical towards 320 (NW), penetrates reservoir rocks of the Tarbert, Ness, Etive, Rannoch and Cook Fms (Fig. 7). Well log correlation from the area identified two faults in the well. A fault with 69

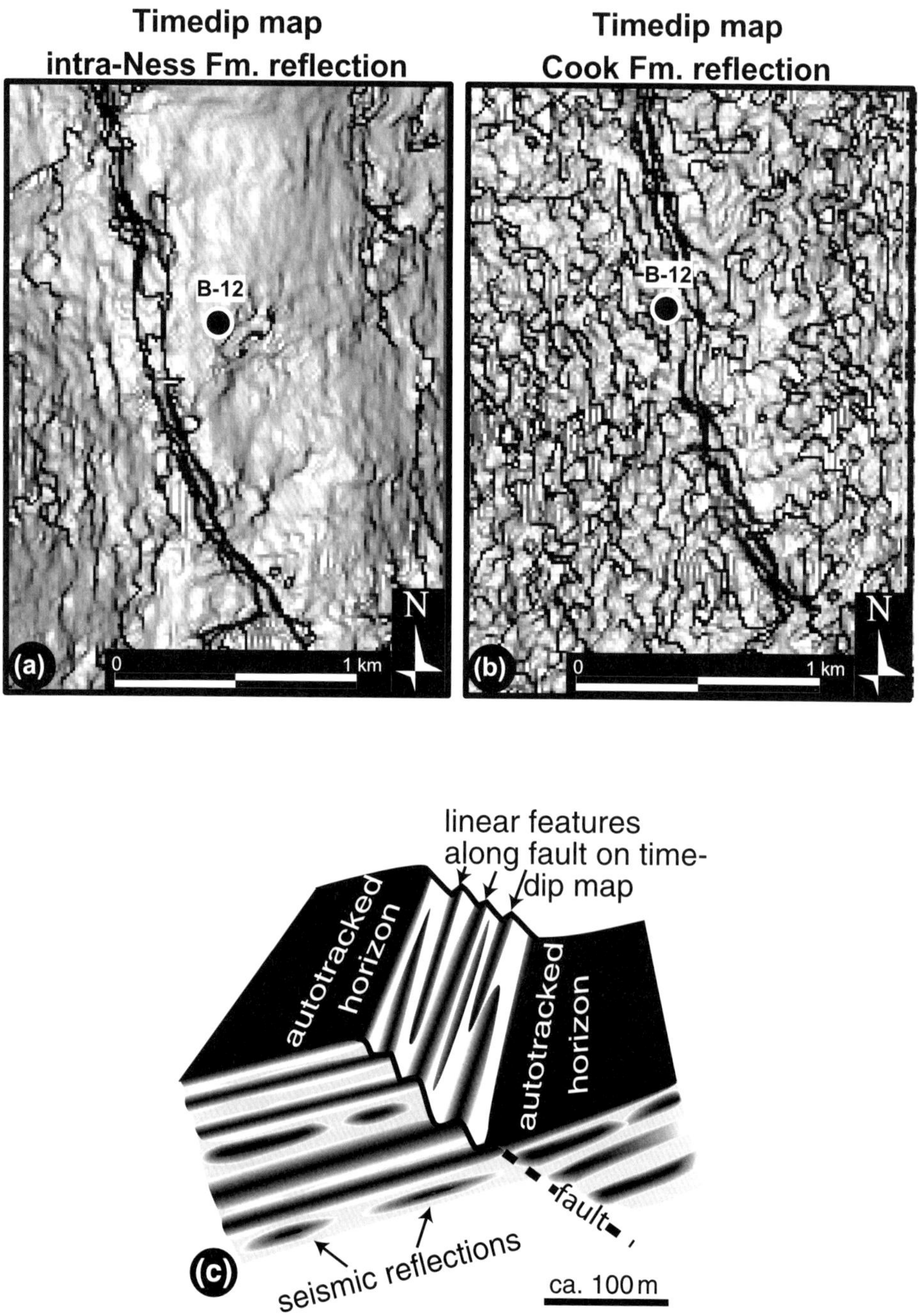

Fig. 5. (**a**) Seismic timedip map of the intra-Ness Fm. reflection in the 1985 seismic survey. The data quality is generally good and the attribute map is little affected by noise. The dark centrelines represent offsets associated with the main faults. (**b**) A timedip map from the Cook Fm. reflection. This seismic reflection is much weaker than the intra-Ness Fm. reflection, and is therefore more affected by (curvi-)linear noise interference features. (**c**) Cartoon illustrating how numerous linear features arise as the autotracked horizon attempts to follow the trace of the fault plane. Where the fault plane reflection is weak or lacking, the autotracked horizon tends to follow bedding plane reflections or multiples in a stepwise manner.

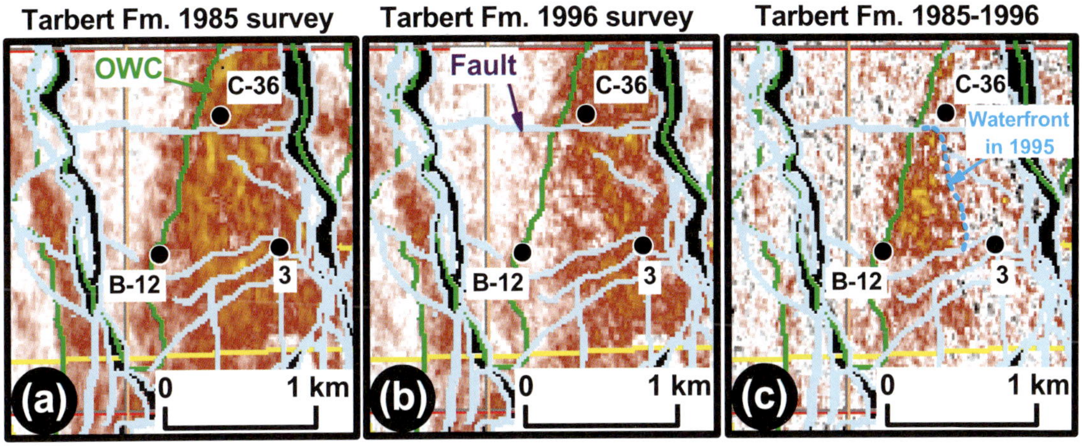

Fig. 6. (**a**) Amplitude map of the Tarbert Fm. from the area around well 34/10–B-12 based on the 1985 seismic survey. (**b**) Amplitude map of the Tarbert Fm. based on the 1996 seismic survey. (**c**) Difference map displaying changes in amplitude values between the two surveys. The changes in reflectivity are likely related to changes in reservoir property caused by injection of water in well 34/10–B-12. The waterfront had almost reached well 34/10–C-36 by the time the C-36 well was drilled and the well experienced water breakthrough shortly after production was initiated. The dark green lines represent the initial oil–water contact.

m missing section is located within the Rannoch Fm. at 2950 m measured depth (MD) (2170 m true vertical depth, TVD) and a fault with 92 m missing section exists in the Drake Fm. at 2983 m MD (2199 m TVD). Due to the large number of wells on the Gullfaks Field, the sedimentologists are generally able to locate minor faults with as little as 4 or 5 m missing section. Furthermore, studies show that more or less all faults on the Gullfaks Field with displacements larger than 10 m are identified from well log correlation (Fossen & Rørnes 1996; Fossen & Hesthammer 2000). The area around well 34/10–B-12 contains several other wells that make detailed control of isopach thickness possible. In particular, correlation against a nearby exploration well (34/10–3) shows that any possible additional faults in well 34/10–B-12 must have displacement less than 3–5 m.

Dipmeter data

Dipmeter data yield information on a scale between seismic data and core data (Rønningsland 1990; Kaya & Norman 1993; Hurley 1994). Well 34/10–B-12 was drilled with oil-based mud. As a result, it was necessary to use an oil-based dipmeter tool to record information on resistivity changes caused by lithological changes (Schlumberger 1986, 1990, 1991; Stuart-Bruges 1984; Dumont *et al.* 1987). Studies from more than 23 km of dipmeter data from the Gullfaks Field demonstrate that, although the data quality from oil-based dipmeter tools is poorer than the quality from water-based tools, it is commonly possible to obtain good control on changes in bedding orientation (Hesthammer & Fossen 1998). Figure 8 shows the results of processing the resistivity curves acquired by the dipmeter tool. In general, the data quality of the resistivity curves indicates good borehole conditions and little tool sticking. However, since the use of oil-based mud causes more scatter of the resulting tadpoles, care must be taken when carrying out structural interpretation.

The general dip of bedding in well 34/10–B-12 changes from 12–14° towards the west in the uppermost parts of the well to subhorizontal and even shallow dips to the east near the faults located at 2950 m MD and 2983 m MD. The gradual decrease in dip of bedding is consistent with observations of large-scale drag towards the east-dipping faults as seen in seismic data (Fig. 4). A sudden change in dip occurs around 2900 m MD. Although this interval contains much scatter, it appears that the dip is much higher than for the intervals above and below. If the change in dip of bedding is caused by drag related to a fault, the fault must dip in a westerly direction since the interval is characterized by steepening drag (Bengtson 1981; Hesthammer & Fossen 1998). The well log correlation, however, shows that no larger faults exist in the interval around 2900 m MD. Any faults must, therefore, have displacement less than 3–5 m. Also, the seismic data show no indications of a westerly

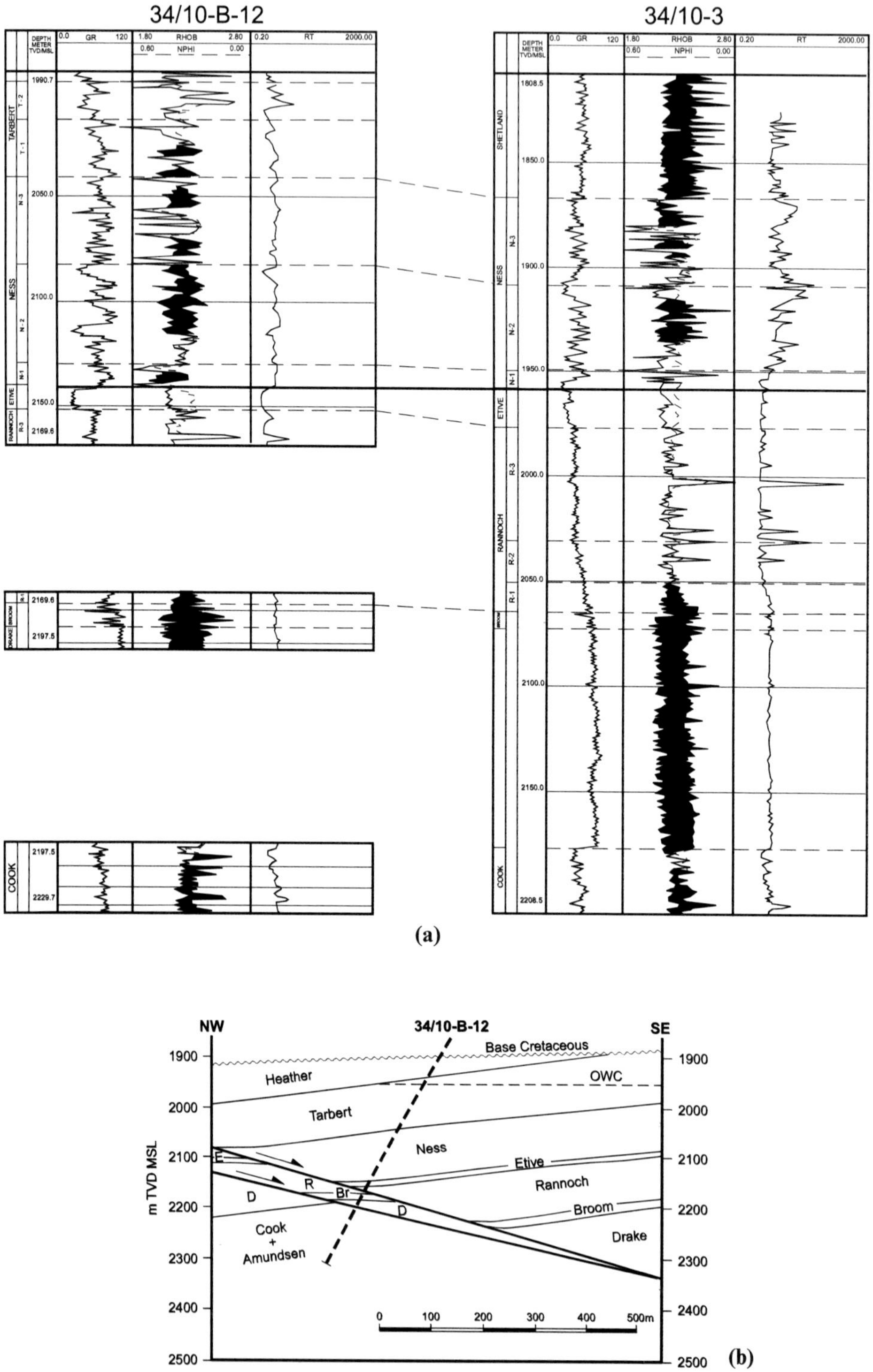

Fig. 7. (**a**) Well log from well 34/10–B-12. See Figure 2 for lithological description. (**b**) Geological profile along well 34/10–B-12.

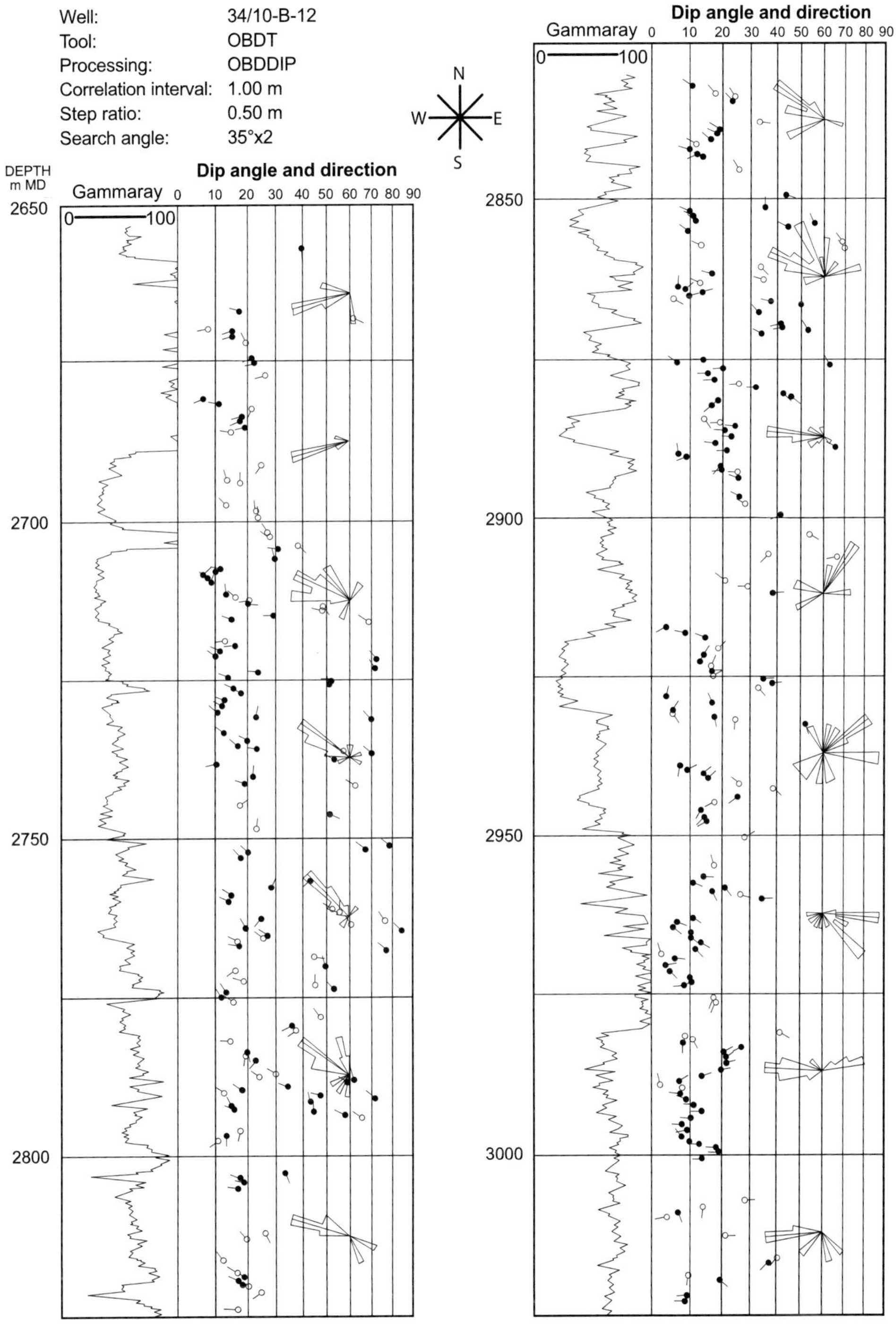

Fig. 8. Dipmeter data from well 34/10–B-12. The well was drilled using oil-based mud. It was therefore necessary to use an oil-based dipmeter tool. This causes more scatter of the processed data than data collected using a water-based dipmeter tool. It is, however, still quite possible to observe general changes in bedding orientation from the oil-based dipmeter data. Dip of bedding decreases with depth and shows easterly dip near the two main faults.

dipping fault, suggesting that any displacement must be below seismic resolution (i.e. 20–30 m). Since there is no obvious relationship between the width of the zone affected by local drag of bedding and the displacement of the faults on the Gullfaks Field, it is quite possible that a minor antithetic fault is located around 2900 m MD.

In the interval from 2950 to 2985 m MD, where the two large-scale faults are located, many of the dipmeter measurements indicate easterly dip. Seismic data clearly demonstrate that the faults must dip in an easterly direction. This suggests that the two faults are associated with flattening drag and that bedding decreases from 12–14° towards the west in the Tarbert Fm. to subhorizontal and easterly dip in the Rannoch and Drake Fms. Dipmeter data alone are not of sufficiently good quality that detailed changes in bedding orientation can be observed, and such data must therefore be used in conjunction with other data types.

The amount of dip of the two faults cannot be determined with certainty from dipmeter data alone. It may, however, be possible to indicate a rough minimum dip of the faults by identifying the maximum dip of the strata at the exact location of the faults. Only if bedding is dragged completely parallel to the faults can studies of bedding dip give the exact dip of the faults. At the location of the faults, there are indications of localized drag, suggesting easterly fault dips in excess of 20–40°. Since the data are processed from an oil-based dipmeter tool, such detailed interpretation must be treated with extreme care and should not be used alone. In the present case, seismic data (Fig. 4) indicate that the faults dip approximately 25–35° towards the east, which is consistent with observations from dipmeter data.

Core data

Core data provide the geologist with the most detailed information about reservoir properties. The reservoir rocks in the hanging wall to the main faults were cored in well 34/10–B-12 to investigate petrophysical, sedimentological and structural properties. Figure 9a displays the frequency of deformation bands per half metre as observed from cores in well 34/10–B-12. Deformation bands are microfaults with less than a few centimetres of displacement. In addition, the bands have not developed distinct slip surfaces and, therefore, represent localized areas of strain hardening rather than the typical weakening of rocks associated with larger-scale faults (e.g. Antonellini *et al.* 1994). It is clear from the diagram that the two main faults located at 2950 m MD and 2983 m MD are associated with abundant deformation bands in a narrow interval that defines the width of the deformation zones. In addition, there is an interval of abundant deformation bands located around 2900 m MD. This is the same location where dip of bedding increased according to dipmeter data. Well log correlation data demonstrate that a possible fault can only be associated with minor (3–5 m) displacement. The width of the damage zone for the possible minor fault located at 2900 m MD is wider than that observed around the 69 m fault located at 2950 m MD. This observation may be surprising, but is consistent with other studies of core data from the Gullfaks Field, which show no simple relationship between the width of a fault's damage zone and the displacement of the fault (Hesthammer & Fossen 2001). The reason for this may be that the deformation bands were created prior to development of a distinct slip surface (Aydin & Johnson 1978, 1983; Fossen & Hesthammer 1998*b*). Once a weak slip surface exists, there is no need to develop more deformation bands unless irregular fault plane geometries cause local areas of fault locking and strain hardening.

A total of 256 m of cores exist for the B-12 well. Of these, 222.6 m were available for analyses. The remaining interval was either missing (6.9 m), characterized by low coherence (22.2 m) or intensely tectonically fragmented (4.3 m). A total of 392 deformation bands were identified, most of which occur in narrow damage zones. The average displacement along each deformation band is 6.1 mm. This gives an accumulated displacement for all deformation bands in the well of 2.4 m. This is much too little to explain the gradual decrease of bedding observed in the hanging wall (large-scale drag) as being caused by deformation bands alone. In addition, well log correlation and dipmeter data demonstrate that there are not enough faults with discrete slip surfaces to account for the change in bedding dip. Since deformation of the Gullfaks Field occurred immediately after deposition of the Brent Group when the sediments were only loosely consolidated, it is likely that much of the sub-seismic deformation occurred by a reorganization of individual grains rather than by discrete fractures or deformation bands (Hesthammer 1999*b*).

Figure 9b shows dip of bedding with respect to depth as observed from core data. Since the well is deviated from vertical, the measurements do not represent the true dip of bedding. The observed gradual decrease in bedding dip towards the two main faults supports observations of hanging wall drag from seismic and dipmeter data. In addition, the high dip of

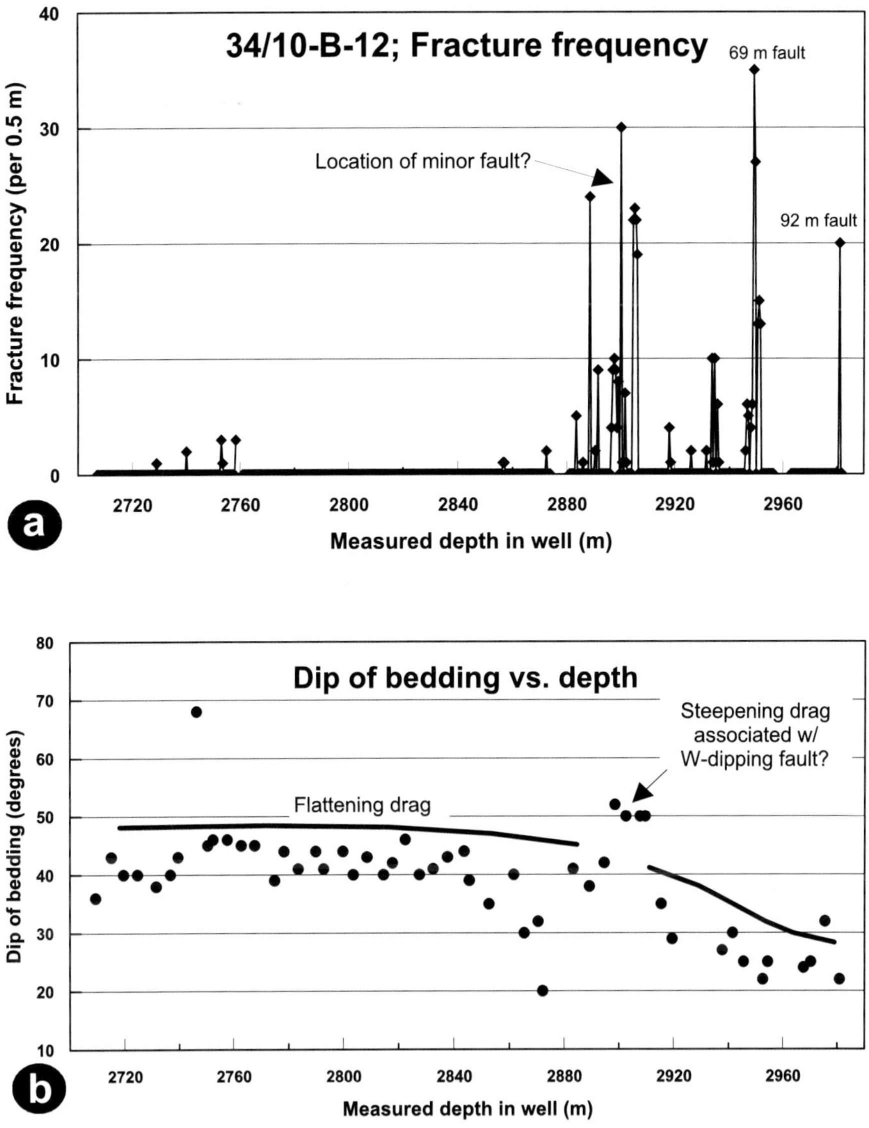

Fig. 9. (**a**) Fracture frequency diagram for well 34/10–B-12. Number of deformation bands per half metre is plotted along the y-axis whereas measured depth is plotted along the x-axis. The two faults associated with 69 m and 92 m missing section in the Rannoch and Drake Fms, respectively, are associated with a narrow deformation zone containing numerous deformation bands. Also, a minor fault (less than 3–5 m offset) may be located around 2900 m MD where another cluster of deformation bands is observed. (**b**) A plot showing change in dip (as observed in core data) versus depth illustrates how the dip of bedding decreases towards the main faults. This is a result of large-scale drag in the hanging wall. A steepening drag associated with a minor fault may be observed at around 2900 m MD. Modified after Hesthammer & Henden (2000).

bedding observed from dipmeter data around 2900 m MD is also clearly observed in the core data. This supports the view that a westerly dipping fault with minor offset is located within this interval.

The core from well 34/10–B-12 was not orientated. It is, therefore, not possible from core data alone to determine the exact bedding orientation. However, observations of dip from the slabbed core (the core is slabbed parallel to the dip direction of bedding) allows calculations of a solution population curve showing all possibilities for bedding orientation (see Hesthammer (1998) for a description of the procedure). Figure 10 shows solution population curves for bedding orientation for the Tarbert Fm. (blue curve), Ness Fm. (green curve), Etive Fm. (yellow curve), Rannoch Fm. (orange curve) and the Drake Fm. (red curve). Seismic data from the Tarbert Fm. indicate a dip of 13° towards 290. This is consistent with observations from dipmeter data. Also, the interpreted orientation from dipmeter and seismic data plot on the solution population curve from core data. This gives a unique orientation of bedding for the Tarbert Fm. Similarly, seismic interpretation of the intra-Ness Fm. reflection indicates that bedding dips 6° towards 295, which is also consistent with observations from dipmeter data. These measurements plot on the solution population curve for bedding orientation measured from core data from the Ness Fm.

Although seismic data and dipmeter data are not conclusive for measurements of bedding orientation within the Etive Fm., core data suggest that bedding dip is subhorizontal. Within the Rannoch and Drake Fms, dipmeter data and core data both suggest that the bedding dips shallowly (2–4°) towards the east.

These observations are consistent with large-scale drag related to a ESE-dipping fault. The 69 m fault located at 2950 m MD is associated with a zone of fault breccia (Fig. 11). Abundant deformation bands appear to be subparallel to the edges of the brecciated zone. It is thus likely that the deformation bands are subparallel to the large-scale fault. Since bedding orientation is determined from seismic, dipmeter and core data, it is possible to estimate the orientation of the deformation bands and faults as observed in the core data and thus estimate the orientation of any larger-scale fault. There are two different approaches that may be applied. One is to use vector calculations and polar coordinates, as described by Hesthammer & Henden (2000).

A different approach is to use stereographic projection. This method does not require the use of mathematics and is easier if only a few orientation data are to be measured. The well is deviated 28° from vertical towards 320, indicated

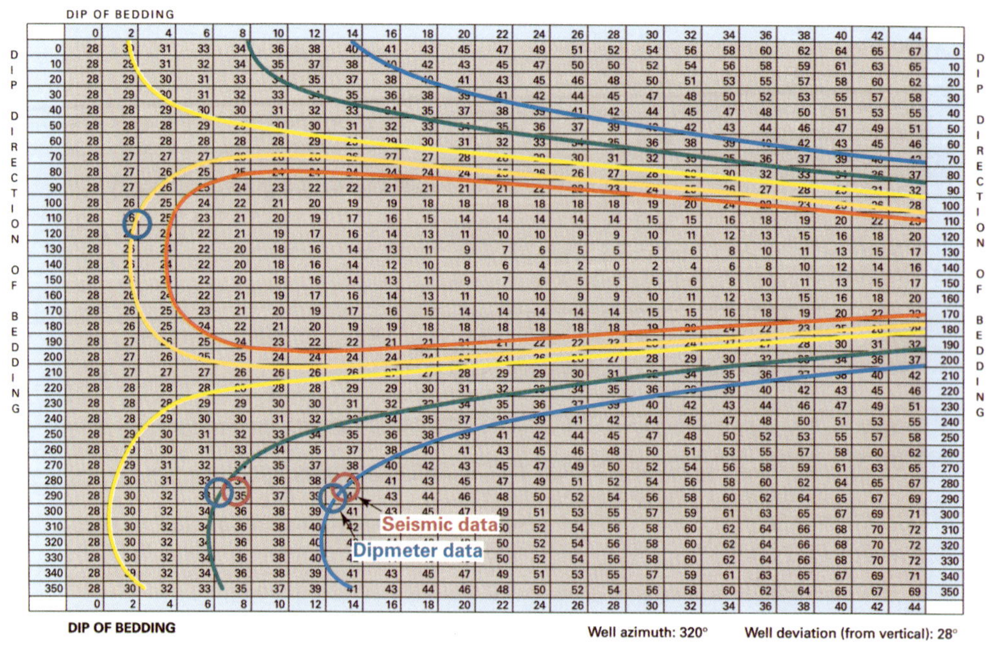

Fig. 10. Solution population diagram for possible orientations of bedding based on analyses of core data from well 34/10–B-12. The dip of bedding decreases from 12–14° towards the west in the Tarbert Fm. (blue curve) to 6° in the Ness Fm. (green curve), via subhorizontal in the Etive Fm. (yellow curve) to easterly dips in the Rannoch and Drake Fms (orange and red curves, respectively). Seismic data and dipmeter data are consistent with core data, thus verifying the gradual change in bedding dip.

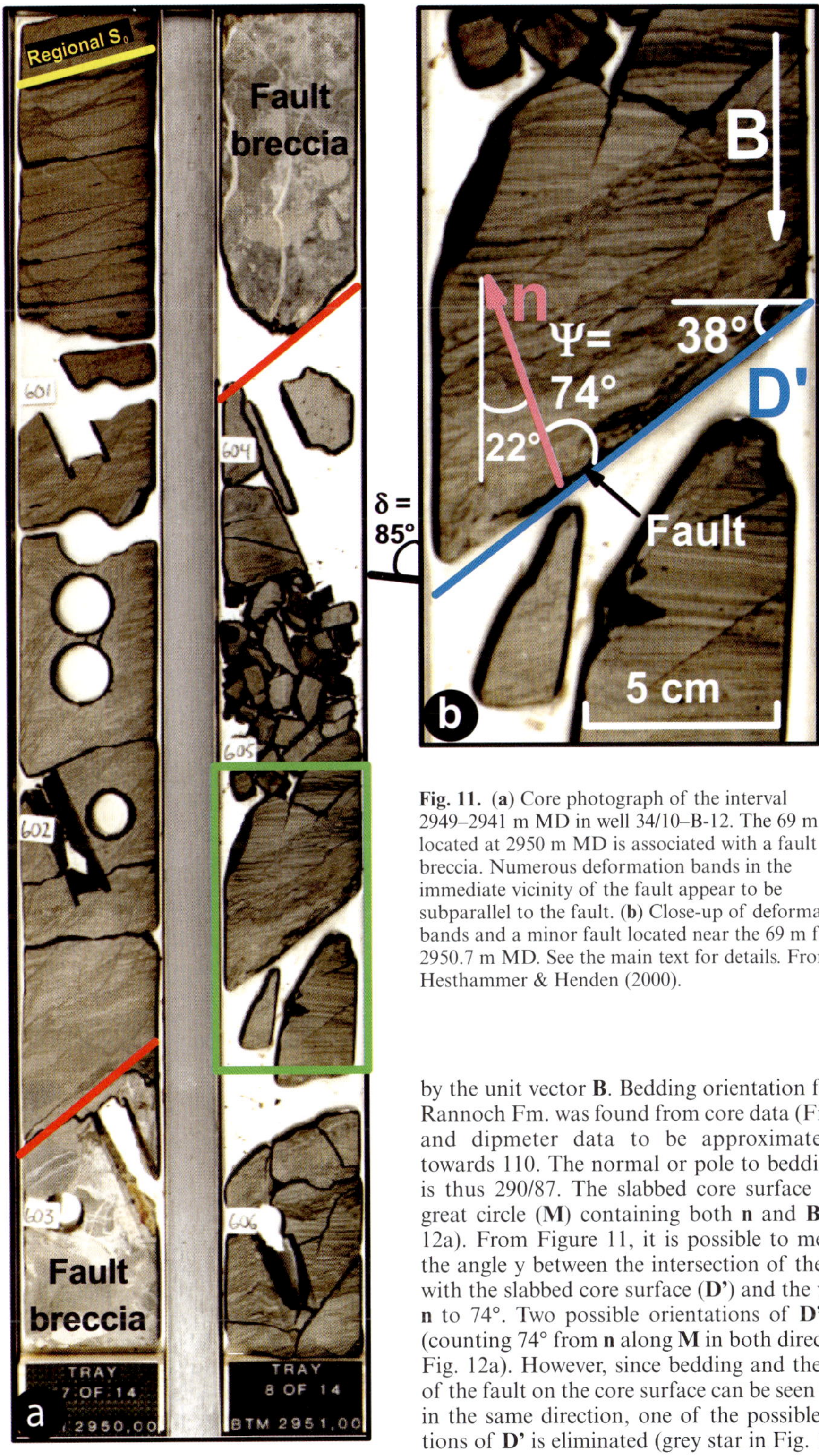

Fig. 11. (**a**) Core photograph of the interval 2949–2941 m MD in well 34/10–B-12. The 69 m fault located at 2950 m MD is associated with a fault breccia. Numerous deformation bands in the immediate vicinity of the fault appear to be subparallel to the fault. (**b**) Close-up of deformation bands and a minor fault located near the 69 m fault at 2950.7 m MD. See the main text for details. From Hesthammer & Henden (2000).

by the unit vector **B**. Bedding orientation for the Rannoch Fm. was found from core data (Fig. 10) and dipmeter data to be approximately 3° towards 110. The normal or pole to bedding (**n**) is thus 290/87. The slabbed core surface is the great circle (**M**) containing both **n** and **B** (Fig. 12a). From Figure 11, it is possible to measure the angle y between the intersection of the fault with the slabbed core surface (**D'**) and the vector **n** to 74°. Two possible orientations of **D'** arise (counting 74° from **n** along **M** in both directions; Fig. 12a). However, since bedding and the trace of the fault on the core surface can be seen to dip in the same direction, one of the possible positions of **D'** is eliminated (grey star in Fig. 12a).

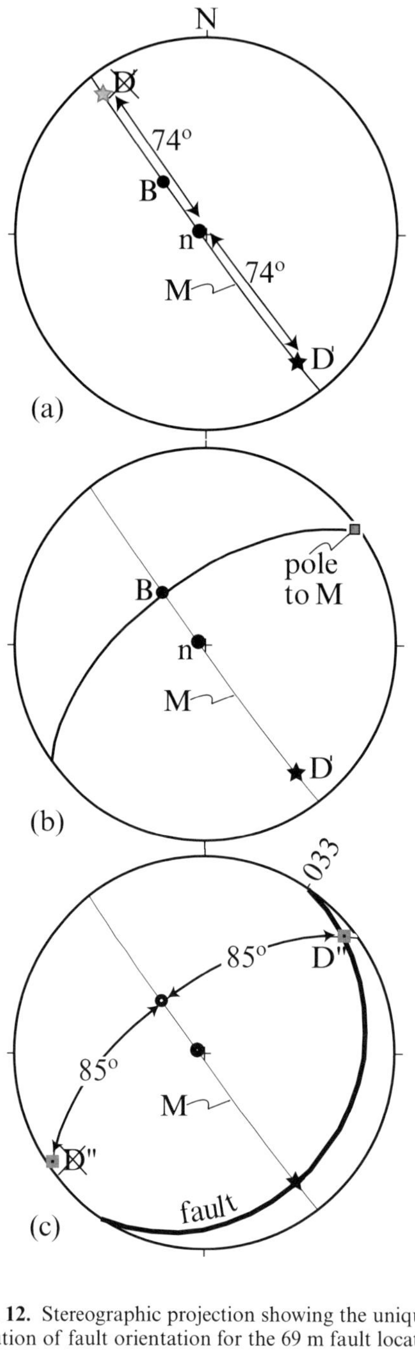

Fig. 12. Stereographic projection showing the unique solution of fault orientation for the 69 m fault located at 2950 m MD in well 34/10–B-12. The diagram shows that the fault must dip 21° towards 123 (SE). This is consistent with observations from dipmeter data and seismic data. M is the slabbed core surface displayed in Figure 11, **n** is the pole to bedding, **D'** is the trace of the fault on **M**, and **D''** is the corresponding trace on the core section perpendicular to **M**. See text for further explanation.

Plotting the pole to **M** and drawing a great circle through this pole and **B** displays the section perpendicular to the main slabbed core surface (Fig. 12b). The trace of the fault (**D''**) in this section makes an acute angle (d) of 85° with **B**. Since the angle can be measured in both directions along the great circle, we are left with two possibilities for **D''** (Fig. 12c). We observe the direction of apparent dip of layering (and **D'**) in the M-section is to the left (Fig. 11). From the stereonet data this means that we are looking at the M-surface from the NE. Knowing from core observation that **D''** is plunging out of the page (Fig. 11), i.e. to the NE, enables us to choose the NE-plunging alternative for **D''** (Fig. 12c). The great circle that contain **D''** and **D'** represents the fault, which is dipping about 21° towards 123 (SE). This result is consistent with observations from seismic data (Figs 3 & 4) and dipmeter data (Fig. 8) and demonstrates how analyses of different data types can be integrated to obtain a coherent structural interpretation.

A similar approach can be applied to the fault located at 2983 m MD. The fault itself is not preserved within the cored section and it is necessary to assume that the nearby deformation bands are subparallel to the fault. Figure 13 shows a core photograph near the fault. Values of y and d from the intersection of a deformation band with the slabbed core surface and the section orientated perpendicular to this were measured to 64° and 100°, respectively. Orientation of bedding was measured from dipmeter data and core data to be 5° towards 110 within the Drake Fm. The well orientation is the same as within the Rannoch Fm. The resulting unique solution based on stereographic projection (Fig. 14) shows that the fault must dip 33° towards 124. This is sufficiently close to the orientation of the fault located at 2950 m MD to assume that they are subparallel, as is indeed indicated by seismic data and dipmeter data.

Conclusions

Integrated, rather than individual, analyses of all available data enable the structural interpreter to confidently understand the characteristics of reservoir geometries. Typical data types that are available for structural interpretation include seismic, well log correlation, dipmeter and core data. The range in scale varies from many kilometres to less than a millimetre. Since the different data types normally overlap in scale, it is possible to control the quality results from the analyses and thereby strengthen the observations or identify possible sources of error.

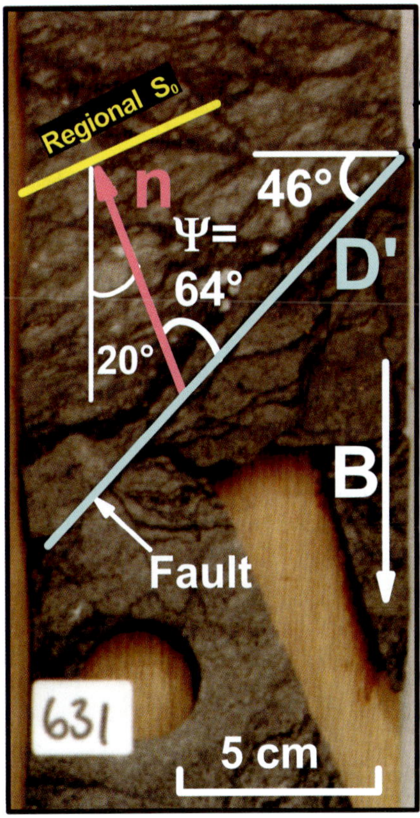

Fig. 13. Core photograph near the 92 m fault located at 2983 m MD in well 34/10–B-12. Several deformation bands are identified. See the main text for detailed discussion. From Hesthammer & Henden (2000).

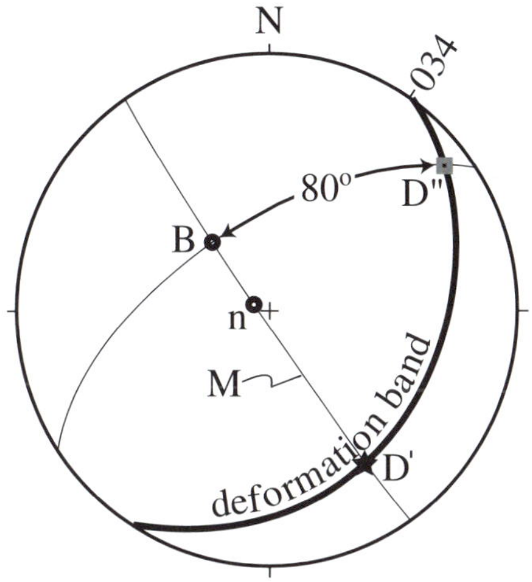

Fig. 14. Stereographic projection showing the orientation of the deformation band located near the 92 m fault at 2983 m MD in well 34/10–B-12. Analyses of core data show that the fault likely dips 33° towards 124 (ESE). This is subparallel to the 69 m fault located at 2950 m MD and consistent with observations from dipmeter data and seismic data. See text for further explanation.

Integrated use of seismic and well data from the area around well 34/10–B-12 on the Gullfaks Field has helped the structural geologist to understand the properties and geometries of the reservoir. In particular, the data analyses lead to the following conclusions:

Seismic data are of reasonably good quality in the area around the B-12 well and show that at least one large-scale east-dipping fault cut through the seismic reflections. The fault is associated with hanging-wall drag, which affects most of the reservoir penetrated by well 34/10–B-12.

Studies of time-lapse seismic identify a waterfront caused by injection of water in well 34/10–B-12. This waterfront was verified by water breakthrough in well 34/10–C-36 located updip of the B-12 well.

Well log correlation identified two large-scale faults located only a few metres apart: a fault associated with 69 m missing section within the Rannoch Fm. and a fault with 92 m missing section in the Drake Fm. Any additional faults must have displacements less than 3–5 m.

Dipmeter data show that the dip of bedding in the 34/10–B-12 well decreases from 12–14° at the uppermost reservoir level to subhorizontal and shallow easterly dip in the lowermost parts of the well. The two large-scale faults are associated with easterly dip of bedding and flattening drag, suggesting that the faults must dip to the east. This is consistent with observations from seismic data.

Core data demonstrate that the two large-scale faults are associated with abundant deformation bands in a narrow deformation zone. The number of deformation bands is too small to account for the change in bedding dip associated with the large-scale drag in the hanging wall. Since well log correlation shows that no larger fault exists in the hanging wall to the main faults, it is likely that most of the deformation was by a widely distributed grain reorganization rather than by faulting.

Core data also show that the dip of bedding decreases towards deeper reservoir levels. Bedding orientation measurements from seismic data and dipmeter data plot on the solution population curves for possible bedding orientation based on core data. This verifies that the different data types are consistent and reliable.

Analyses of fault orientation from core data show that the 69 m fault located at 2950 m MD must dip 23° towards 120, whereas the 92 m fault located at 2983 m MD dip 32° towards 125. This is consistent with observations of fault orientation from seismic data and dipmeter data.

The authors would like to thank Statoil and Norsk Hydro for permission to publish this paper.

References

ANTONELLINI, M., AYDIN, A. & POLLARD, D. D. 1994. Microstructure of deformation bands in porous sandstones at Arches National Park, Utah. *Journal of Structural Geology*, **16**, 941–959.

AYDIN, A. & JOHNSON, A. M. 1978. Development of faults as zones of deformation bands and as slip surfaces in sandstones. *Pure and Applied Geophysics*, **116**, 931–942.

AYDIN, A. & JOHNSON, A. M. 1983. Analysis of faulting in porous sandstones. *Journal of Structural Geology*, **5**, 19–31.

BAHORICH, M. & FARMER, S. 1995. 3–D seismic discontinuity for faults and stratigraphic features: the coherence cube. *Leading Edge*, **14**, 1053–1058.

BENGTSON, C. A. 1981. Statistical curvature analysis techniques for structural interpretation of dipmeter data. *AAPG Bulletin* **65**, 312–332.

BRENNEKE, J. C. 1995. Analysis of fault traps. *World Oil*, **216**, 63–71.

DALLEY, R. M., GEVERS, E. C. A., STAMPFLI, G. M., DAVIES, D. J., GASTALDI, C. N., RUIJTENBERG, P. A. & VERMEER, G. J. O. 1989. Dip and azimuth displays for 3D seismic interpretation. *First Break*, **7**, 86–95.

DUMONT, A., KUBACSI, M. & CHARDAC, J. L. 1987. The oil-based mud dipmeter tool. *28th Annual SPWLA Logging Symposium*, **2**.

FOSSEN, H. & HESTHAMMER, J. 1998*a*. Structural geology of the Gullfaks Field. *In*: COWARD, M. P., JOHNSON, H & DALTABAN, T. S. (eds) *Structural Geology in Reservoir Characterization*. Geological Society, London, Special Publications, **127**, 231–261.

FOSSEN, H. & HESTHAMMER, J. 1998*b*. Deformation bands and their significance in porous sandstone reservoirs. *First Break*, **16**, 21–25.

FOSSEN, H. & HESTHAMMER, J. 2000. Possible absence of small faults in the Gullfaks Field, northern North Sea: implications for downscaling of faults in some porous sandstones. *Journal of Structural Geology* **22**, 851–863.

FOSSEN, H. & RØRNES, A. 1996. Properties of fault populations in the Gullfaks Field, northern North Sea. *Journal of Structural Geology*, **18**, 179–190.

GEORGE, D. 1996. Coherence cube reveals stratigraphic features more readily than traditional 3D seismic slice plots. *Offshore International*, **56**, 22–23.

HESTHAMMER, J. 1998. Integrated use of well data for structural interpretation of seismic data. *Petroleum Geoscience*, **4**, 97–109.

HESTHAMMER, J. 1999*a*. Improving seismic data for detailed structural interpretation. *The Leading Edge*, **18**, 226–247.

HESTHAMMER, J. 1999*b*. *Analysis of fault geometry and internal fault block deformation in the Gullfaks region, northern North Sea*. PhD thesis, University of Bergen, Norway.

HESTHAMMER, J. & FOSSEN, H. 1997*a*. Seismic attribute analysis in structural interpretation of the Gullfaks Field, northern North Sea. *Petroleum Geoscience*, **3**, 13–26.

HESTHAMMER, J. & FOSSEN, H. 1997*b*. The influence of seismic noise in structural interpretation of seismic attribute maps. *First Break*, **15**, 209–219.

HESTHAMMER, J. & FOSSEN, H. 1998. The use of dipmeter data to constrain the structural geology of the Gullfaks Field, northern North Sea. *Marine and Petroleum Geology*, **15**, 549–573.

HESTHAMMER, J. LANDRØ, M. & FOSSEN, H. 2001. Use and abuse of seismic data in reservoir characterisation. *Marine and Petroleum Geology*, **18**, 635–655.

HESTHAMMER, J., & FOSSEN, H. 2001. Structural core analysis from the Gullfaks area, northern North Sea. *Marine and Petroleum Geology*, **18**, 411–439.

HESTHAMMER, J. & HENDEN, J. O. 2000. Information on fault orientation from unoriented cores. *AAPG Bulletin*, **84**, 472–488.

HESTHAMMER, J. & LØKKEBØ, S. M. 1997. Combining seismic surveys to improve data quality. *First Break*, **15**, 103–115.

HOETZ, H. L .J. G. & WATTERS, D. G. 1992. Seismic horizon attribute mapping for the Annerveen Gasfield, the Netherlands. *First Break*, **10**, 41–51.

HURLEY, N. F. 1994. Recognition of faults, unconformities, and sequence boundaries using cumulative dip plots. *AAPG Bulletin* **78**, 1173–1185.

KAYA, M. A. & NORMAN, T. N. 1993. Verifying a geological structure by applying SCAT (statistical curvarure analysis techniques) method on dipmeter data at the Umurca Field in the Turkish Thrace. *4th Annual SPE and Archie Conference Proceedings*, 67–72.

RØNNINGSLAND, T. M. 1990. Structural interpretation of dipmeter results in the Gullfaks Field. *In*: HURST, A., LOVELL, M. A. & MORTON, A. C. (eds) *Geological Applications of Wireline Logs*. Geological Society Special Publications, London **48**, 273–286.

SCHLUMBERGER 1986. *Dipmeter Interpretation*. Schlumberger Limited, Texas.

SCHLUMBERGER 1990. *Data Services Catalog*. Schlumberger Educational Services, New York.

SCHLUMBERGER 1991. *Wireline Services Catalog*. Schlumberger Educational Services, Texas.

STUART-BRUGES, W. P. 1984. A dipmeter for use in oil based muds. *9th International SAID (division of SPWLA) Formation Evaluation Symposium*.

TOLLEFSEN, S., GRAUE, E. & SVINDAL, S. 1994. Gullfaks development provides challenges. *World Oil* (April), 45–54.

VOGGENREITER, W. R. 1991. Kapuni 3D interpretation – imaging Eocene paleogeography. *New Zealand Oil Exploration Conference Proceedings*, **1**, 287–298.

Fault and stress magnitude controls on variations in the orientation of *in situ* stress

BY DAVID P. YALE

ExxonMobil Upstream Research Company, PO Box 2189, Houston, TX 77252

Abstract: There has been a great deal of interest in mapping regional trends in stress orientation and their relationships to crustal deformation processes over the last quarter century. Less emphasis has been placed on explaining local variations in the stress field. However, it is variations in the local stress field which are most critical to characterizing hydrocarbon reservoirs. Fracture orientation, well stability, well orientation, and permeability anisotropy are all strongly affected by variations in the local stress field.

Using stress orientation data from a number of fields in different tectonic environments, we have tried to determine some of the tectonic and geological controls on variations in in situ stress orientation. We find that nearness to faults, fault structure, and magnitude of tectonic stress play primary roles in determining whether the regional stress field will be perturbed in a given reservoir and whether small-scale variations in the stress field can be expected.

We find that high tectonic stress environments (large horizontal differential stress) lead to a much more consistent local stress field than more tectonically quiescent areas. Faults can play a large role in rotating the local stress field. The smaller the difference between the maximum and minimum horizontal stress magnitude (i.e. lower tectonic stress), the larger a fault's zone of influence is. Large-scale faulting with segmentation of the reservoir into discrete fault blocks can lead to significant stress orientation variations across the reservoir even in areas of large differential stress.

There has been a great deal of interest in mapping the orientation of *in situ* stress in the upper crust across the globe over the last quarter century (Zoback & Zoback 1980; Zoback 1992). This effort has improved our understanding of crustal deformation processes and the influence of plate tectonics on these processes. Most of this effort, however, has been aimed at understanding large-scale, regional patterns in the upper crust's stress field, not smaller-scale variations which are influenced by local structure (Zoback 1992).

A number of other studies have looked in more detail at these small scale variations (Gough & Bell 1982; Plumb *et al.* 1984; Griffin 1985; Bell and Babcock 1986; Mount & Suppe 1987; Plumb & Cox 1987; Yale *et al.* 1991, 1992, 1994; Aleksandrowski *et al.* 1992; Cowgill *et al.* 1992; Evans & Bereton 1990; Laubach *et al.* 1992;). Fault structure, topographic loading, depth, sedimentary structure, disconnection with basement and geologicalstructure (anticlines, salt domes, joints) have all been shown to have some influence on local variations in the measurement of *in situ* stress orientation.

Local variations in *in situ* stress orientation are especially critical in the development of hydrocarbon reservoirs. Stress orientation controls the direction of hydraulically induced fractures, the direction of open natural fractures, permeability anisotropy, and the optimal orientation of stable horizontal wells. Extrapolation of stress orientation measurements from a small number of wells in one part of the field to other parts of the field or inference of local stress orientation from published regional trends can result in incorrect reservoir development decisions if there are significant variations in the local stress field.

In this paper, we look at the variations in stress orientation across several fields in different tectonic environments, where we have conducted *in situ* stress orientation measurements. Comparing these results with some published studies of local stress field variability (Griffin 1985; Bell & Babcock 1986; Aleksandrowski *et al.* 1992) leads to certain criteria which can help discriminate those fields which are most likely to exhibit significant variations in local stress orientation.

Stress orientation methods

A wide variety of techniques are currently used to determine the orientation of *in situ* stress in the upper crust. Some of the more common methods include: dipmeter analysis of wellbore ellipticity (Babcock 1978; Gough & Bell; 1982); anelastic recovery strain analysis of core samples

From: AMEEN, M. (ed.) *Fracture and In-Situ Stress Characterization of Hydrocarbon Reservoirs.* Geological Society, London, Special Publications, **209**, 55–64. 0305–8719/03/$15.00
© The Geological Society of London 2003.

(Teufel 1983), shear acoustic anisotropy analysis of core samples (Yale & Sprunt 1989; Yale *et al.* 1992); orientation of hydraulic fractures via borehole televiewer (Strubhar *et al.* 1975), via tiltmeters (Evans 1983) or via borehole microseismic (Sorrels & Mulcahy 1986), earthquake focal mechanisms (Zoback & Zoback 1980); and overcoring methods (Plumb *et al.* 1984). For petroleum reservoirs between 2000 and 7000 m depth, the first four techniques are the most useful and are the methods utilized by the studies described in this report.

The first three techniques (wellbore ellipticity, shear acoustic anisotropy and anelastic recovery strain) utilize the rock's response to the drilling process to infer the stress orientation. In the case of wellbore ellipticity, the stress distribution around the open cavity of the wellbore can result in the formation of 'breakouts' normal to the maximum horizontal stress direction if the tangential stress in the borehole is greater than the compressive stress of the rock forming an elliptical wellbore. This ellipticity of the wellbore can be measured by either caliper logs or borehole televiewer logs. For both shear acoustic anisotropy and anelastic recovery strain, the relief of the *in situ* stress field on a drilled, orientated core sample results in a measurable expansion and anisotropy of shear acoustic wave propagation which can be used to determine the orientation of the maximum horizontal stress. The fourth technique (orientation of hydraulic fractures) is considered the most 'direct' technique as the hydraulic fracture will always propagate in the direction of the maximum horizontal stress. However, since all the current methods (borehole televiewers, tiltmeters and borehole microseismic) 'infer' the hydraulic fracture orientation from geophysical measurements, their accuracy and success is similar to the three indirect methods described previously. More details on all of the techniques can be found in the references cited in the previous paragraph.

Variations in stress orientation

As mentioned above, a number of tectonic and geological structures appear to influence the orientation of *in situ* stress on a local level. In this paper, we will look mainly at the influence of faults, fault blocks, and stress magnitude on stress orientation. All the examples cited are from 2000 to 4000 meters in depth, which in general should remove near-surface influences such as topographic loading (Plumb *et al.* 1984). In addition, we will concentrate on analysing stress variations across petroleum reservoirs within and immediately adjacent to the producing reservoir, therefore removing the variations some have observed over larger depth intervals (Aleksandrowski *et al.* 1992). The map scale of the stress variations we will be examining is between 100 and 500 km^2.

Northern North Sea, Quad 15

Figure 1 shows a 10 km by 10 km map view of a field in the northern part of the British North Sea. The field is in Quad 15 and is in the Witch Ground Graben (Cowgill *et al.* 1992). The Witch Ground Graben is an extensional basin trending WNW by ESE between Quads 14 and 15. The data presented in Figure 1 are taken from Yale *et al.* (1994). The data are taken from acoustic anisotropy measurements on core from the Piper Sands and wellbore elongation data from the overlying Kimmerage shale and the Piper Sands.

The WNW trending faults in Figure 1 are interpreted as normal faults and lie roughly parallel to the major normal faults defining the Witch Ground Graben. The NNE trending fault through the centre of the field is interpreted as a wrench fault. The throw on faults is generally between 250 and 400 m.

The regional stress trend in this area is considered to be maximum compression in a NNW direction (Klein & Barr 1986; Muller *et al.* 1992). However, the data in Figure 1 show significant variations from this general trend and significant variations between individual fault blocks. The wells on the west side of the NNE wrench fault show a NNE trend in maximum horizontal stress whereas those in the NE quadrant show a more northwesterly trend. In the southern part of the field there is more variation, but most of the results suggest maximum horizontal stresses parallel to the local fault structure. The data from acoustic anisotropy results agree very well with the data from wellbore elongation measurements.

Maximum horizontal stress parallel to the strike of the faults is consistent with the normal faulting seen in the field. However, the 30 to 50° variation between the regional stress field and the local stress field strongly suggests a decoupling or rotation of the stress field by the local fault structure. The very large throws on these faults and the strong segmentation of the field into self-contained fault blocks may be the cause of the decoupling between the local and regional stress fields in this area.

Yale & Sprunt (1989) suggested that there is a correlation between the magnitude of horizontal *in situ* stress differences and the magnitude of the acoustic anisotropy. The larger the difference between the maximum and minimum horizontal

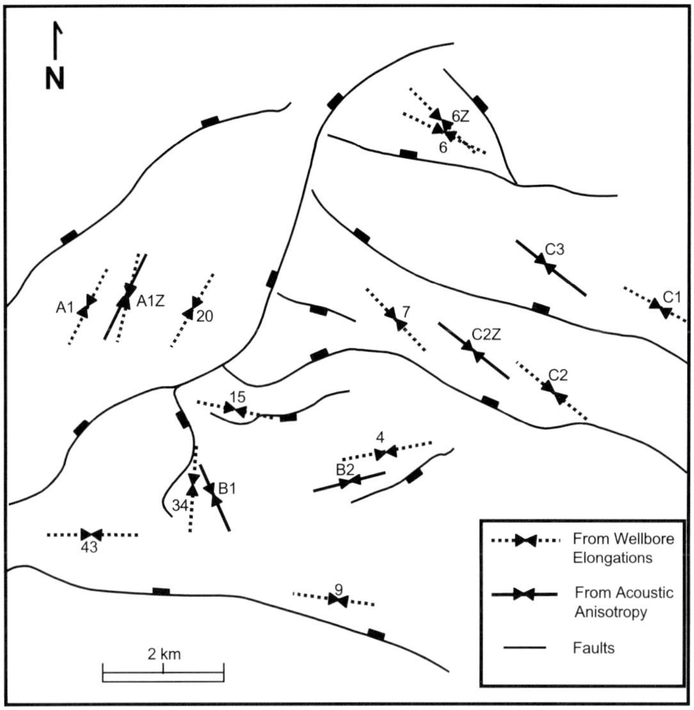

Fig. 1. Orientation of maximum horizontal stress, northern North Sea, Quad 15 (adapted from Yale *et al.* 1994).

stresses (i.e. the larger the magnitude of the tectonic stress), the larger the difference in shear acoustic velocities between the maximum and minimum horizontal stress directions (i.e. the larger the shear acoustic anisotropy magnitude). For the Quad 15 wells, the average magnitude of the shear acoustic anisotropy was 3.4%. Our experience suggests that this level of anisotropy is indicative of a fairly high differential horizontal stress from a moderate to high tectonic stress environment. The strong segmentation of the reservoir by the faulting, however, appears to control the local stress field despite this apparently large differential horizontal stress.

Southern North Sea, Quad 48

Figure 2 shows the fault structure and maximum horizontal stress orientation in a field in the southern North Sea Basin. The field is in Quad 48, off the coast of Yarmouth. The data consist of acoustic anisotropy results from core and wellbore elongation measurements from calipers. The figure is taken from Yale *et al.* (1991).

The fault structure here is quite different from the previous field in Quad 15. Most of the faults have a NW trend and are low-angle growth faults. The age of the faulting is significantly older than in the Quad 15 field and the throw on the faults is significantly less. There is evidence in the area of inversion structures on older normal faults due to the NNW trending compressive stress from the relative motion of the African and European plates (Klein & Barr 1986; Lake & Karner 1987; Muller *et al.* 1992).

The overall orientation of the maximum horizontal stress is much more consistent in this field, varying from slightly east of north to NNW. The average orientation agrees well with the regional stress field (Klein & Barr 1986; Muller *et al.* 1992) and there is good agreement between the acoustic anisotropy results and the wellbore elongation data. However, the variations seen are more likely due to fault influences than random scatter in the data. Two interesting and unusual orientations occur in wells E and J. In well E, there are two major trends in the data, one perpendicular to the nearby fault and one roughly parallel to the trend of the majority of the wells. In well J, the stress orientation is perpendicular to the fault. The zones tested in both of these wells are very close to the faults, within 200 m in the case of well E and within 75 m in well J. The rotation of

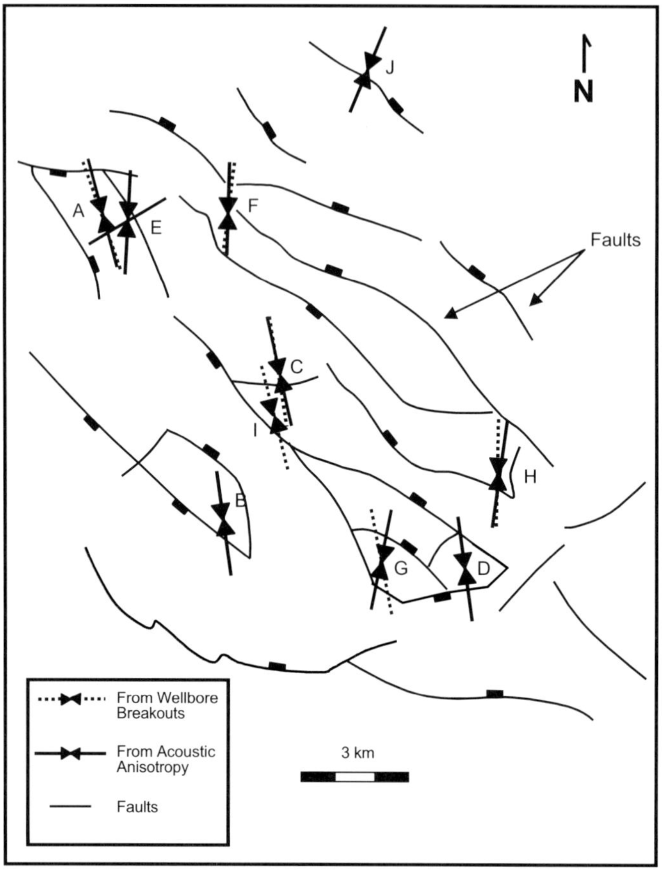

Fig. 2. Orientation of maximum horizontal stress, Quad 48, southern North Sea (adapted from Yale *et al.* 1991).

the local stress field very close to faults has been noted by others (Sbar *et al.* 1981; Mount & Suppe 1987; Barton & Zoback 1993). The small variation in stress orientation that is observed between wells may correlate with the orientation of nearby faults.

The average magnitude of the shear acoustic anisotropy was 3.9% for this field. Like the Quad 15 data, this is indicative of a moderate to high tectonic stress in the area. Unlike the Quad 15 data, the stress orientations are very consistent in this area (except very close to faults). This may be explained by the more consistent nature of the fault trends in Quad 48 versus Quad 15 or the lower degree of segmentation of the reservoir by the faults in Quad 48 versus Quad 15. In addition, the difference in the age of the faulting between the Quad 48 (older) and Quad 15 (younger) examples suggests that the Quad 15 faults may still be active enough to perturb the regional stress field.

Western Wyoming

Figure 3 shows the maximum horizontal stress orientations in the western part of Wyoming. The area is slightly north and west of the area studied by Laubach *et al.* (1992) and is within the Green River Basin. As shown by Zoback & Zoback (1989) the area is on the edge of a large regional trend of SW mid-plate compression and Cordilleran extension leading to a more NNW maximum horizontal stress orientation. The stress field is further complicated by structural features such as the thrust faults shown in Figure 3. Acoustic anisotropy, wellbore elongation, and anelastic strain recovery data were used in the various wells to determine stress orientation.

As shown, wells A, B, and D showed two trends in stress orientation. The NW trend is dominant in wells A and B and secondary in well D, and the east to NE trend is dominant in well D and secondary in wells A and B. The most

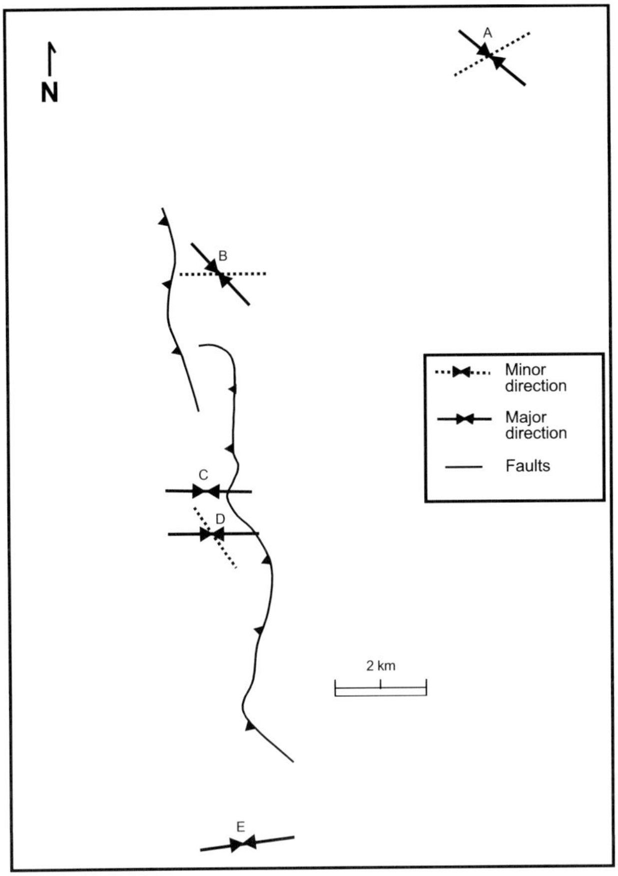

Fig. 3. Orientation of maximum horizontal stress, western Wyoming.

striking difference between the five wells, however, is the switch from a NW dominant stress orientation east of the fault to an east–west direction west of the fault. The east–west trend of maximum horizontal stress is consistent with the north–south trending thrust faults and the NW trend in consistent with the regional stress field trend (Zoback & Zoback 1989).

Laubach *et al.* (1992) show a significant amount of scatter in stress orientation data just to the south and east of the map area in Figure 3. They see scatter between wells and among different stress orientation techniques. Laubach *et al.* (1992) conclude that the scatter in results and the lack of wellbore breakouts may be indicative of small differences between the maximum and minimum horizontal stresses. This small difference would lead to a difficulty in measuring the stress orientation due to low signal-to-noise ratios in the measurements and more variations between wells as small-scale features can influence the stress field more easily.

This is further supported by our shear acoustic anisotropy measurements. The magnitudes of the shear acoustic anisotropy were small in most of the samples tested from this field (average anisotropy only 1.9%). This suggests a fairly small differential horizontal stress environment which could lead to more variation in apparent stress orientation.

Supporting studies

The three studies detailed above show three stress regimes which we often encounter in determining stress orientation across petroleum reservoirs. The Quad 48 example shows the effects of moderate to high tectonic stress (and resulting high differential horizontal stress) leading to a consistent stress orientation except where the wells are very close to faults. The Quad 15 example represents an area of moderate to high tectonic stress but where aspects of the faulting appear to decouple the local stress field from the regional stress field. The faults in this area are younger (and therefore may be more active) and appear to more fully segment the reservoir than in Quad 48 supporting this decoupling of local and regional stress fields. The Wyoming example shows the effects of low tectonic stress leading to higher variability in stress orientation due to local structure. There are a number of other published reports as well as some of our own field experience which can be divided into these three groupings.

Figures 4 and 5 show two fields in the United States (South Texas and Kansas–Oklahoma respectively) where tectonic regime and shear acoustic anisotropy magnitude (only 1.8% in the South Texas field and 2.1% in the Kansas field) suggest fairly small differences between the principal horizontal stresses (Yale & Ryan 1994). In both cases the predominant orientation of maximum horizontal stress in the fields is aligned with the regional stress field (north–south in South Texas and slightly south of east in Kansas; Zoback 1992). However, nearby faulting appears to be able to rotate the local stress field even though the faults are several miles away. In all three wells where the apparent stress direction has rotated, the maximum stress orientation is parallel to the nearby normal faults. In the Kansas example (Yale & Ryan 1994), acoustic anisotropy, anelastic recovery strain, and hydraulic fracture orientation (from tiltmeters and borehole seismic data) were used to orientate the stress field. In South Texas, acoustic anisotropy and anelastic recovery strain were used.

These results are similar to work by Griffin (1985) in northern Alaska. In Figure 6, Griffin (1985) shows significant variability in the orientation of maximum horizontal stress as determined from wellbore breakouts. Again the local fault structure appears to influence the stress orientation. Griffin (1985) used acoustic well log anisotropy and differential strain analysis to predict the stress orientation. In five of the nine wells where all three techniques were used, all three gave the same orientation within

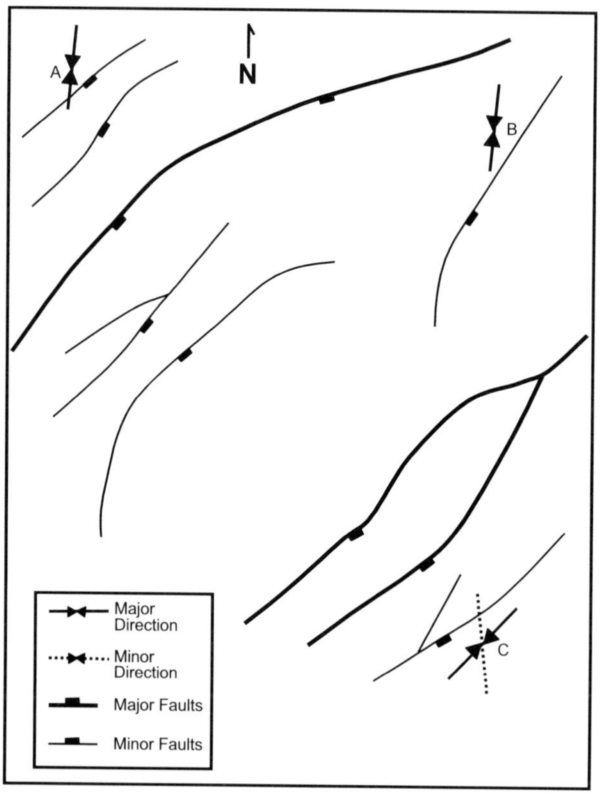

Fig. 4. Orientation of maximum horizontal stress, South Texas.

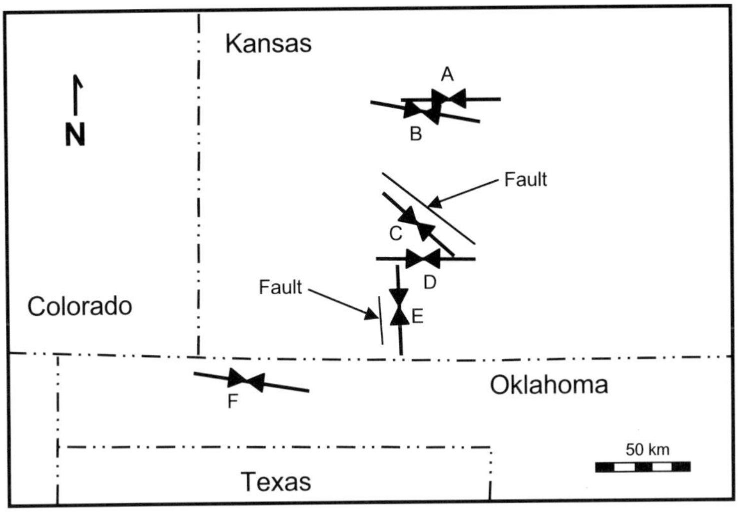

Fig. 5. Orientation of maximum horizontal stress, Kansas – Oklahoma area (from Yale & Ryan 1994).

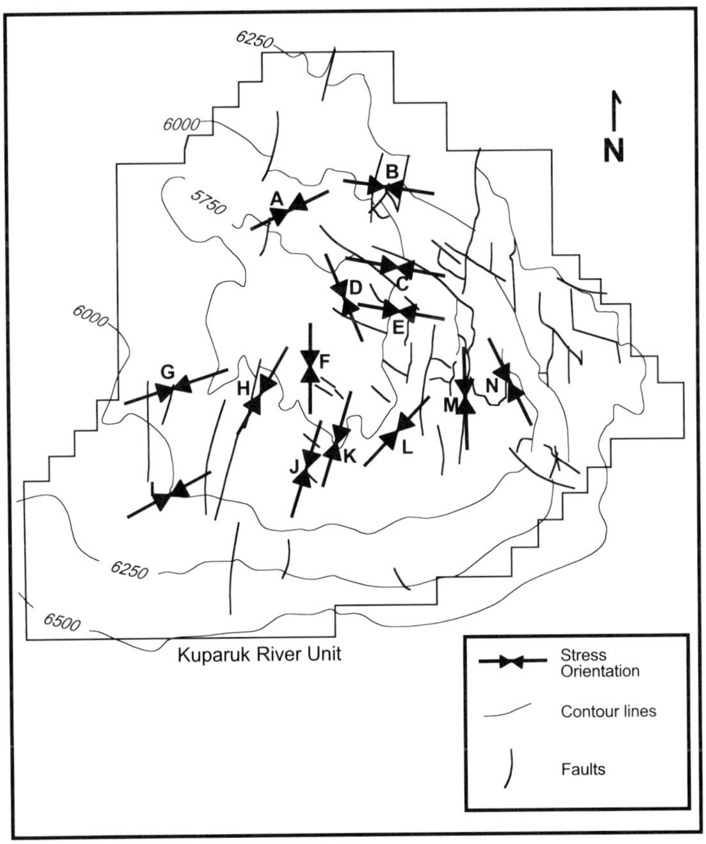

Fig. 6. Orientation of maximum horizontal stress, northern Alaska (adapted from Griffin 1985).

25°. However, in four of the nine wells, the stress orientations from the different techniques were up to 90° different. As in Laubach *et al.* (1992), this suggests that the variability in the orientation of maximum horizontal stress may correlate with small horizontal stress differences.

These three fields are in direct contrast to Figure 7 which shows the widely consistent orientation of maximum horizontal stress in the area of western Canada just east of the Rocky Mountains (Fig. 7 shows wellbore breakout direction which is orthogonal to the maximum horizontal stress orientation). This is similar to our results in Quad 48 of the southern North Sea. Bell & Babcock (1986) and Bell & Price (1987) have shown that the orientation of maximum horizontal stress is NE–SW throughout the oil-producing zones of Alberta and parts of British Columbia and Saskatchewan. This orientation is perpendicular to the trend of the Rocky Mountains and suggests that large tectonic stresses (and the resulting large differential horizontal stress) are leading to a very consistent stress orientation across a large area, similar to the situation we found in the Quad 48 field. Our own studies, within the area of 'Detail C' in Figure 7, show similarly consistent stress orientations in five different wells. Our results, from acoustic anisotropy measurements, show the same NE–SW trend as Bell & Price's (1987) wellbore elongation data. The shear acoustic anisotropy results support the existence of a strong differential horizontal stress field (and therefore high tectonic stress environment) as the magnitude of the shear acoustic anisotropy is large, averaging 4.6%.

Work by Aleksandrowski *et al.* (1992) supports the type of local stress regime found in our Quad 15 example. As shown in Figure 8, Aleksandrowski *et al.* (1992) finds similar variability in local stress orientation in a field in Quad 30 of the Norwegian North Sea as well as in Njord field in the central part of the Norwegian North Sea. Well B in Figure 8 shows a maximum horizontal stress orientation nearly parallel to the regional stress trend but well A shows an orientation nearly 60° from the regional trend. Aleksandrowski *et al.* (1992) interpret the stress orientation in well A as illustrative of the maximum stress being rotated parallel to the normal faults forming the local fault block. In the Njord field, Aleksandrowski *et al.* (1992) find a similar rotation of the regional stress field in a heavily normal-faulted area.

Conclusions

The trends observed in the orientation of the maximum horizontal *in situ* stress in the nine

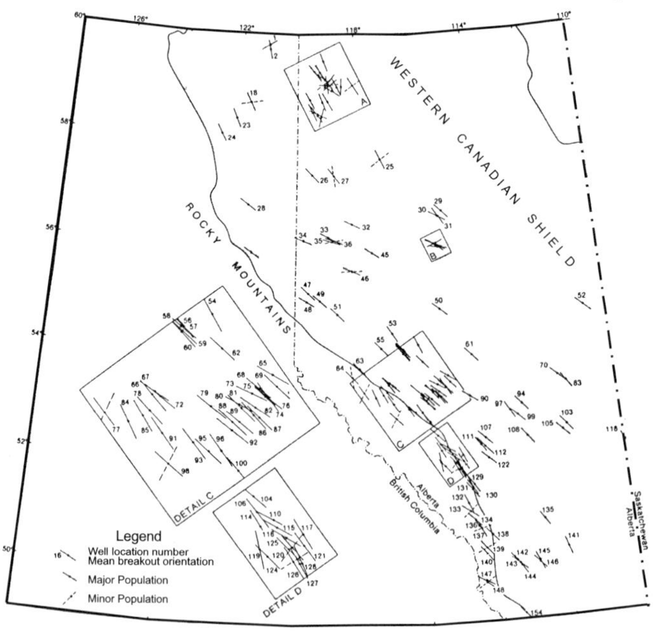

Fig. 7. Wellbore breakout orientations (minimum horizontal stress), western Canadian Basin (adapted from Bell & Price 1987)

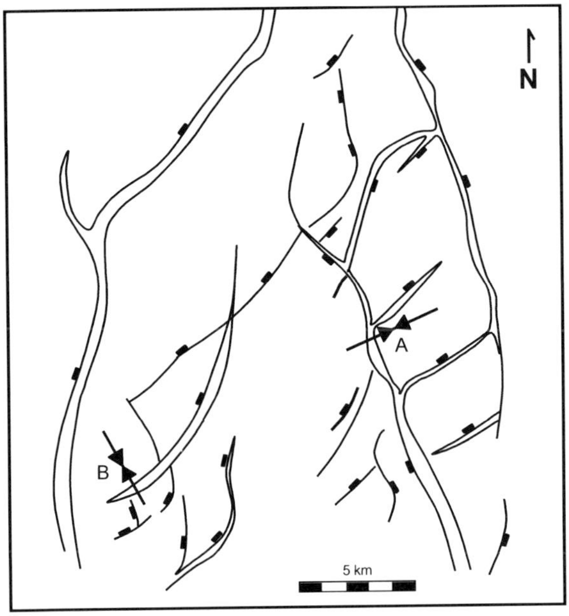

Fig. 8. Orientation of maximum horizontal stress, Norwegian North Sea, Quad 30 (adapted from Aleksandrowski *et al.* 1992).

fields discussed above suggest the following guidelines could be helpful in determining when local stress orientation is likely to deviate significantly from regional trends.

In areas with large differential horizontal stress (which in general correlates with high tectonic stress), the orientation of the maximum horizontal stress is most likely to be fairly consistent and parallel to regional trends. Small-scale faulting can affect the regional trends but generally only in an area within a few hundred meters of the fault.

It appears that large-scale faulting, which separates areas of the field into distinct fault blocks, can rotate the regional stress field significantly and lead to large variations in stress orientation from fault block to fault block, even in areas of moderate to large tectonic stresses. The presence of younger, active faulting may be a key element in this rotation.

Areas with small differential horizontal stress (low tectonic stress) have a greater chance of having geological and tectonic structures rotating the regional stress trend even if those features are several kilometres away. Measurement of *in situ* stress orientation is further complicated in these areas due to the lack of some stress indicators (wellbore elongations) and the low signal-to-noise ratio of other methods (shear acoustic anisotropy and anelastic recovery strain). More direct methods (hydraulic fracture orientation), however, work well even under these conditions.

The magnitude of *in situ* stresses, especially the difference between the maximum and minimum horizontal stress, are the least understood yet most critical data for determining where variations in stress orientation are likely to occur. Shear acoustic anisotropy magnitude, anelastic strain recover, and hydraulic fracturing tests can give estimates of the degree of differential horizontal stress and therefore tectonic stress under the right conditions. However, better stress magnitude measurement techniques are needed to resolve these issues.

The author would like to thank Mobil Technology Company for permission to publish this paper as well as J. M. Rodriguez for all his help with the laboratory and field measurements and J. Nieto, D. Blaisdell, S. Austin, T. Blanton, B. Thompson, T. Ryan, B. Sneed, and B. Scribbins for their help with the field studies.

References

ALEKSANDROWSKI, P., INDERHAUG, O. H. & KNAPSTAD, B. 1992.) Tectonic structures and wellbore breakout orientation. *In* TILLERSON, J. R. AND WAWERSIK, W. R. (eds) *Proceedings of the 33rd U.S. Rock Mechanics Symposium*, Balkema, Rotterdam.

BABCOCK, E. A. 1978. Measurement of subsurface fractures from dipmeter logs, *AAPG Bulletin*, **62**, 1111–1126.

BARTON, C. & ZOBACK, M. D. 1993. Stress perturbations associated with active faults penetrated by boreholes: Evidence for near complete stress drop and a new technique for stress magnitude measurement. Presented at *34th U. S. Symposium on Rock Mechanics, Madison, WI, June 1993.*

BELL, J. S. & BABCOCK, E. A. 1986. The stress regime of the Western Canadian Basin and implications for hydrocarbon production. *Bulletin of Canadian Petroleum Geology,* **34**, 364–378.

BELL, J. S. & PRICE, P. R. 1987. In situ *stresses in Western Canada: Implications for the oil industry.* Geologic Survey of Canada, Handout No. **87**, Institute of Sedimentary and Petroleum Geology.

COWGILL, S. M., MEREDITH, P. G., MURRELL, S. A. F. & BERETON, N. R. 1992. *In situ* stress orientations in the Witch Ground Graben, North Sea revealed by borehole breakouts: preliminary results. *In*: HURST, A., GRIFFITHS, C. M. & WORTHINGTON, P. M. (eds) *Geological Applications of Wireline Logs II.* Geological Society, London, Special Publications **65**, 179–184.

EVANS, K. 1983. On the development of shallow hydraulic fractures as viewed through the surface deformation field: Part I – Principles, *Journal of Petroleum Technology* (February), 406–410.

EVANS, C. J. & BERETON, N. R. 1990. *In situ* crustal stress in the United Kingdom from borehole breakouts. *In*: HURST, A., LOVELL, M. A. & MORTON, A. C. (eds.) *Geological Applications of Wireline Logs*. Geological Society, London, Special Publications **48**, 327–338.

GOUGH, D. I. & BELL, J. S. 1982. Stress orientations from borehole wall fractures with examples from Colorado, east Texas, and northern Canada. *Canadian Journal of Earth Science*, **19**, 1358–1370.

GRIFFIN, K. W. 1985. Induced fracture orientation determination in the Kuparuk Reservoir, presented at *60th Annual Technical Conference of Society of Petroleum Engineers, Las Vegas, NV, September 1985,* SPE 14261.

KLEIN, R. J. & BARR, M. V. 1986. Regional state of stress in Western Europe. *In*: *Proceedings of the International Symposium on Rock Stresses and Rock Stress Measurement.* Lulea: CENTEK, 33–44.

LAKE, S. D. & KARNER, G. D. 1987. The structure and evolution of the Wessex Basin, southern England: an example of inversion tectonics. *Tectonophysics,* **137**, 347–378.

LAUBACH, S.E., CLIFT, S. J., HILL, R. E. & FIX, J. 1992. Stress directions in Cretaceous Frontier Formation, Green River Basin, Wyoming. *In*: MULLEN, C. E. (ed.) *Wyoming Geological Association Forty-Third Field Conference Guidebook,* Pioneer Printing, Cheyenne.

MOUNT, V. S. & SUPPE, J. 1987. State of stress near the San Andreas Fault: Implications for wrench tectonics. *Geology*, **15**, 1143–1146.

MULLER, B. ZOBACK, M. L. *et al.* 1992. Regional patterns of tectonic stress in Europe. *Journal of Geophysical Research,* **97**, 11783–11803.

PLUMB, R. A. & COX, J. W. 1987. Stress directions in Eastern North America determined to 4.5 km from borehole elongation measurements, *Journal of Geophysical Research,* **92**, 4805–4816.

PLUMB, R. A., ENGELDER, T. & YALE, D. P. 1980. Near surface *in situ* stress – 3. Correlation with microcrack fabric within New Hampshire granites. *Journal of Geophysical Research,* **89**, 9350–9364.

SBAR, M. L., ENGELDER, T., PLUMB, R. & MARSHAK, S. 1979. Stress pattern near the San Andreas fault, Palmdale, California from near-surface *in situ* measurements. *Journal of Geophysical Research,* **84**, 156–164.

SORRELLS, G. G. & MULCAY, C. C. 1986. Advances in the microseismic method of hydraulic fracture azimuth estimation, presented at the SPE Unconventional Gas Technololgy Symposium, Louisville, Kentucky, May 1986, SPE paper no. 15216.

STRUBHAR, M. K., FITCH, J. L. & GLEN, E. E. 1975. Multiple, vertical fracture from an inclined wellbore – field experiment. *Journal of Petroleum Technology,* **27**, 641–647.

TEUFEL, L. W. (1983) Determination of *in situ* stress from anelastic strain recovery measurements of oriented core. Presented at *SPE/DOE Symposium on Low Permeability, Denver, Colorado*, SPE/DOE 11649.

YALE, D. P. & RYAN, T. 1994. Comparison of *in situ* stress and hydraulic fracture direction in the Mid-Continent are, U. S. *In*: NELSON P. P. AND LAUBACH, S. E. (eds), *Rock Mechanics: Models and Measurements, Challenges from Industry.* Balkema, Rotterdam, 393–400.

YALE, D. P. & SPRUNT, E. S. 1989. Prediction of fracture direction using shear acoustic anisotropy. *The Log Analyst*, (March-April), 65–70.

YALE, D. P., NIETO, J. A., RODRIGUEZ, J. M. & MOORE, R. D. 1991. *In situ* stress direction from shear acoustic anisotropy: A case history from the Southern North Sea. *In*: *Proceedings of the 14th Annual European SPWLA THAMES Symposium.* Society of Professional Well Log Analysts, London.

YALE, D. P., STRUBHAR, M. K. & EL RABAA, A. W. 1992. A field comparison of techniques for determining the direction of hydraulic fractures. *In*: TILLERSON, J. R. AND WAWERSIK, W. R., *Rock Mechanics Proceedings of the 33rd U. S. Symposium*, Balkema, Rotterdam, 89–98.

YALE, D. P., RODRIGUEZ, J. M., MERCER, T. B. & BLAISDELL, D. W. 1994. *In situ* stress orientation and the effects of local structure – Scott Field, North Sea. In: *Rock Mechanics in Petroleum Engineering.* Balkema, Rotterdam, 945–952.

ZOBACK, M. L. 1992. First- and second-order patterns of stress in the lithosphere: The World Stress Map Project. *Journal of Geophysical Research,* **97**, 11703–11728.

ZOBACK, M. L. & ZOBACK, M. D. 1980. State of stress in the conterminous United States. *Journal of Geophysical Research,* **85**, 6113–6156.

ZOBACK, M. L. & ZOBACK, M. D. 1989. Tectonic stress field of the continental United States. *In*: PAKISER, L. C. AND MOONEY, W. D. (eds), *Geologic Framework of the Continental United States.* Geological Society of America, Memoirs, **172**, 523–539.

Seismic volume attributes for fracture analysis

H. TRAPPE & C. HELLMICH

Trappe Erdöl Erdgas Consultant, Burgwedeler Strasse. 89, D-30916 Isernhagen, Germany (e-mail: Trappe@TEEC.de)

Abstract: Seismic volume attributes are introduced for the characterization of subtle fault patterns. These patterns are important in understanding the flow behaviour in reservoirs. The seismic attributes are based on neighbourhood analysis of the seismic signal and take 2D or 3D information into account. In the case of fracture systems, attributes such as coherency, that quantify change in the seismic signal are used. Case studies are presented to show the impact of the attribute analysis on the description of reservoirs.

Knowledge of fracture systems is of great importance in understanding reservoir behaviour. Fractures and faults can have either positive or negative impacts on the flow of hydrocarbons. While fractures may open the pathways for hydrocarbons due to increased fracture permeability, they can also be responsible for building flow barriers. Acid waters from nearby layers can propagate through fracture zones and modify the reservoir quality by diagenetic processes.

In all cases the knowledge of location, orientation, and length distribution of the fracture system helps in understanding and modelling the flow paths, or, on the contrary, the sealing effects of faults. Such an analysis requires a characterizing effort on all length scales from the microscopic world of core measurements to general geological trends. While, with no doubt, seismic measurements can laterally resolve the range of tens of metres, the challenge for 3D seismic methods lies in the ability to resolve features as subtle as possible.

To achieve these goals we propose the use of seismic multi-trace filtering to compute 2D and 3D seismic volume attributes. This includes advanced seismic processing algorithms with extensive use of higher order statistics and feature extraction methods. These methods are equally applicable in fields such as medical image processing or material science, where direct measurements of desired properties are either costly, or impossible.

Case studies presented will show the benefit of using 2D and 3D volume attributes in exploration and reservoir description.

3D seismic volume attributes in fractured reservoirs

For the discussion of volume attributes, it is assumed that the data are processed to achieve the best available image of the zone of interest. As in other advanced seismic techniques like AVO (amplitude versus offset), or seismic inversion, a processing flow that preserves true-amplitude information is of advantage.

A well known approach to fault characterization is the building of a structural model, with the interpretation of seismic horizons and fault patterns as the first step. Faults and fracture systems are then highlighted through fault mapping and through related seismic attributes, like time-dip maps. Problems can arise when approaching the limits of seismic resolution. Noise and an ambiguous signal character make the manual interpretation difficult and time consuming. Even if small faults and lineaments are directly visible in the data, the great number of traceable events makes it difficult to interpret them manually. Examples from many fields have shown that subtle faults can easily slip through the interpretation process. Time-dip maps, on the other hand, take only time information into account, and not the seismic signal character. They are generally applied when horizons have already been interpreted.

Regularly used seismic attributes like complex trace attributes, are mainly derived by single trace processes. The complex trace attributes, like amplitude envelope, phase or frequency, are computed by taking the seismic samples in moving time windows into account (e.g. Taner *et al.* 1979).

Extending this concept to more than one trace leads to multi-trace filters. These filters can be used to calculate simple and higher-order statistics, or other characteristic measures of the seismic data. A variety of possibilities exist that include 2D operators on time or horizon slices or on seismic sections and 3D filters operating on a seismic volume. The goal of all these multi-trace filters is to quantify points or parts of the seismic

From: AMEEN, M. (ed.) *Fracture and In-Situ Stress Characterization of Hydrocarbon Reservoirs*. Geological Society, London, Special Publications, **209**, 65–75. 0305–8719/03/$15.00
© The Geological Society of London 2003.

volume. Some examples of measures are given next.

The simplest quantification of a seismic subvolume is the mean value:

$$\overline{S} = \frac{1}{N}\sum_i S_i \tag{1}$$

where s_i is the seismic sample, and N the total number of samples. Even this filter may render valuable information. Examples are the discrimination of high energy zones, and the detection of processing-induced amplitude scaling. This operator comes in a large variety of forms, depending on the definition of the subvolume, and on a possible further reduction of the data within this volume. Samples may be rejected for being close to faults, or of uncertain data quality. 2D and 3D computations may be constrained along constant times, or interpreted horizons.

For seismic data, the mean is often substituted by energy related measures like taking absolute values, or the root mean square amplitude:

$$S_{RMS} = \sqrt{\frac{1}{N}\sum_i S_i^2} \tag{2}$$

The standard deviation is a starting point for designing quantities that measure the change inside the subvolume. Taking the absolute values of the differences, instead of the squares, produces an operator known in image processing as dispersion:

$$d = \sum_i \left| S_i - \overline{S} \right| \tag{3}$$

In general, all edge detection and derivative filters can be adapted for the use in seismic data characterization. This includes the wide field of texture classification methods (Haralick 1979). Correlation and cross-correlation based measures are another way of expressing the change in signal character (see Yilmaz (1987); for a number of different methods). In seismic velocity analysis the semblance (Yilmaz 1987, Rabbel *et al.* 1991) is a widely used operator for expressing the correlation between traces. Semblance may take the wave character of the seismic trace into account, if a summation over a time window t is included:

$$s = \frac{1}{N}\frac{\sum_i \left(\sum_i S_i(t)\right)^2}{\sum_i \sum_i \left[S_i(t)\right]^2} \tag{4}$$

Directionally sensitive filters are necessary in order to account for local dip and azimuth. Almost all operators can be modified for this task, generally at the cost of more extensive computations. The output of these filters consists of the general filter response, and of additional attributes describing the orientation of the seismic events. A careful calibration and honouring of the chosen operator is necessary to translate this into a reasonable geological orientation.

The heterogeneity of the seismic data can also be determined with measures such as entropy or fractal dimension. The entropy is based on the histogram of the samples

$$e = \sum_{bins} i \ln i \tag{5}$$

with i being the contribution of each bin in the histogram. In the literature (Mandelbrot 1982), many different methods, such as the box counting method exist to compute the fractal dimension. One of these methods is the scaled range analysis, that estimates the Hurst coefficient H. For the 2D dimensional case, the Hurst coefficient is related to the fractal dimension FD by $H=3-FD$. Looking at the frequency spectrum of a subsample in relation to the wave number, the Hurst coefficient is obtained by the regression in the log/log plot of the two quantities. For both methods valid formulas exist in 2D and 3D, which can be adapted taking the wave character of the seismic data into account. Alternative algorithms based on Fourier or wavelet transforms, are more commonly used in seismic facies characterization of reservoir zones.

In many cases, advanced measures are necessary to capture the complexity of the data. The description by volume attributes naturally leads to a stochastic view of the data. Studies have shown (Trappe *et al.* 1996; Trappe & Hellmich 1997) that volume attributes can be linked to depositional environments and can be used in the discrimination of facies and lithology. This reflects the view of the authors, that a complex geological setting cannot be understood through the analysis of single seismic events.

The detection and mapping of faults and fracture zones commonly require multi-trace filters, that take neighbourhood relations into account. In the case of fractures, all kinds of correlation measures and filters that detect change are of interest. Custom-made filters allow measurement of the length and orientation of seismic features. This includes coherency which has gained popularity lately (Bahorich & Farmer

1995; Marfurt *et al.* 1998), but many other methods exist to characterize seismic data for this purpose.

One of the alternatives is the binary cube (Trappe *et al.* 2000). By comparison to a threshold amplitude, the seismic data set is converted to a binary volume, containing only the values 0 and 1. Similarity measures are calculated from the binary data within subvolumes. The binary data can be processed very efficiently. Similarity calculations with binary values require much less computing time than similarity measures for real amplitudes. Moreover, most of the similarity measures for binary data have a simple structure. One possibility that is demonstrated in this paper, is to count the number n_1 of ones in a subvolume, and divide them by the total number of samples N:

$$B = \frac{n_i}{N} \quad (6)$$

Pattern recognition and feature extraction are further methods used to generate seismic attributes, and handle the great amount of data contained in 3D surveys. They open the possibility to automatically map features which otherwise can only be obtained by time consuming hand-picking. Through these methods the orientation and length distribution of fault and fracture systems can be determined. In combination with geostatistical simulation techniques, it is then possible to build stochastic reservoir models that honour the determined distribution. The combination of volume attributes with the results of AVO and inversion studies is a powerful tool to gain more information from the seismic data (Trappe & Hellmich 1999).

The following case studies delineate the contribution of 3D volume attributes to the investigation of low permeability reservoirs. Since fractures are mainly detected by the changes in seismic signal character, the following examples use a semblance attribute according to Equation 4, and the binary cube according to Equation 6, respectively.

Case study 1: re-evaluation of an existing oil field

The study was conducted as part of a re-evaluation study of an existing oil field. An area of about 100 km^2 was extracted from an existing 3D seismic survey for detailed investigations with 3D volume attributes. An interpretation of key horizons, and well data were available. Both types of additional information, however, were not incorporated in the investigations, but only used for a quality control for the attribute generation and subsequent structural interpretation. Figure 1 is a random line taken from the 3D seismic survey. It shows that the area is highly faulted, and that data quality is an issue in some areas.

To map subtle faults and to identify fracture zones that may impact reservoir behaviour, a volume attribute was used that highlights discontinuities in the seismic signal. For each time sample a subvolume was extracted from the 3D survey. A coherency attribute was chosen according to the semblance algorithm of Equation 4. It was computed along dipping data planes. In addition to the similarity of the signals, the orientation parameters for the optimal alignment of the filter are stored, leading to three parameters for each examined sample point: coherency attribute (Fig. 2a), dip (Fig. 2b), and azimuth (Fig. 2c) of local strike.

All three values can be combined in a so-called intensity-hue-saturation (IHS) display, where the coherency attribute controls the intensity, the azimuth determines the colour (hue), and the dip defines the saturation of the image. Figure 2d is one such IHS display of the corresponding section shown in Figure 1. While major faults are visible in Figure 1, smaller faults

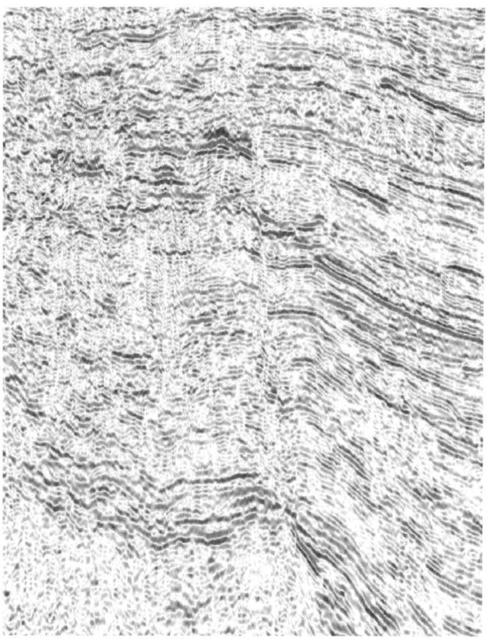

Fig. 1 Random seismic line over the prospect of case study 1.

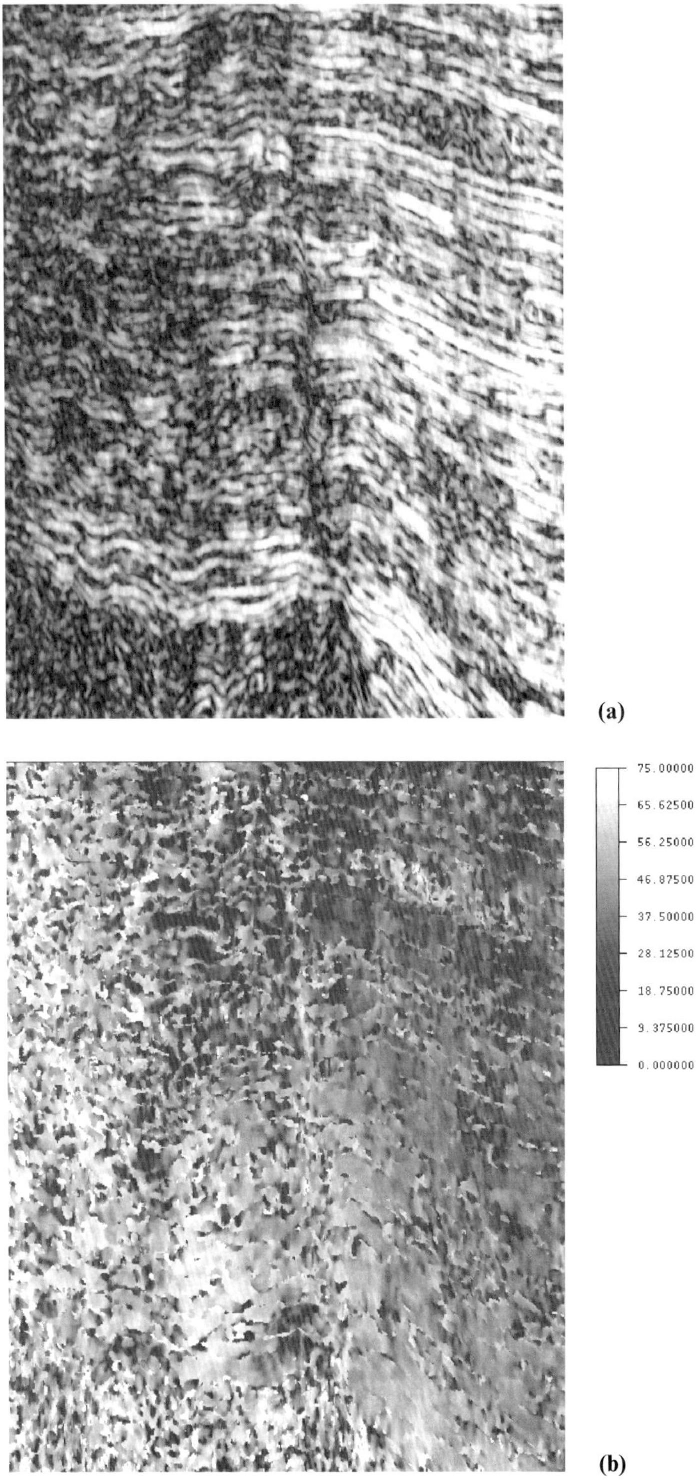

Fig. 2. (**a**) Semblance processing of seismic line in Figure 1 to highlight faults and fracture zones. (**b**) Dip angle estimated by directional filter.

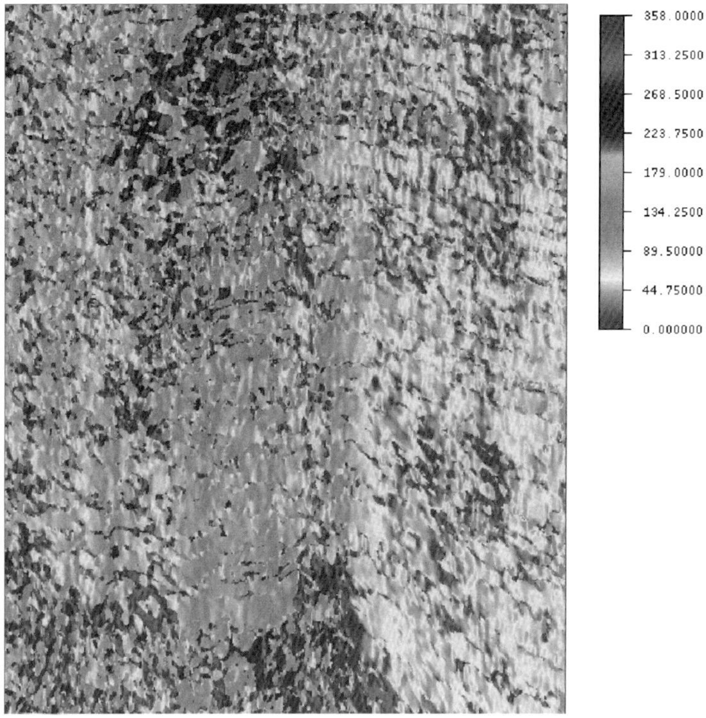

(c)

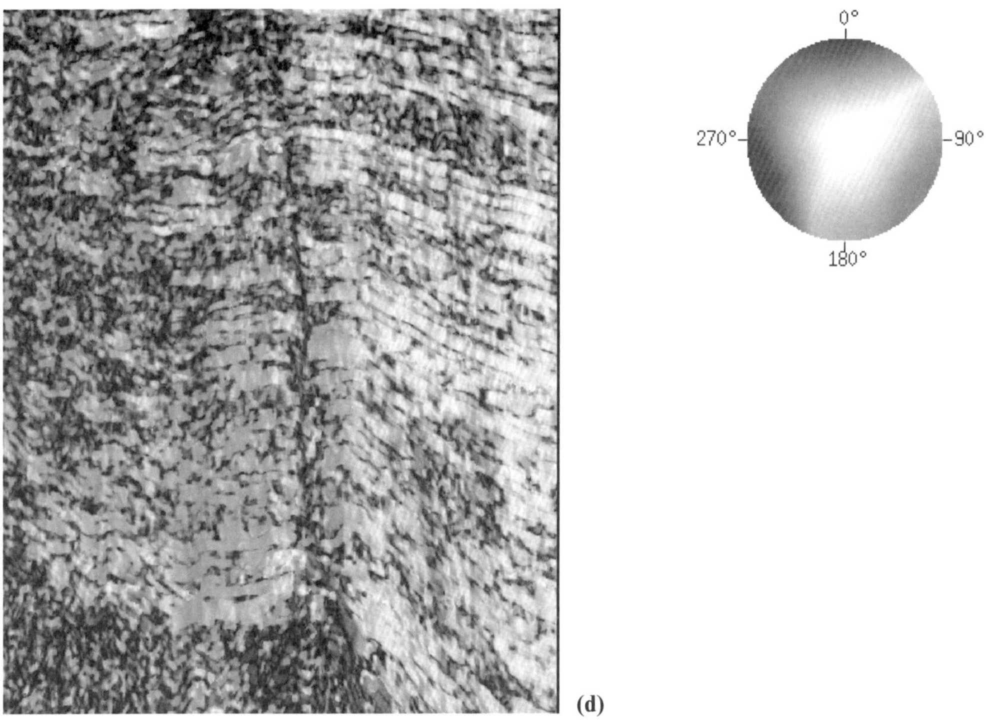

(d)

Fig. 2. (**c**) Azimuth direction estimated by directional filter. (**d**) Combination of (a), (b) and (c) into IHS view of semblance-related attribute, dip and azimuth.

and fracture zones are better resolved in Figure 2d due to the directional sensitivity of the filter. Especially noteworthy are the small features that are distinguished by their colours only (Fig. 2b and c), but otherwise exhibit the same coherency values as their surroundings (Fig. 2a).

Figure 3 is a time slice through the volume showing the coherency, azimuth, and dip as IHS display as another aid for interpreting the subtle features. The structure indicated by the arrow was interpreted as a reef, and gave one of the reasons to conduct this study. For this potential reef structure, four time slices were merged together in Figure 4. The internal structure is clearly defined, as it evolves in two-way traveltime.

In this example, a directional filter was necessary to obtain the presented results. The multi-trace filter takes the neighbourhood of a trace into consideration and assigns true 3D information from this environment to each considered data point. Whereas the simultaneous view of inlines and crosslines allows the experienced interpreter to build an understanding of the signal distribution in space, the combined display of dip, azimuth and coherency immediately shows the spatial orientation of subsurface elements.

Case study 2: diagenetic effects on reservoir quality

This case study shows how different disciplines help to shape the understanding of the mechanism responsible for the reservoir quality. The target is formed by Rotliegend sandstones at a depth of about 4000 m. A 3D seismic survey, as well as some sparse reference wells are available for calibration of seismic and petrophysical parameters.

A random seismic line crossing one of the wells in the study area is presented in Figure 5. Major fault zones in the depth interval from Zechstein to Carboniferous are visible, and can easily be defined with the help of an interpretation system. They are classified as belonging to a post-Zechstein fault system. The well indicated on the seismic section was drilled to test the amplitude anomalies in the Rotliegend. Amplitude mapping and modelling as well as seismic inversion techniques had been applied, suggesting

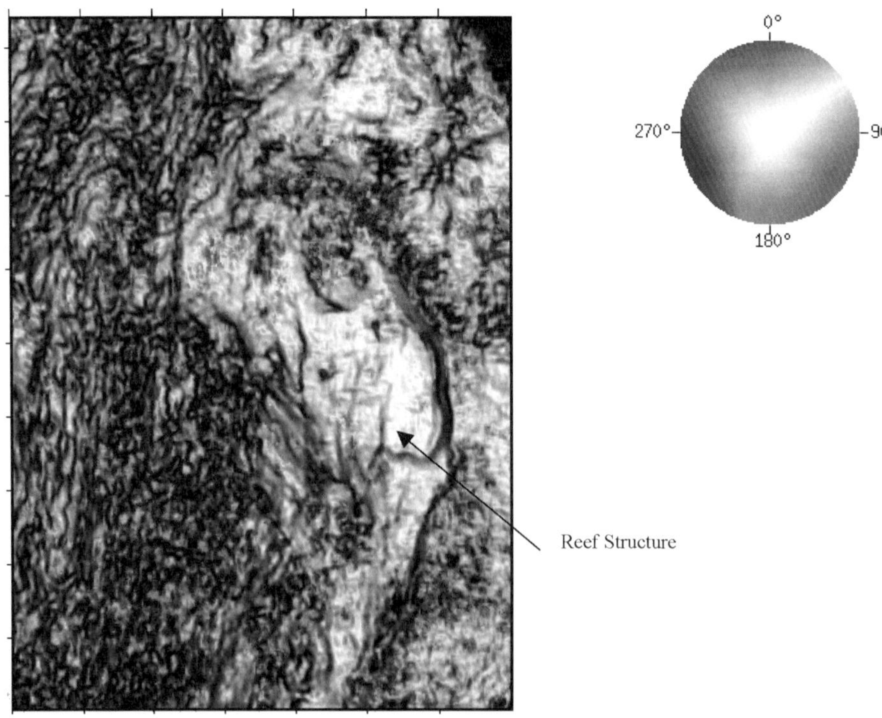

Fig. 3. Time slice showing semblance related attribute, dip, and azimuth as IHS display. The size of the area shown is *c*.100 km2.

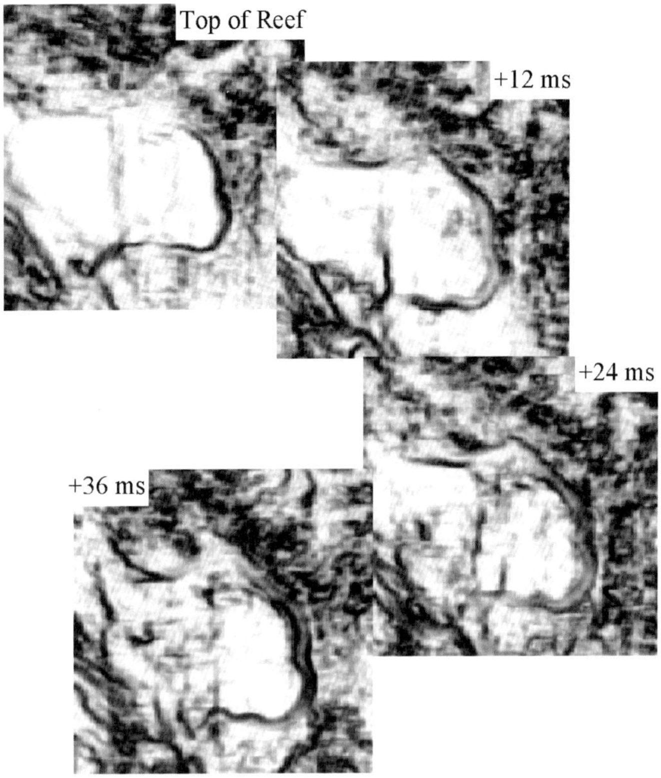

Fig. 4. Time slices of a reef structure.

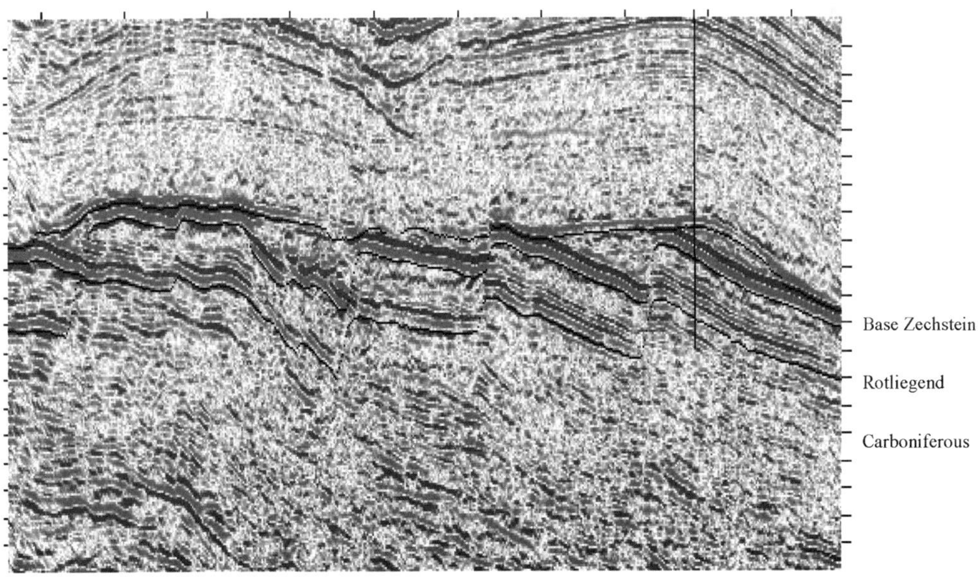

Fig. 5. Random seismic line over potential prospect of case study 2.

that these anomalies were related to Rotliegend sandstone reservoirs of increased porosities. The drilling result supported this prediction. However, despite these good porosities, the reservoirs were characterized by poor permeabilities.

Sedimentological and petrophysical investigations revealed the importance of diagenetic influences on the reservoir quality. Diagenetic processes seemed to be responsible for the poor permeabilities. In Figure 6 the intergranular pore space filled with secondary Illite minerals (pore-bridging Illite) between quartz grains is shown. The Illite minerals were identified as one of the controlling factors for reduced permeabilities in the area. A possible explanation for the occurrence of this Illite meshwork is seen in ascending waters from deeper layers of the Carboniferous. The pathways of such waters are often faults, which connect the Carboniferous and Rotliegend layers. In most of the cases the waters show a high acidity. This causes feldspar leaching, and therefore kaolinite, and at higher depth of burial Illite cementation within the Rotliegend layers (Gaupp *et al.* 1993).

As a consequence high porosities could be mapped using traditional seismic attributes. However, the flow of hydrocarbons is strongly hindered by the illite meshwork, which is not directly mappable using seismic attributes. Thus, the investigations focused on the identification of fracture systems, which could be possible pathways for ascending waters between Carboniferous and Rotliegend layers.

Therefore, a seismic attribute volume was produced, in order to highlight possible fracture zones in addition to the faults already addressed in the seismic amplitude section. The multi-trace filter was chosen according to the binary classification method, which has been introduced in Equation 6.

A result of these computations is presented in the time slices of Figure 7a and b. These binary attribute maps display the heterogeneity of the seismic data along the Rotliegend reservoirs. The fault system seen on the seismic profile (Fig. 5) is also visible in the time slices. In addition to the known fault system, fracture zones can be identified which are hardly visible on the seismic section. These zones are interpreted as belonging to a second, pre-Zechstein fracture system. The position of the reference well discussed before is indicated on the time slices. It is located in the area of a fracture zone of pre-Zechstein age.

A second seismic line is shown in Figure 8. It was chosen perpendicular to the line of Fig. 5, with the intersection of both lines being at the

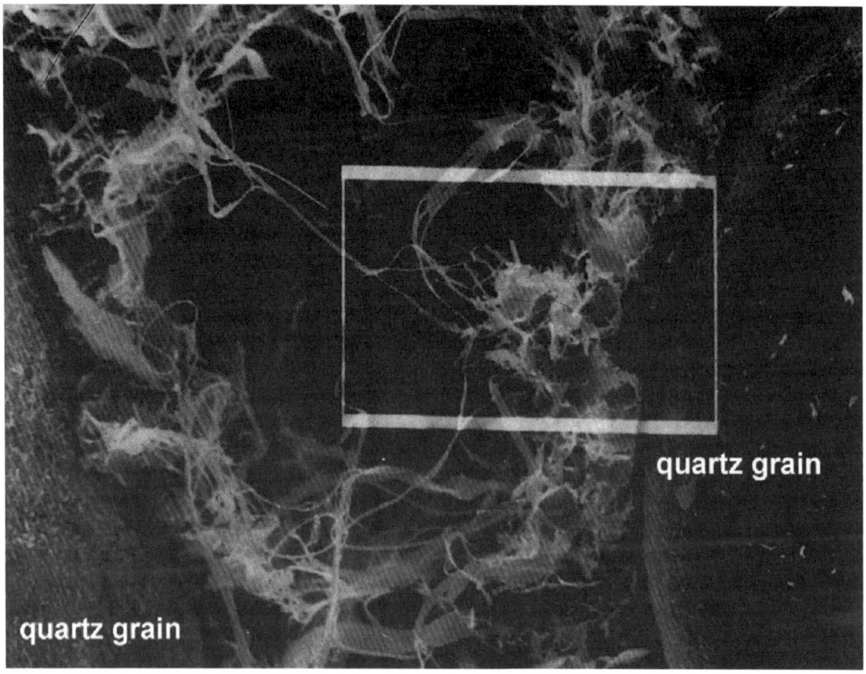

Fig. 6. Intergranular pore space filled with secondary illite minerals (Source: University of Clausthal, Germany).

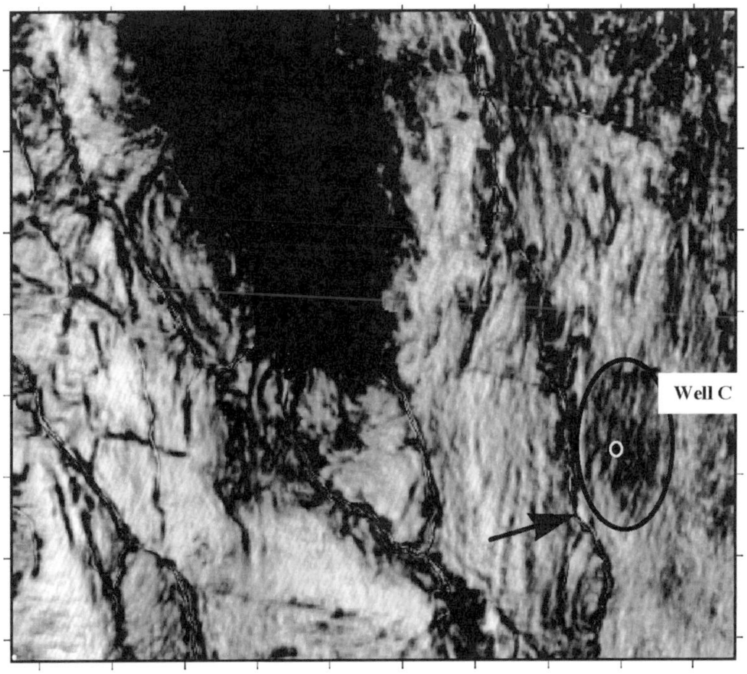

(a)

arrow: fault system known from Fig. 5, *circle*: Pre-Zechstein fracture zone.

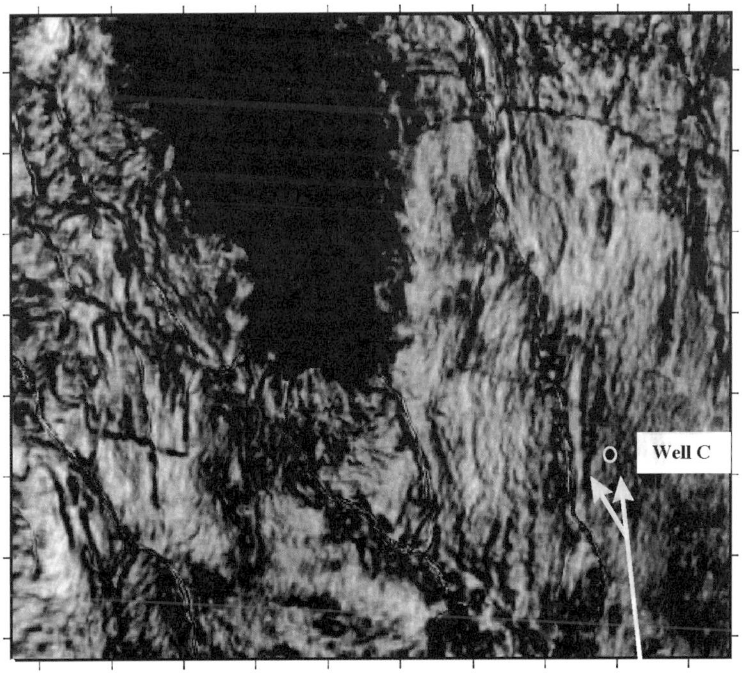

(b)

Pre-Zechstein fractures

Fig. 7. (**a**) Time slice showing the lateral appearance of fracture systems inside the Rotliegend. (**b**) Time slice 12 ms below.

well location. The result of the binary classification along this second line is given in Figure 9. From this vertical slice, and from the horizontal time slices of Figures 7a and b, it is obvious that the reference well has been drilled in a pre-Zechstein fracture zone, and that this fracture zone extends from the Rotliegend down to the Carboniferous. The results support the assump-

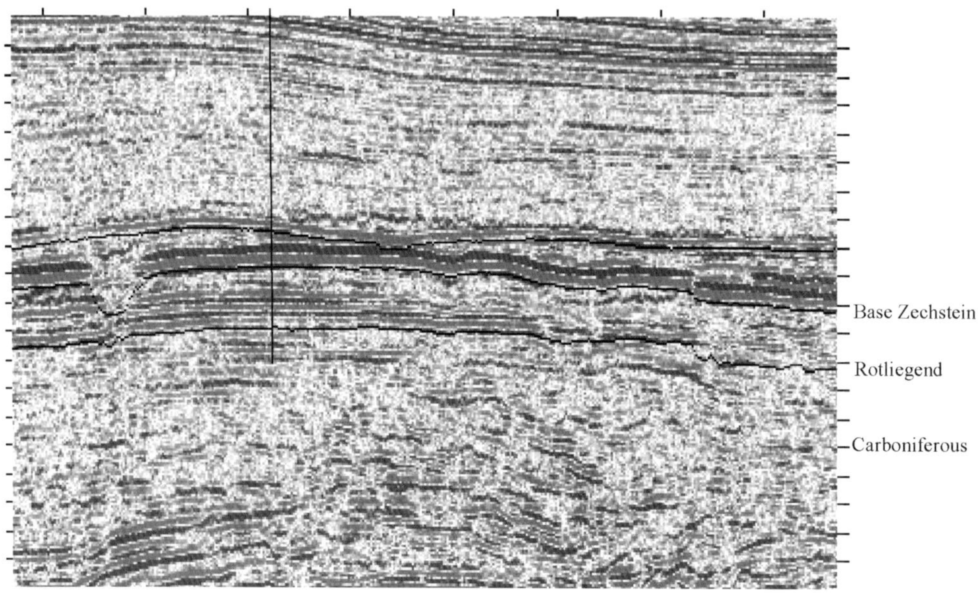

Fig. 8. Seismic line perpendicular to Figure 5, intersecting the well.

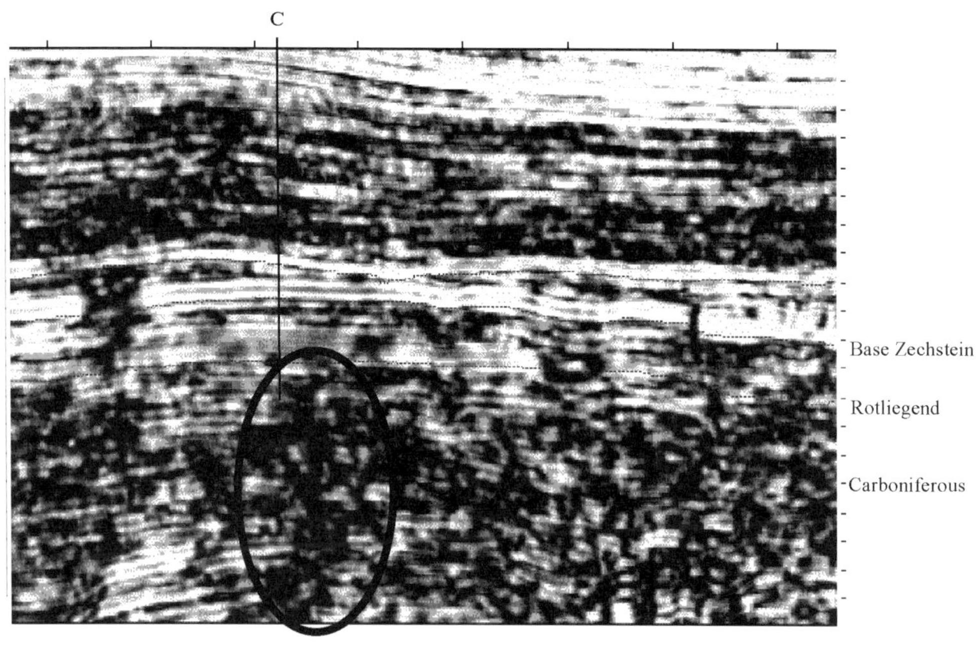

Fig. 9. Same line as Figure 8. Binary classification highlights fracture zones.

tion that locally poor reservoir conditions of Rotliegend sandstones may be related to fault and fracture zones which connect Carboniferous and Rotliegend layers. These faults and fractures represent a possible conduit for formation waters, which can cause diagenetic alterations.

Conclusion

3D volume attributes are means for a detailed characterization of seismic data. The use of multi-trace filters and feature extraction methods open the field for a variety of seismic attributes. Their impact on reservoir characterization has been shown in two examples, which cover the detection of subtle faults, possible fracture zones, and tectonic style. The understanding of the reservoir is improved by the simultaneous use of several attributes, in combination with a reliable reservoir model.

The generated attributes have to be thoroughly examined for their physical and geological meaning to avoid misinterpretation. Well control, geological information, and production data are suited to calibrate the seismic attributes to rock properties. This paper is devoted to 2D and 3D post-stack methods, but the approach can be equally successfully applied to acoustic impedance data, or pre-stack data.

References

BAHORICH, M. & FARMER, S. 1995. 3-D Seismic coherency for faults and stratigraphic features. *The Leading Edge*. **14**, 1053–1058.

GAUPP, R., MATTER, A., PLATT, J., RAMSEYER, K. & WALZEBUCK, J. 1993. Diagenesis and fluid evolution of deeply buried Permian (Rotliegende) gas reservoirs, Northwest Germany. *AAPG Bulletin*, **77**, 1111–1128.

HARALICK, R. M. 1979. Statistical and Structural Approaches to Texture. *Proceedings of the IEEE*, **67**, 768–804.

MANDELBROT, B. B. 1982. *The Fractal Geometry of Nature*. Freeman, San Francisco.

MARFURT, K., KIRLIN, R., FARMER, S. & BAHORICH, M. 1998. 3-D seismic attributes using a semblance-based coherency algorithm. *Geophysics*, **63,** 1150–1165.

RABBEL, W., BITTNER, R. & GELCHINSKY, B. 1991. Seismic mapping of complex reflectors with the common-reflecting-element method (CRE method). *Physics of the Earth and Planetary Interiors*, **67**, 200–210.

TANER, M., KOEHLER, F. & SHERIFF, R. 1979. Complex seismic trace analysis. *Geophysics*, **44,** 1041–1063.

TRAPPE, H. & HELLMICH, C. 1997. Image processing and geostatistics upper carboniferous reservoir description. *EAGE 59th Conference and Exhibition 1997 Extended Abstract.*

TRAPPE, H. & HELLMICH, C. 1999. Carboniferous case study : combining 3D seismic inversion and 3D heterogeneity results for optimal well and production planning. *EAGE 61st Conference and Exhibition 1999 Extended Abstract.*

TRAPPE, H., HELLMICH, C. & FERTIG, J. 1996. Bildverarbeitung seismischer Attribute und Geostatistik im Oberkarbon. *DGMK Conference volume 9601.*

TRAPPE, H., HELLMICH, C. & FÖLL, M. 2000. Potential Power in the Application of Seismic Volume attributes. *First Break*, **18**, 397–402.

YILMAZ, Ö. 1987. *Seismic Data Processing*. Society of Exploration Geophysicists, Tulsa.

Microseismic monitoring of geomechanical reservoir processes and fracture-dominated fluid flow

ANDREW J. JUPE, ROBERT H. JONES,
STEPHEN A. WILSON & JOHN F. COWLES

ABB Offshore Systems, Rosemanowes, Penryn, Cornwall, TR10 9DU, UK
(e-mail: andrew.j.jupe@gb.abb.com)

Abstract: Microearthquakes (microseismic events) are induced during hydrocarbon and geothermal fluid production operations in naturally fractured reservoirs. They typically result from shear-stress release on pre-existing faults and fractures due to production/injection induced perturbations to the effective stress conditions. These stress changes may be due to reservoir depletion, flooding or stimulation operations. Over a number of years it has been shown that microseismic monitoring has the potential to provide valuable time-lapse 3D information on the geomechanical processes taking place within a reservoir. These processes include the distribution of fluid flow and pressure fronts within naturally fractured systems, production-related compaction and the reactivation of faults. With the advent of permanent reservoir monitoring systems (e.g. intelligent wells), microseismic monitoring has the potential to become a practicable means of time-lapse imaging of hydrocarbon reservoir processes remote from production/injection boreholes. This paper illustrates some of the ways in which microseismic monitoring can contribute to the development and management of hydrocarbon reservoirs through the presentation of examples from both hydrocarbon and geothermal reservoirs.

Throughout the life of a reservoir the hydrocarbon production process induces pressure changes within the reservoir. These changes result in perturbations to the *in situ* stress conditions that propagate through the reservoir. One of the key drivers for the development of 4D (3D time-lapse) reflection seismic data has been the operator's desire for time-lapse images of where and when these changes are taking place, and also their magnitude and impact on production. This information then feeds into the reservoir management process, particularly through integration into reservoir simulation and in the longer term through coupled hydro-geomechanical reservoir models.

An additional and direct consequence of these pressure and stress changes is the possibility of re-activation of pre-existing reservoir structures such as faults and fractures, which manifests itself as small earthquakes, referred to as micro-earthquakes or microseismic activity. In mature fields and fracture-dominated reservoirs the pressure and stress changes tend to be more severe and the seismic activity more intense. The following are three situations in which induced microseismic activity is likely to occur.

Reservoir depressurization can lead to localized reservoir compaction, or on a reservoir scale to compaction-assisted or compaction-driven production. In several North Sea fields where compaction is taking place there are associated problems of wellbore stability and integrity. Large-scale depressurization near fault-sealed or fault-compartmentalized reservoirs will lead to fault reactivation and the potential for pressure breakthrough into neighbouring reservoir compartments or even fields.

Enhanced recovery, particularly in mature fields, may require massive injection and flooding programmes aimed at pressure maintenance. Even in moderately fractured reservoirs the efficiency of these operations is known to be strongly affected by anisotropic or heterogeneous flood patterns.

Reinjection is also becoming an increasingly common approach to the disposal of waste from development and in-fill drilling, from increased solids and water production, and ultimately it may be used in seafloor waste recovery and disposal prior to decommissioning. The question being asked by operators and regulators is how can this process of reinjection be managed and the size and shape of disposal domains monitored.

This paper aims to demonstrate how induced microseismic activity can be used for continuous 3D time-lapse monitoring of these hydro-geo-mechanical processes and assist in the reservoir

From: AMEEN, M. (ed.) *Fracture and In-Situ Stress Characterization of Hydrocarbon Reservoirs.* Geological Society, London, Special Publications, **209**, 77–86. 0305–8719/03/$15.00
© The Geological Society of London 2003.

development and management. This is achieved through the presentation of data sets from three hydrocarbon reservoirs (Clinton County, USA, and Ekofisk and Valhall, North Sea) and also an example from the geothermal industry where the technique is already well established.

The origin of microseismic events

Microseismic events are basically small earthquakes, usually Richter magnitude (M_L) <0, and they have been detected and located at distances of over 1 km from the monitoring well in some hydrocarbon reservoirs, depending on seismic attenuation and complexity of the velocity structure. They occur because the Earth stresses acting within the reservoir are anisotropic. This causes shear stresses to build up on naturally occurring fracture surfaces that under normal conditions are locked together (Fig. 1). When the *in situ* stresses are perturbed by reservoir production activity, such as changing fluid pressures, the fractures can shear producing small earthquakes (Fig. 1). The seismic signals from these microseismic events can be detected and located in space using high bandwidth borehole geophones. Microseismic activity has been successfully detected and located in rocks ranging from unconsolidated sands, to chalks to crystalline rocks (e.g. Keck & Withers 1994; Ilderton *et al.* 1994; Deflandre *et al.* 1995; Wallroth *et al.* 1996; Maxwell *et al.* 1998; Warpinski *et al.* 1999; Snell & Close 1999; Bell *et al.* 2000; Maxwell & Urbancic 2001). Other potential sources of microseismic activity include the generation of new fractures through tensile splitting (e.g. conventional hydraulic fracturing) or pore collapse associated with compaction, that might produce implosion events. However, shearing mechanisms are generally believed to be the most common and also most energetic failure mechanism.

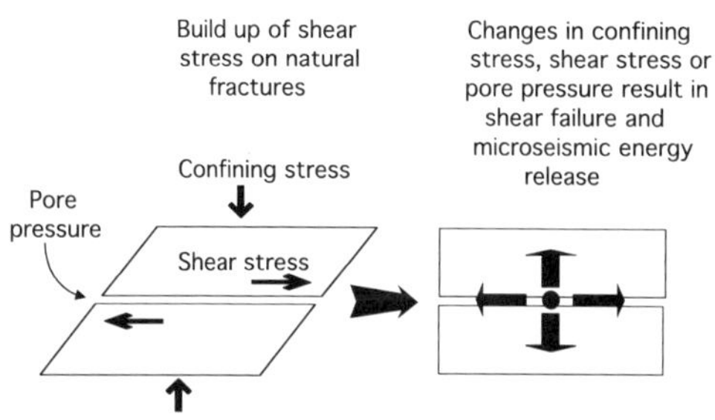

Fig. 1. Schematic of the main microseismic shear failure mechanism.

The examples presented in this paper illustrate the potential of microseismics to:

- identify fault structures that can result in reservoir compartmentalization or act as flow channels and routes for premature water breakthrough;
- image flow anisotropy associated with production from fracture-dominated reservoirs;
- provide real-time 3D monitoring of fluid pressure front movement, such as water flood fronts;
- assist in targeting new producer/injector wells;
- identify areas of reservoir compaction and potential wellbore instability;
- Provide input and condition permeability grids for reservoir simulation.

Fracture zone controlled water production (Clinton County, Kentucky, USA)

Clinton County, Kentucky is located within the Cumberland Saddle of the Cincinnati Arch, immediately west of the Grenville Front. The reservoir in this field consists of a low porosity carbonate within which fracture storage and permeability are suggested by isolated, high-volume production wells which subsequently produced brine (Phillips *et al.* 1996). Initial production rates as high as 400 barrels per hour and cumulative production of 100000 barrels from a single well have been reported.

Basement-controlled wrench-fault structures have been associated with oil production from shallow (135–180 m) carbonate reservoirs 65 km west of Clinton County. Operators have based drilling programmes on fracture/lineament patterns delineated from radar images and interpreted as being associated with wrench faulting of an east–west trending basement fault.

Several programmes of microseismic monitoring have been carried out by personnel from the Los Alamos National Laboratory during production to help identify active and possibly producing fault structures (e.g. Rutledge *et al.* 1994; Phillips *et al.* 1996). Microseismic results reported by Phillips *et al.* (1996) defined three low-angle, reverse slip fracture zones (Fig. 2), where the seismic activity was clearly related to the production of over 8100 barrels of oil from well HT1. Rutledge *et al.* (1994) also reported similar structures from other parts of the field. The mapped fractures fell above and below the HT1 production interval, but intersected or could be mapped to old production well intervals (GT1, GT2 and GT4). A cumulative volume of 4600 barrels of oil was extracted from these three wells in the nine months preceding monitoring.

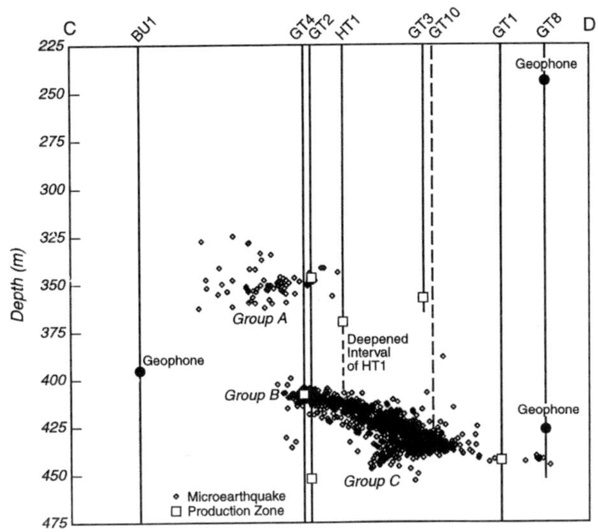

Fig. 2. Cross-sectional view of microseismic activity at Clinton County (after Phillips *et al.* 1996). Reproduced with the kind permission of the SPE.

Well GT10 was subsequently drilled into the main mapped fracture and produced brine. Later well HT1 was deepened, encountering brine within 3 m of the same mapped fracture. The interpretation is that the microseismic monitoring defined oil-bearing producing fractures that subsequently filled with brine, presumably from an active, but in this case poorly connected water drive.

The presence of these low-angle, oil-bearing faults has implications for field development. Drilling horizontal or deviated wells does not increase the probability of intersecting productive fractures in Clinton County. However, interwell correlation and mapping of the conductive fractures could allow better planning of plug and abandonment operations, so as to avoid premature contamination of the main pay zone with water. Pressure maintenance operations are also being considered once the fracture zones between the wells have been mapped.

The results obtained in Clinton County are a clear illustration of how microseismic imaging of active fractures can assist in identifying fracture-controlled water breakthrough and assist planning reservoir development.

Monitoring fractured and compacting chalk reservoirs (Ekofisk and Valhall, North Sea)

In 1997/98 microseismic monitoring trials were conducted in the Ekofisk and adjacent Valhall fields, operated by Phillips Petroleum and BP-Amoco respectively. Both fields comprise highly productive naturally fractured chalk reservoirs that are undergoing varying levels of production-induced compaction.

In April 1997 an 18–day microseismic monitoring trial was conducted in the Ekofisk field (Maxwell *et al.* 1998; Dangerfield *et al.* 1999). The monitoring system consisted of a six-level triaxial VSP wireline tool deployed within the reservoir, in an observation well (2/4–C11a) located near the crest of the field. Approximately 2100 microseismic events were recorded, corresponding to roughly five events per hour at distances of at least 1 km from the observation well. The majority of the events were located in the upper part of the reservoir, predominantly within low-porosity layers overlying relatively porous layers which are undergoing water flooding and compaction (Fig. 3). Events were most accurately located near the monitoring borehole, with the location of events further away being more uncertain. Within 300 m of the well most events were precisely located to an accuracy of better than 30m.

Locations of the events close to the monitoring well (C11A) were obtained with a 3D-velocity model, which is particularly accurate in this region due to the large number of wells. Figure 4 shows that the events around C11A cluster along discrete linear bands. These linear clusters are interpreted to result from induced movements on pre-existing faults. They have the same orientations as the majority of the faults

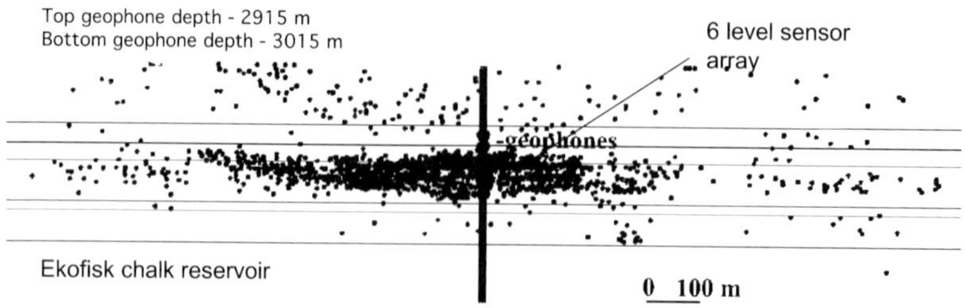

Fig. 3. Vertical cross-section showing the location of the microseismic events and major reservoir layers at Ekofisk (after Maxwell *et al.* 1998).

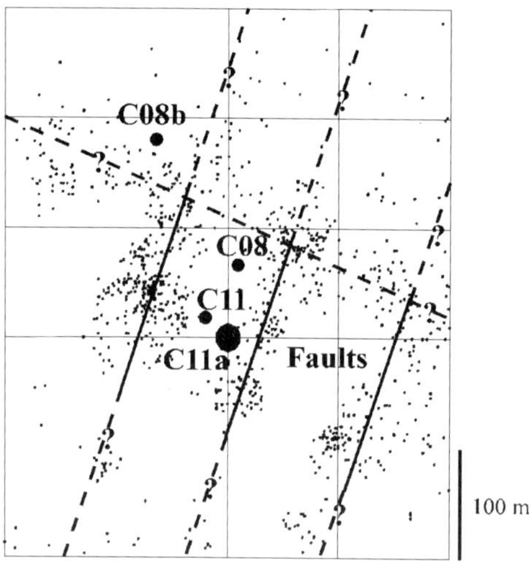

Fig. 4. Map view of events showing distinct lineations attributed to induced movement on pre-existing faults at Ekofisk (after Maxwell *et al.* 1998). Well C11A is the monitoring well.

within the reservoir (NNE-SSW and NW-SE). These results are particularly useful because this portion of the field is not imaged by the normal surface seismic methods due to the presence of crestal gas cloud.

A comparison was made between a numerical waterflood simulation and the positions of the induced microseismicity (Maxwell *et al.* 1998). Only events above magnitude –2 were included, corresponding to the magnitude range where all events are detected independent of distance. Comparisons with simulations of the waterflood front indicated that the saturated zones around the injection wells were aseismic. However, the position of the waterflood front was not known accurately enough to test if any of the clusters of microseismicity were associated with it. Nevertheless the results are encouraging despite the relatively short monitoring period. The compaction of the chalk layer may be strongest at the waterfront. If a significant proportion of the total acoustic activity is caused in association with the waterfront, then time-lapse microseismic monitoring could represent a useful tool for tracking the position of the water flood.

In summary, microseismicity was successfully recorded and processed using a VSP sonde in a single borehole in Ekofisk. Events were accurately located and found to be concentrated in the low-porosity sublayers. Close to the monitoring well the microseismicity clustered in specific lineations in plan view, believed to be reactivation of pre-existing faults. The dynamic relationships between fault movements, reservoir compartmentalization compaction and waterflood are clearly complex. It will require longer test to see if these mechanisms can be isolated; however, this initial study was highly encouraging.

The Valhall field, which is adjacent to Ekofisk, has been producing oil since 1982 (Dyer *et al.* 1999). It currently produces in excess of 100 000 barrels of oil per day. Estimated original oil in place is 2.4 billion barrels, of which 640 million barrels can be recovered under the current plan. Development drilling is still ongoing. The field is recognized as one of the most challenging in the North Sea. The reservoir is relatively thin, structurally complex and covers a 50 km^2 area. It was formed as a slightly asymmetric anticline by inversion along the major Skrubbe fault that lies west of the field.

While 3D seismic data play an essential role in assessing depletion of the structures flanks, gas charging of the overburden has severely distorted the seismic image in the crestal part, as in Ekofisk. The reservoir rock is weak, highly porous Cretaceous chalk that undergoes significant compaction when depleted. Fracture permeability and compaction compensate for generally low chalk-matrix permeability. Horizontal wells have significantly improved recovery.

Although compaction of the chalk adds to oil recovery, it also results in the overburden deformation. This implies seafloor subsidence (currently 3.5 m) and a negative impact on well life. Overburden deformation associated with compaction calls for special wells and casing design, in both overburden and reservoir zones. Still one or two wells per year are lost due to permanent failure or casing collapse.

Most of the horizontal wells are completed by multiple proppant fracturing with up to five or six fractures in one well. The ability to monitor and qualify these stimulations is currently limited to using overall production data. Additional observations that could be used to delineate the magnitude and direction of fractured zones would be very valuable in optimising wellbore stimulation.

A microseismic monitoring trial (Dyer *et al.* 1999) was conducted to see if the technique could provide information on:

- deformation mechanisms;
- poor seismic control in the crestal area;
- wellbore stability issues associated with subsidence;
- dominant stress orientation;
- feasibility for monitoring proppant fracturing in the reservoir.

As with Ekofisk a single six-level VSP wireline tool was deployed in a monitoring well. In addition to its location in the crestal part of the reservoir it was only a few hundred metres away from a well that was being sidetracked during the monitoring period, with hydraulic fracturing of the reservoir planned when drilling ended. The unfortunate limitation was a clay plug in the bottom hole section of the monitoring well which caused the geophone array to be located 250 m shallower than planned and above the top reservoir. This increased seismic attenuation and significantly reduced the number of events detected.

Frequency of microseismic event rates ranged from none to ten per day. During the 57-day monitoring period 572 events were detected, of which 324 could be reliably located. The event rate was about 10% that observed in the Ekofisk field. This appears to be due to the high attenuation of the low velocity sequence between the geophones and reservoir, and the fact that the sensor string could not be deployed as close to the reservoir as it had in Ekofisk, where the deepest geophones were within the reservoir. This means that only nearby or more energetic events were detected.

All located events except those related to drilling activities were located within a 50 m thick zone directly above the Top Balder reservoir formation. The limited vertical and lateral distribution of the microseismic events indicated that the microseismicity was related to reservoir production rather than random background seismicity. This is a major zone where wellbore stability problems are experienced. From event locations two potential structures were interpreted (Fig. 5). These events are shown as ellipsoids representing one standard deviation of the event location uncertainty. Analysis of the focal mechanisms of events in these lineaments indicated a significant normal-faulting component. Although the direction of the lineaments does not necessarily indicate fault strike it is likely to be the case in Valhall. These lineaments indicated two sets of fault directions: one roughly NW and SE and the other roughly SW to NE. This agrees with the current fault pattern model for Valhall.

As with Ekofisk this survey proves a high level of microseismic activity in the Valhall field that is

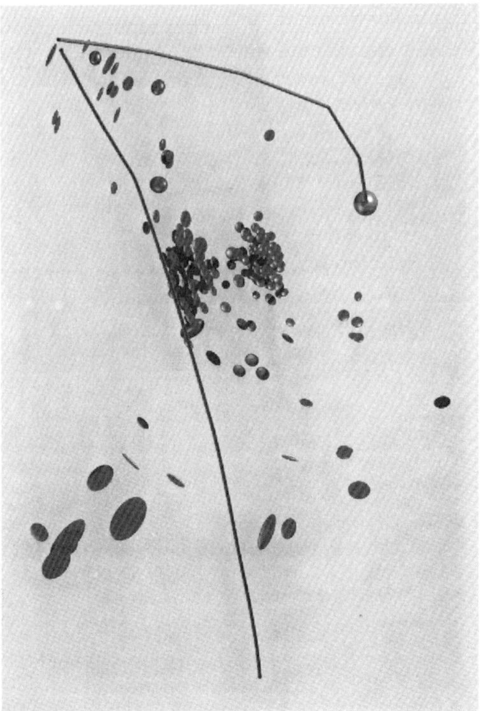

Fig. 5. Plan view of one standard deviation confidence ellipsoids for event locations in Valhall (after Dyer *et al.* 1999). Note clustering of event locations into two planar structures. Green sphere indicates location of geophones.

most likely related to reservoir production. The observed microseismic activity agrees with general fault trends and occurs in a zone with significant wellbore stability and integrity problems.

Geothermal reservoir development and management (Soultz-sous-Forets, Alsace)

The Soultz European Hot Dry Rock (HDR) experimental site (Baria *et al.* 1999) is located in the Rhine Graben *c.* 50 km north of Strasbourg. The geology comprises *c.* 1.5 km thick sedimentary cover overlying granitic basement. The geothermal reservoir itself is 2.5 to 4 km in depth in a faulted horst block, with rock temperatures of >160°C at 4 km depth. The system has been developed progressively since 1992, through a series of reservoir stimulation and characterization operations. In 1995 this led to the targeting and drilling of a second deep (4 km) borehole, creating a circulating HDR system. The first major development phase took place in 1993 with the large-scale stimulation (massive hydraulic fracturing, MHF) of a single deep borehole GPK1 (Fig. 6), over a 700 m open-hole length (2800 to 3500 m). This operation comprised the injection of 45 000 m^3 of water at flowrates up to $50\,l\,s^{-1}$ and overpressures of *c.* 10 MPa. The operation was aimed at opening up the existing fracture system, thus reducing flow impedance.

The operation was monitored using a network of three permanent four-component downhole accelerometers (1.5 km depth) and a single hydrophone, deployed in observation boreholes (Jupe *et al.* 1999). During the course of the operation 18 000 microseismic events were detected and located (Fig. 6). These formed a roughly ellipsoidal cloud with a NW–SE to north–south alignment (Fig. 7), a vertical extent of *c.* 1.2km and horizontal dimensions of *c.* 800 m and 150 m (Fig. 6).

At shallower depths (<3000 m) it was found that the near-wellbore growth direction was more north–south (close to the maximum horizontal stress direction σ_H) and believed to be controlled by the tensile opening (jacking) of fractures subparallel to σ_H. At greater depths, where the overpressure was lower compared to σ_H, the growth direction was more NW–SE; indicative of shear-failure-dominated growth along pre-existing fractures.

The hypothesized opening up of near-wellbore fractures at shallower depth was further confirmed by flow logging, where *c.* 60% of the flow left the borehole in the first 100 m of openhole, and by the microseismic depth distribution, with *c.* 70% of the microseismic events occurring at shallower depths than 3000 m.

Following the 1993 stimulation a series of reservoir characterisation operations were undertaken in 1994, at which time the microseismic activity was used to target a second deep (4 km) borehole GPK2. The objectives with this borehole were to achieve a hydraulic connection to GPK1 over the deepest (hottest) sections of the reservoir, thus maximizing recovery temperatures. The anisotropy of the microseismic cloud clearly dictated that the borehole should be located either NW or SE of GPK1, but that the deepest connection (3500 m) would be achieved by drilling to the S–SE, also satisfying a second criterion of a well separation of *c.* 350 m.

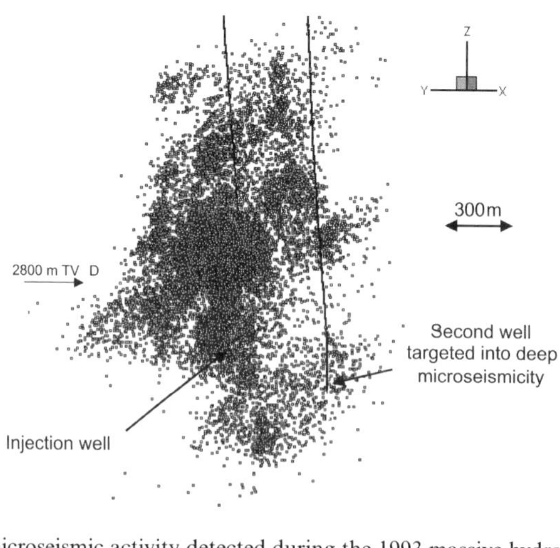

Fig. 6. Side elevation of microseismic activity detected during the 1993 massive hydraulic fracturing (MHF) of GPK1, Soultz. Well separation is approximately 350 m.

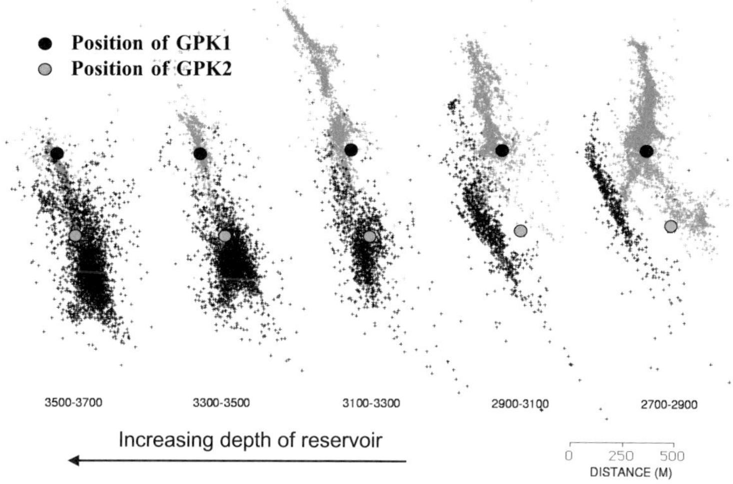

Fig. 7. Plan view of 300 m thick depth slices through the 1993 (GPK1) and 1995/96 (GPK2) MHF data sets.

In 1995/96 the second borehole GPK2 was stimulated at flowrates up to 80 l s^{-1}, reaching downhole overpressures of *c.* 12 MPa. Over 12 000 microseismic events were detected (Fig. 8), forming a NW–SE trending structure, encroaching on the deepest sections of the GPK1 microseismic cloud and extending as far as the GPK1 borehole (Fig. 7). The GPK2/GPK1 reservoir was then successfully tested through a series of circulation operations. Long-term reservoir circulation (injection/production) started in 1997, accompanied by production logging, hydraulic and inert tracer tests.

The results of production logging during circulation indicated that flow exits were distributed along the entire openhole length of GPK2, but that flow entry into GPK1 occurred at two discrete intervals; at the bottom of GPK1 and in the upper section of GPK1 close to the casing shoe. The results of the microseismic monitoring (Fig. 8) demonstrated a strong correlation with this flow path distribution, indicating intersections with GPK2 over the entire openhole length and two apparent intersections with GPK1. This observation prompted an investigation into whether the microseismic event distribution from the Soultz reservoir could be used to construct a quantitative fluid flow model for reservoir performance prediction and use in reservoir development decisions.

Simulator grid construction

The microseismic data from the two MHF treatments has been analysed in terms of event spatial density and connectivity. This information is then mapped to a grid of reservoir simulator cells. The degree of grid connectivity and the scaling of cell hydraulic properties (i.e. pipe conductivity) is based on the analysis of microseismic attributes for the cells (i.e. density, cumulative energy release, cumulative shear slip) in conjunction with the measured hydraulic properties of the system (i.e. flow logs). Once constructed the grid of simulator cells is then input into a steady-state flow model and the simulation results compared with observations from the water circulation (i.e. flow logs, fluid recovery and inert tracer results). Figure 9 presents the fluid pressure field and flow distribution obtained from simulations of the circulation conducted in 1997 between GPK1 (injector) and GPK2 (producer). Figure 9 shows the most direct flow path between the two boreholes, in which flow leaves the bottom of GPK1 and is recovered in GPK2. This flow path represents the shortest breakthrough time of inert tracer (*c.* 4 days). A second significant flow path is observed in the upper section of GPK1, however this is seen to move out of the plane between the two wells. Fig. 9 demonstrates that this is in fact a hydraulic connection between the two wells, but is more tortuous than the deeper connection. This movement of fluid out of the plane of the two wells results in a longer tracer recovery time than the deeper connection. Both the deep and more shallow flow exits correlate well with the results of flow exits/entries in production logs.

Both the spatial distribution of the major water flow paths remote from the wellbore, and reasonable estimates of tracer transport times

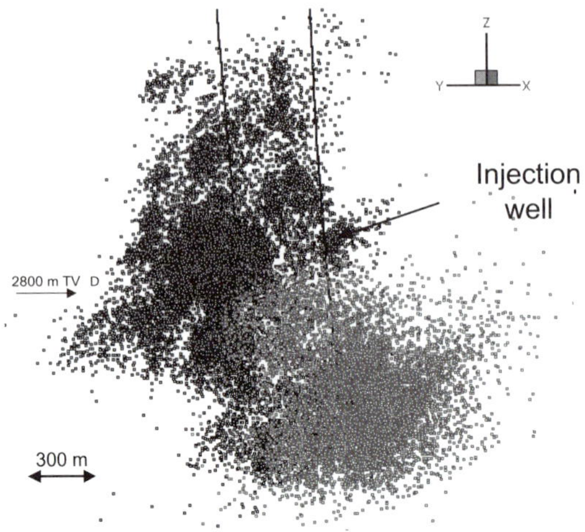

Fig. 8. Side elevation of microseismic activity detected during the 1993 MHF of GPK1 (left) and 1995/96 MHF of GPK2 (right). Well separation is approximately 350 m.

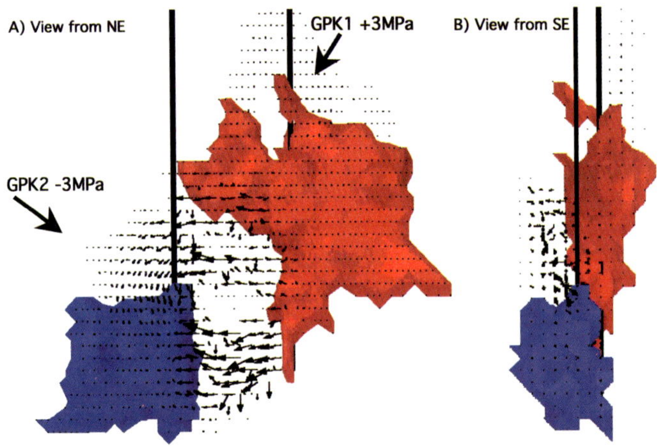

Fig. 9. 3D steady-state fluid pressure iso-surfaces (is ±3 MPa) and flow vectors calculated for circulation between GPK1 and GPK2. The length of the flow vector arrows represents the magnitude of the flow velocity. Note that the flow vectors also indicate discrete flow paths that are out of the plane between the two wells. (**a**) View from NE; (**b**) view from SE.

between boreholes can be reproduced from this analysis of connectivity. This work confirms the importance of flow path geometry (i.e. anisotropy and heterogeneity) in controlling the flow behaviour of naturally fractured systems and highlights the significance of microseismic imaging for reservoir characterization and monitoring.

In summary, the key contributions to the development and management of this naturally fractured geothermal reservoir have been the following:

- Microseismic monitoring was successfully used to monitor the growth direction and extent of a series of Stimulation (massive hydraulic fracturing) operations at the Soultz site.
- Microseismicity has been used to identify discrete flow paths and quantify the flow/pressure field within the reservoir.
- Microseismicity has been used successfully to target a water injector well at a depth of around 4 km.

- Microseismicity has been used to assess 3D reservoir volume and shape, and hence quantify the success and efficiency of reservoir development operations.
- Results of the initial GPK1 stimulations directly influenced the parameters of the later GPK2 hydraulic operations.
- Microseismicity has been successfully used to construct a reservoir simulator grid for modelling the long-term behaviour and development of the geothermal system.

Summary

Microseismic imaging is well developed in geothermal applications and is directly applicable to the monitoring and management of hydrocarbon reservoirs, where it offers a high resolution means of continuous monitoring of the hydro- and geomechanical processes taking place within an evolving hydrocarbon reservoir.

The examples illustrate the potential of microseismics to:

- identify fault structures that can result in reservoir compartmentalization or act as flow channels and routes for premature water breakthrough;
- image flow anisotropy associated with production from fracture-dominated reservoirs;
- provide real-time 3D monitoring of fluid pressure front movement, such as waterflood fronts ;
- assist in targeting new producer/injector wells
- identify areas of reservoir compaction and potential wellbore instability;
- provide input and condition permeability grids for reservoir simulation.

Despite the potential of the technique there have been barriers preventing its widespread uptake by the industry. However, the industry is now more open to new ideas and 'step changes' in technology. In particular 4D seismic has succeeded in moving the focus from static reservoir models to understanding more about long-term reservoir behaviour and production processes in the inter-well region. The remaining technological difficulty is the deployment of permanent seismic sensors in production well completions; however, this barrier is also disappearing through the emergence of permanent borehole sensor installations and intelligent well completion technology, with subsurface sensor installations, dedicated telemetry systems and well control devices. Service companies are now introducing permanently deployed seismic systems into their product ranges (e.g. Deflandre *et al.* 1995; Hottman & Curtis 2001) and operators are beginning to trial these systems in intelligent well deployments (e.g. Bell *et al.* 2000). Soon it will be possible to fully exploit the 'grunts and groans' induced by hydrocarbon production.

The Soultz European HDR Programme is part of the 'Community Research Programme' of the European Commission. Funding for the European programme was provided by DGXII of the European Commission (Brussels), ADEME, BRGM and CNRS (France), BMBF and KFA (Germany) and other national and private sources.

The authors also acknowledge the funding support provided for aspects of the reported work by the NEDO (Japan) funded MTC (More Than Cloud) and MURPHY Projects managed by Tohoku University, by the UK DTI-OSO/LINK programme, by the European Community DGXVII Thermie Programme and by Phillips Petroleum and BP-Amoco.

References

BARIA, R., BAUMGARTNER, J., GERARD, A., JUNG, R. & GARNISH, J. 1999. European HDR research programme at Soultz-sous-Forets (France) 1987–1996. *Geothermics* **28**, 655–669.

BELL, M., KRAAIJEVANGER, H. & MAISONS, C. 2000. Integrated downhole monitoring of hydraulically fractured production wells. *Proceedings of 2000 SPE European Petroleum Conference, Paris, 24–25 October*. SPE Paper 65156.

DANGERFIELD, J., YOUNG, R. P. & MAXWELL, S. C. 1999. Microseismicity – so what? *First Break* (May), 159–161.

DEFLANDRE, J. P., LAURENT, J., MICHON, D. & BLONDIN, E. 1995. Microseismic surveying and repeated VSPs for monitoring an underground gas storage reservoir using permanent geophones. *First Break* (13 April), 129–138.

DYER, B. C., JONES, R. H., COWLES, J. F., BARKVED, O. & FOLSTAD, P. G. 1999. Microseismic survey of a North Sea reservoir. *World Oil* (March), 74–78.

HOTTMAN, W. E. & CURTIS, M. P. 2001. Borehole seismic sensors in the instrumented oil field. *The Leading Edge* (June), 630–634.

ILDERTON, D. C., PATZEK, T. W., RECTOR, J. W. & VINEGAR, H. J. 1994. Passive imaging of hydrofractures in the South Belridge Diatomite. *Proceedings of 1994 SPE Annual Technical Conference and Exhibition, New Orleans, 25–28 September*. SPE Paper 28383.

JUPE, A. J., JONES, R. H., DYER, B. & WILSON, S. 1999. Monitoring and management of fractured reservoirs using microearthquake activity. *Proceedings of the 1998 SPE/ISRM Eurock '98, Trondheim, Norway, 8–10 July*. SPE Paper 47315.

KECK, R. G. & WITHERS, R. J. 1994. A field demonstration of hydraulic fracturing for solids waste injection with real-time passive seismic monitoring. *Proceedings of 1994 SPE Annual Technical Conference and Exhibition, New Orleans, 25–28 September*. SPE Paper 28495

MAXWELL, S. & URBANCIC, T. 2001. The role of passive emicroseismic monitoring in the instrumented oil field. *The Leading Edge* (June), 636–639.

MAXWELL, S. C., YOUNG, R. P., BOSSU, R., JUPE, A. J. & DANGERFIELD, J. 1998. Microseismic logging of the Ekofisk reservoir. *Proceedings of the 1998 SPE/ISRM Eurock '98, Trondheim, Norway, 8–10 July*. SPE Paper 47276.

PHILLIPS, W. S., RUTLEDGE, J. T., FAIRBANKS, T. D., GARDNER, T. L., MILLER, M. E. & SCHUESSLER, B. K. 1996. Reservoir fracture mapping using microearthquakes: Austin Chalk, Giddings Field, Tx and 76 Field Clinton Co., Ky. *Proceedings of 1996 SPE Annual Technical Conference and Exhibition, Denver, 6–9 October*. SPE Paper 36651

RUTLEDGE, J. T., SCOTT PHILLIPS, W., ROFF, A., ALBRIGHT, J., HAMILTON-SMITH, T., JONES, S. K. & KIMMICH, K. C. 1994. Subsurface fracture mapping using microearthquakes detected during primary oil production, Clinton County, Kentucky. *Proceedings of 1994 SPE Annual Technical Conference and Exhibition, New Orleans, 25–28 September*. SPE Paper 28384.

SNELL, J. S. & CLOSE, A. D. 1999. Yates field steam pilot applies latest seismic and logging monitoring techniques. *Proceedings of 1999 SPE Annual Technical Conference and Exhibition, Houston, 3–6 October*. SPE Paper 56791.

WALLROTH, T., JUPE, A. J. & JONES, R. H. 1996. Characterisation of a fractured reservoir using microearthquakes induced by hydraulic injections. *Marine and Petroleum Geology*, **13**(4), 447–455.

WARPINSKI, N. R., WRIGHT, T. B., UHL, J. E., ENGLER, B. P., DROZDA, P. M., PETERSON, R. E. & BRANAGAN, P. T. 1999. Microseismic mapping of the B-sand hydraulic fracture experiment at the DOE/GRI Multisite Project. *SPE Journal* **4**(3), 242–250.

Ongoing compression across South American plate: observations, numerical modelling and some implications for petroleum geology

CLAUDIO LIMA

Petrobras Research Center – Cenpes, Ilha do Fundão Q-7, 21949–900–Rio de Janeiro, Brazil
(e-mail: claudioc@cenpes.petrobras.com.br)

Abstract: The South American plate is now in horizontal compression and shortening. This is shown by stress data compilations, intraplate stress field numerical models and space-based geodetic results. Consistent with the compressive scenario, analyses of leak-off and hydraulic fracturing data indicate that the maximum principal stress is horizontal for most Brazilian basins. The observed compression/shortening is probably mostly due to the convergence of the South American and the Nazca plates and the divergence of the South American and the African plates.

Plate-wide deformation related to the Andean tectonics has been evidenced by analyses of integrated visualizations of available plate-scale information on tectonics, continental geology, topography/bathymetry, seismicity, stresses, active deformation, residual isostatic anomalies, fission track analyses, and seismically derived Moho depths and P and S wave velocity anomalies. Here, some results of these analyses are presented and some implications of the ongoing compression for petroleum exploitation and exploration are discussed.

A conceptual model for the plate-wide deformation is presented and numerically tested using elasto-plastic rheologies. The model states that in response to the compression, the lithosphere as a whole (or only the crust if thermal gradients are high enough) tends to fold and fracture. This tendency is stronger during peaks of the Andean orogeneses. The forming antiforms are responsible for uplift along the erosional basin borders, whereas the forming synforms are sites of continental sedimentation, at basin centres. The denudation of sedimentary covers promotes the exhumation of deeper and deeper rocks, cropping out at the foot of retreating scarps. Consequently, the erosional borders of the basins form local topography highs with respect to the adjacent basement.

In exploration, neotectonics is usually simply disregarded, at least with regard to the exploratory activities developed in 'passive' margin basins. Nevertheless, we infer that neotectonics has an important role in the distribution and preservation of petroleum accumulations, since: (i) accumulations are ephemeral in a geological time scale, being strongly dependent on the seals's fine geometry and biodegredation; (ii) a strong positive correlation exists between permeability anisotropy and maximum horizontal stress; (iii) the source rocks of the most important Subandean and intraplate petroleum systems are still in the oil generation window; and (iv) the disruption of 'kitchens' of generation is a positive factor to primary migration. This inference seems to be confirmed, since the most important South American petroleum accumulations are found along the actively deforming border between the South American and Caribbean plates. Indeed, the most important accumulations in marginal basins are found in the southeastern Brazilian margin (the Campos basin), which has been deformed the most with respect to other margins during the Cenozoic, and continues to be the most seismically active.

The South American plate is now in horizontal compression and shortening, as shown by stress data compilations (Assumpção 1992; Lima *et al.* 1997) and space-based geodetic results (Lima 1999). This tectonic compression can be simulated by intraplate stress field numerical models (Coblentz & Richardson 1996; Meijer 1995). Thrust regimes prevail along the southeastern Brazilian margin, in central Brazil and in the Amazon region; strike-slip regimes prevail along the northeastern Brazilian margin (Assumpção & Suarez 1988; Assumpção, 1998; Ferreira *et al.* 1998). All but one of the available base lines from the French DORIS geodetic space-based system are shortening. Base lines that cross the Andes towards the midplate are shortening 13–20 mm a^{-1}, and midplate shortening reaches 7 mm a^{-1} over the Brazilian craton (Fig. 1; L. Soudarin, CNES, Toulouse, written communication, 1998). Geodetic results have also confirmed that the

From: AMEEN, M. (ed.) *Fracture and In-Situ Stress Characterization of Hydrocarbon Reservoirs.* Geological Society, London, Special Publications, **209**, 87–100. 0305–8719/03/$15.00
© The Geological Society of London 2003.

Andean deformation belt is still propagating eastwards, thrusting the Brazilian craton (Norabuena *et al.* 1998). Cenozoic deformation is continent-wide (Cobbold *et al.* 1996; Lima 1999). Analyses of leak-off and hydraulic fracturing data indicate that the maximum principal stress is predominantly horizontal for most Brazilian basins (Lima Neto 1998, 1999; Lima Neto & Beneduzi 1998). The observed compression/shortening is probably mostly due to the convergence of the South American and Nazca plates and the divergence of the South American and African plates (Lima 1999).

In western Europe, ongoing NW-trending compression is inducing regional tectonic deformation (Illies & Greiner 1979; Illies *et al.* 1979). On the basis of *in situ* stress determinations, focal mechanisms, geodetic measurements and Quaternary volcanic activities, these authors have demonstrated that since Pliocene times the current stress field has controlled faulting, patterns of uplift and deposition through a 800 km long belt of seismotectonic activity affecting the Rhine graben, the Rhenish shield, the lower Rhine embayment and the Zuideer depression. These phenomena have been termed 'incipient inversion', an intraplate deformation attributed to Alpine tectonics (Ziegler 1989).

Similarly, plate wide deformation related to the Andean tectonics was highlighted by integrated visualizations of available plate-scale information on tectonics, continental geology, topography/bathymetry, seismicity, stresses, active deformation, residual isostatic anomalies, fission track analyses, and seismically derived Moho depths and P and S wave velocity anomalies (Lima 1999).

Here, firstly, we present some results of analyses of composite images and geotransects of plate topography and bathymetry, geology, gravity, and seismicity made by Lima (1999), which provide evidence for a number of observations related to the ongoing compression/shortening. Secondly, we summarize some results

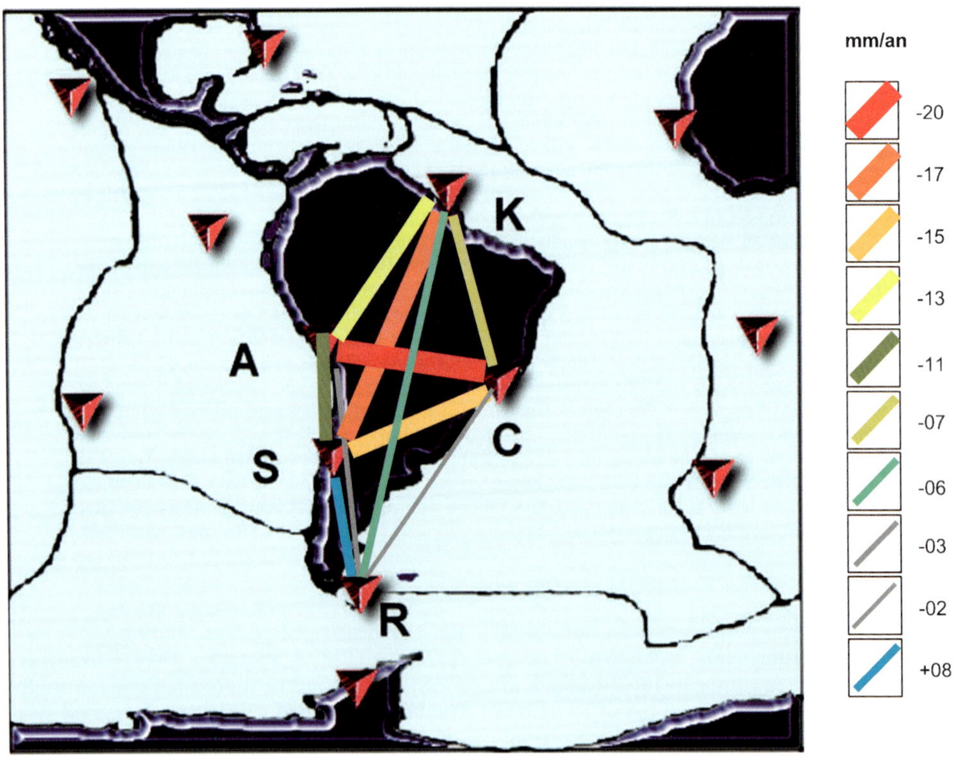

Fig. 1. South America base lines of French space-based geodetic system *DORIS* (Soudarin 1995). Results from four years of observation (1993–96, 48 world-wide distributed stations) indicate that all but one of the base lines are in shortening. The stations are: A, Arequipa; C, Cachoeira Paulista; K, Kourou; R, Rio Grande; S, Santiago. Source: L. Soudarin, CNES-Toulouse, 1998, written communication. Modified from Lima (1999).

from other studies in order to discuss some implications of the ongoing compression for petroleum exploitation and exploration.

Some structural and topographic expressions related to the ongoing compression

Previous work (Gephart 1994) has shown that the topography of the Central Andes and the geometry of the underlying subducting slab of the Nazca plate have the same bilateral symmetry. Moreover, the calculated best plane of symmetry is roughly coincident with the Euler equator of the Tertiary convergence between the South American and the Nazca plates, implying an intimate relationship between the kinematics of the convergence and plate deformation along the Andean margin.

Lima (1999) observed that the same plane is also roughly coincident with a chain of gravity positive isostatic anomalies (Ussami *et al.* 1993) that cross the entire midplate, from the Central Andes salient up to northeastern Brazil (Figs 2 & 3).

This chain concentrates seismicity, is the major plate divide and separates the plate into two different geotectonic, gravity and topographic domains. High topography and negative residual isostatic anomalies prevail southeastwards, where rocks mostly belong to the Brasiliano domain

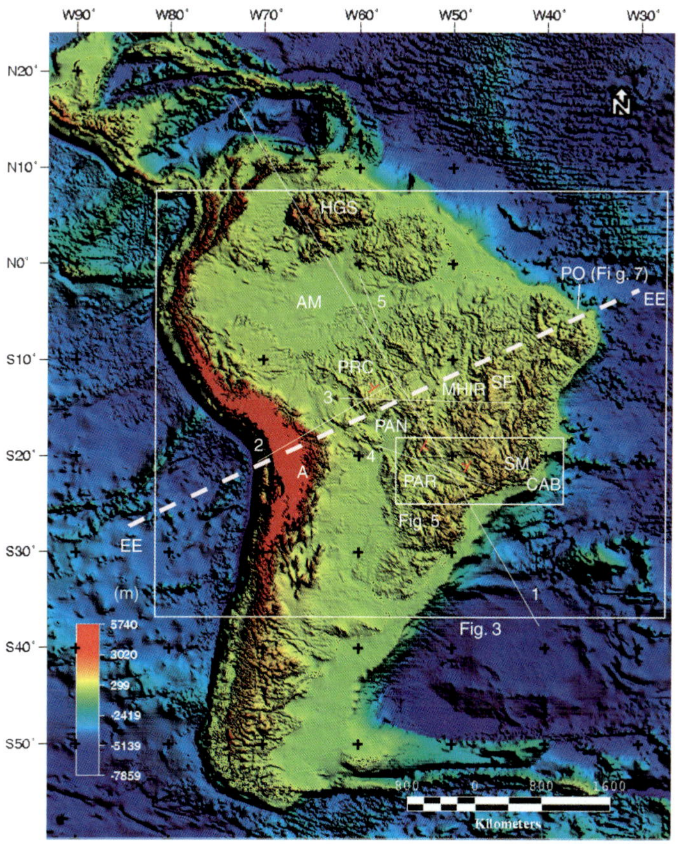

Fig. 2. Topography of South America. Shaded image containing South American topography and bathymetry, built using the DTM3 data base, compiled by the Leeds University in a regular 3–minute arc grid. Lines 1–5 are geotransects (Fig. 6); A, Andes; AM, Amazonia; CAB, Campos (off-shore) basin; EE, Euler equator for Mid-Tertiary convergence between the South American and Nazca plates and best plane for bilateral symmetry of topography and the underlying slab (Gephart 1994); HGS, highlands of the Guyanna shield; MHIR, transcontinental chain of positive residual isostatic anomalies (Fig. 3); PAN, Pantanal basin (Neogene); PAR, Parana basin (Palaeozoic–Tertiary; PO, Potiguar basin (Mesozoic–Tertiary; PRC, Parecis basin (Palaeozoic–Tertiary); SF, São Francisco basin (Mesozoic–Tertiary); SM, southeastern Brazilian Coastal Ranges (Serra do Mar and Serra da Mantiqueira).

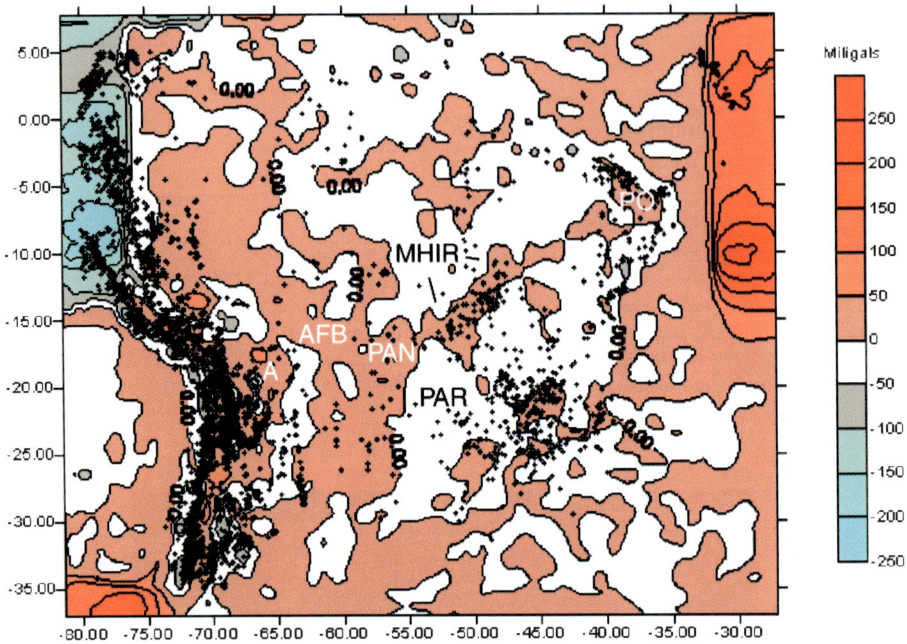

Fig. 3. Isostatic residual anomalies (Ussami *et al.* 1993) and seismicity (crosses) in South America. Negative anomalies are shown in white and bluish hues; positive anomalies are in reddish hues. See the transcontinental chain of isostatic positive anomalies associated with seismicity (MHIR) crossing the plate from the Andean flexural bulge (AFB), where the Pantanal basin (PAN, Neogene) is found, towards the NE continental edge of the plate (Potiguar basin, PO, Mesozoic–Tertiary). A, Andes; PAR, Parana basin (Palaeozoic–Tertiary).

(450–700 Ma; Brito Neves 1991). On the other hand, northwestward, rocks mostly belong to the Tranzamazonico domain (*c.* 2000 Ma; Brito Neves 1991), positive residual isostatic anomalies prevail and high topography is found only near the contact between the South American and Caribbean plates (Figs 2–5).

The implication is that strong mechanical links exist between the convergence of South America and Nazca, the Andean deformation and the intraplate deformation, and the plate-wide deformation is dependent on older Precambrian major structures, namely on a major Neoproterozoic suture (Brito Neves 1991).

In front of the Central Andes salient, the western edge of the Brazilian craton has been uplifted along a line that mimics the structural lines of the Andean deformational belt (Figs 4 & 6, geotransect 2). This surface uplift is due to a flexural bulge (Fig. 3) induced by vertical and horizontal loads associated with the dynamics of the chain (Shiraiwa & Ussami 1993). In response to the uplift, denudation of a Tertiary and older sedimentary cover has been taking place, producing sedimentary scarps that retreat towards the Brazilian craton, exhuming the Precambrian shield and producing large areas of Neogene sedimentation, such as the great Pantanal basin (Figs 4 & 6, geotransect 2; see also Fig. 9).

Seismological data available for the Parana basin eastwards to the Pantanal indicate that the Moho is depressed under the centre of the basin (where there is a thin belt of Neogenic sedimentation) and raised towards its erosional borders (Assumpção *et al.* 1997; Fig. 5). Moreover, estimates of effective elastic thickness (T_e) for the Parana basin (Vidotti 1997) indicate that greater values (about 60 km) are found along the axes of the basin, whereas lower values (about 20–30 km) are found at the erosional and raised eastern and western borders. These observations suggest a continental-scale lithospheric deformation comprising an antiform at the Pantanal and a synform at the Parana basin. The sedimentary bedding tends to follow those plate-scale structures, denoting a breached antiform at the Pantanal and a synform at the Parana basin (Fig. 2).

Processes similar to those described above relating seismicity, gravity anomalies and retreating of sedimentary scarps seem to be operating at different scales in all continental basins. All

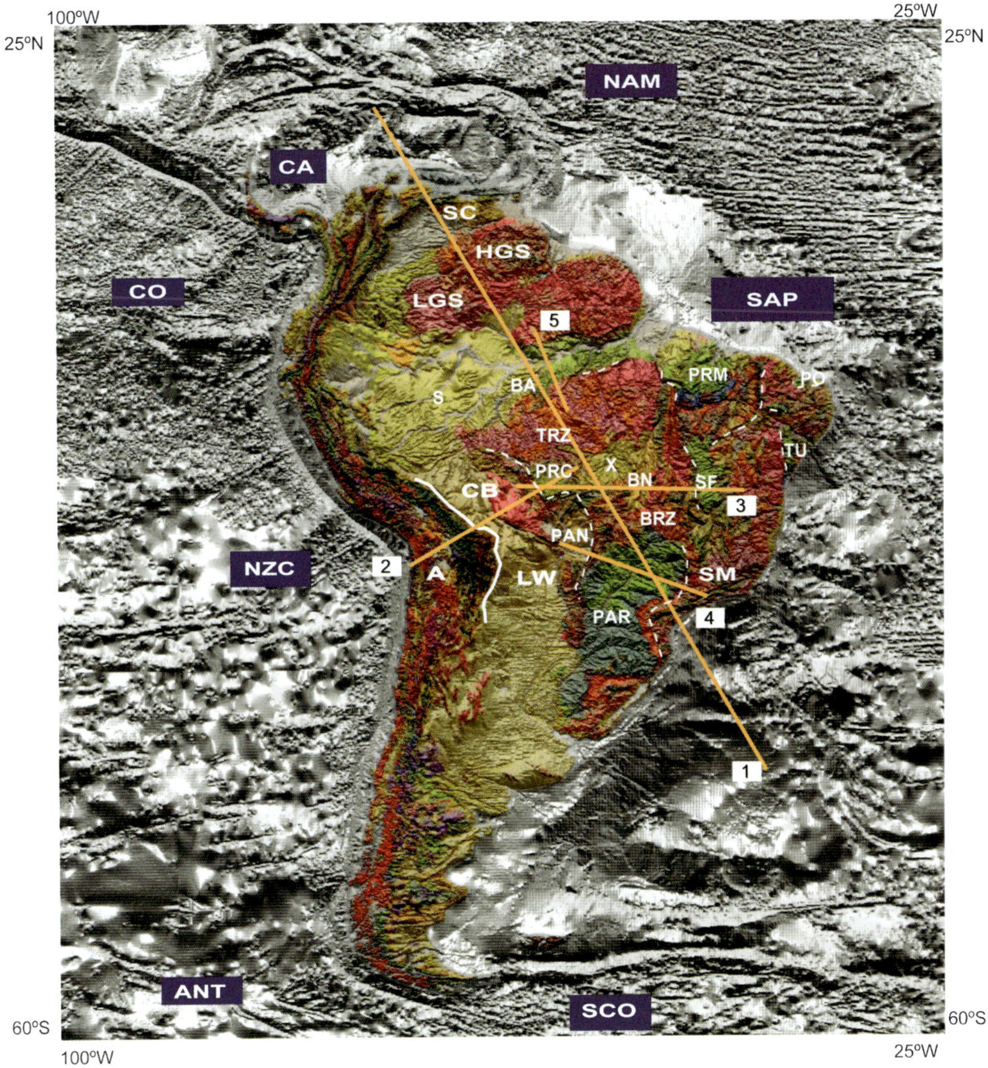

Fig. 4. Geology and Topography of South America. Overlay of a shaded digital terrain model (DTM) of the topography and bathymetry and the Geological Map of South America (UNESCO 1981). The shaded image was built using the DTM3 data base, compiled by the Leeds University in a regular 3-minute arc grid. Dashed lines indicate retreating scarps formed by erosional edges of sedimentary basins. A, Andes; ANT, Antarctica Plate; BA, Amazon basin (Palaeozoic–Mesozoic); BN, Bananal basin (Neogene); BRZ, Brasiliano domain; CA, Caribbean plate; CB, Brazilian craton; CO, Cocos plate; HGS, highlands of the Guyana shield; LGS, lowlands of the Guyana shield; LW, lowlands; NAM, North America plate; NZC, Nazca plate; PAN, Pantanal basin (Neogene); PAR, Parana basin; PO, Potiguar basin; PRC, Parecis basin; PRM, Pará-Maranhão basin; S, Solimões basin (Palaeozoic–Neogene); SAP, South American plate; SC, South Caribbean fold belt; SCO, Scotia plate; SF, São Francisco basin; SM, southeastern Brazilian Coastal Ranges; TRZ, Transamazonico domain; TU, Reconcavo–Tucano basins; X, Xingu basin. Transects 1–5 are shown in Figure 6. Modified from Lima (1999).

these basins are actually remnants of basins whose erosional edges are associated with positive isostatic anomalies, and frequently with seismicity (Figs 4 & 6). They are in incipient inversion; their borders are being uplifted and eroded, whereas their axes, along with major river flows (*e.g.* the Parana and the Amazon) are subsiding, acting as narrow belts of Neogene sedimentation. Along the raised borders, Cretaceous and Devonian marine sediments may be found

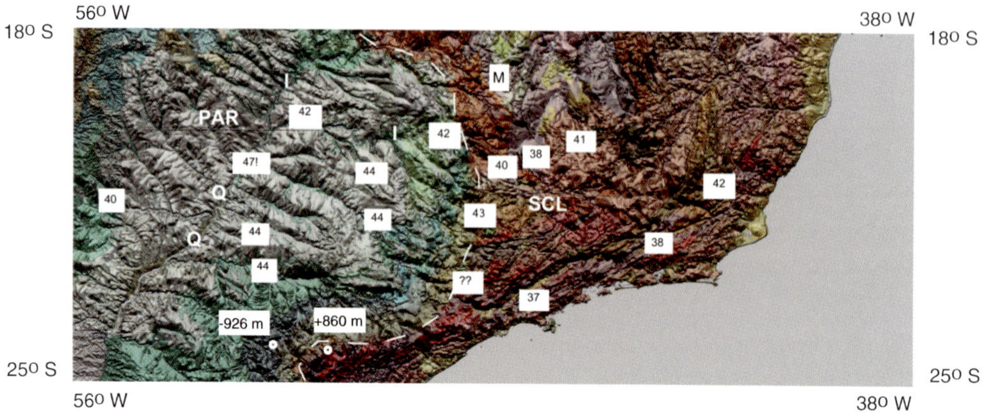

Fig. 5. Crustal thickness and the Parana basin structure in southeastern Brazil. Overlay of a shaded DTM built with the TOPO30 data base and an area of the Geological Map of Brazil (DNPM, 1981). Crustal thicknesses (in km) indicated by numbers in the white squares were determined by seismological studies (Assumpção *et al.* 1997), using data provided by delays between arrivals of directed P and Ps waves, the latter generated at the Moho discontinuity (Receiver Function Analysis). The Moho makes a large wavelength lithospheric synform at the Parana basin (PAR, in greenish hues), accompanying its structure. M, remnants of raised Aptian–Albian marine sediments; SCL, Precambrian basement (in reddish hues). Devonian marine sediments cropping out at +860 m were found by drilling at –926 m (see text). Modified from Lima (1999).

several hundreds of metres in height. The same marine sediments have been found by drilling, several hundred of metres deep at the centres of the basins (Petri 1948; Daemon *et al.* 1967; Lange & Petri 1967; Lima 1999; Fig. 5). As a result of the retreating process, Precambrian crystalline rocks are exhumed adjacent to basins that are in incipient inversion, cropping out at lower topographic levels (Fig. 7).

A conceptual model for the observed wide-scale intraplate deformation

In order to explain this scenario, Lima (1999) proposed that, since the Upper Cretaceous, in response to the compression, the lithosphere as a whole (or only the crust if thermal gradients are high enough) tends to fold and fracture (Fig. 8).

The forming antiforms are responsible for uplift along the erosional basin borders, whereas the forming synforms are sites of continental sedimentation, at basin centres. Actually, during the Upper Cretaceous, a continental sedimentation whose remnant is 250 m thick was deposited along the axis of the Parana basin (Milani 1997). This sedimentation is concomitant with the Upper Cretaceous Peruvian orogenesis. The advection of mantle materials towards the forming antiforms, associated with denudation, promotes adiabatic decompression in the rising materials, facilitating fusion, and hence, the observed Upper Cretaceous and Tertiary alkaline-magmatism. This tendency is stronger during peaks of the Andean orogeneses. The denudation of sedimentary covers promotes the exhumation of deeper and deeper rocks, cropping out at the foot of retreating scarps. Consequently, the erosional borders of the basins form local topography highs with respect to the adjacent basement. The basement low topography is frequently associated with great Neogene basins (e.g. the Pantanal). Seismicity, always associated with positive residual isostatic anomalies, is concentrated about inflection points between forming synforms and antiforms, along the erosional borders of basins, probably absorbing most of the shortening. The tendency to fold and fracture is controlled by the previous lithospheric/crustal structure. This is a key element in the conceptual model. An example of such a control is the Moho topography under the Paraná basin: the extensive basalt flooding at the centre of the basin is supposed to be responsible for underplating and hence for the initial synform of the Moho.

Results of a comprehensive study of the deformational history of Andean sedimentary basins were used to postulate the idea that the evolution of this belt would have been dominated by a compressional tectonic continuum since the Upper Cretaceous (Noblet *et al.* 1996). This is consistent with the intraplate compressional

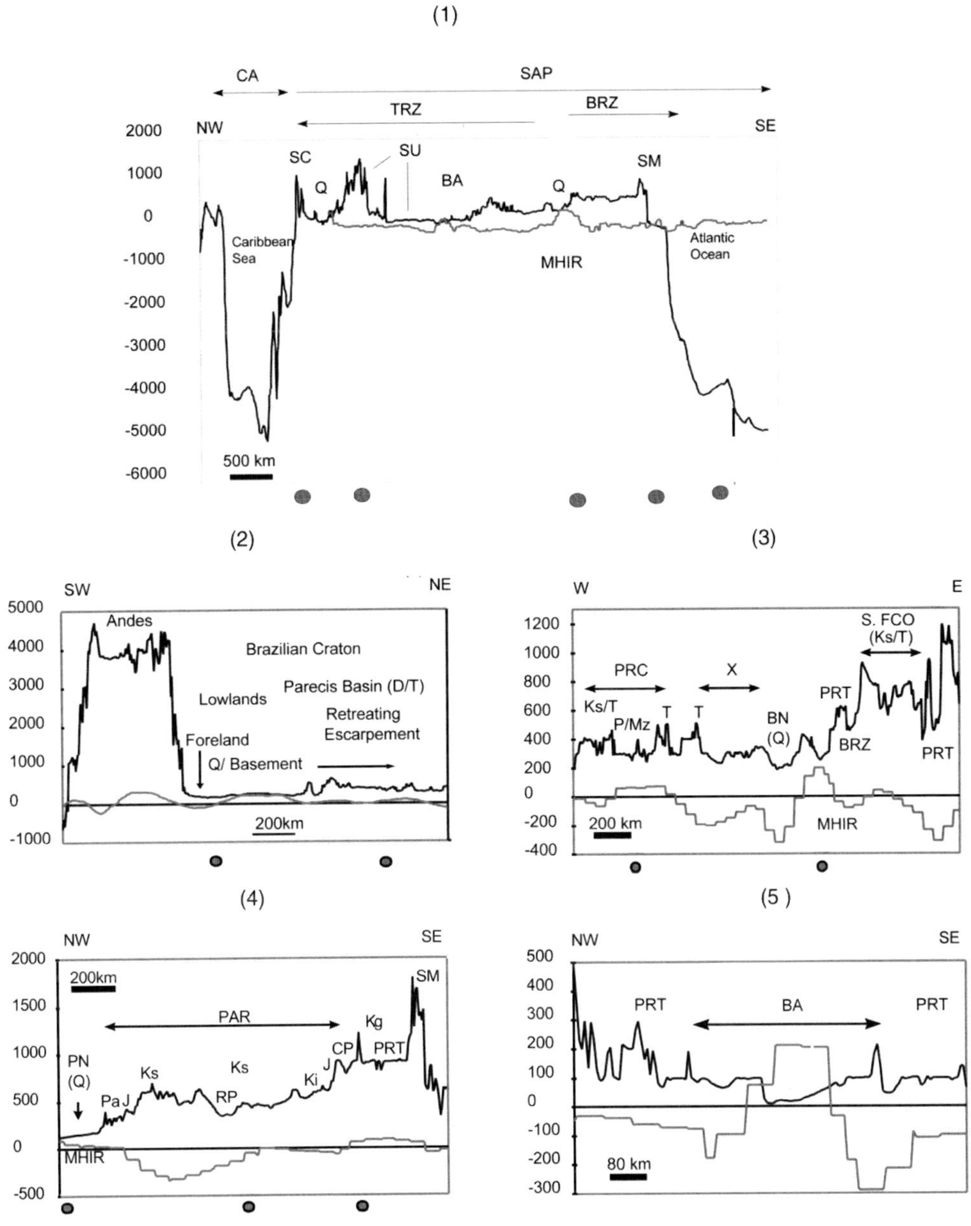

Fig. 6. Geotransects. Circles below boxes containing the transects indicate clusters of epicentres found along them, not the depth of hypocentres. Numbers at the vertical axes of the geotransects indicate both topography, in metres (black line) and residual isostatic anomalies, in 10 times miligals (grey line). BA, Amazon basin (mostly Palaeozoic); BN (Q), Bananal basin (Neogene); BRZ, Brasiliano domain; C, Carboniferous; CA, Caribbean plate; J, Jurassic; Ki, Lower Cretaceous (tholeitic basalts); K(g) Upper Cretaceous alkaline volcanic rocks; Ks, Upper Cretaceous; HGS, highlands of the Guyana shield; LGS, lowlands of the Guyana shield; MHIR, transcontinental chain of positive isostatic anomalies; Mz, Mesozoic; P, Permian; Pa, Palaeozoic; PAR, Parana basin (mostly Palaeozoic); PN (Q), Pantanal basin (Neogene); PRT, Proterozoic; Q, Quaternary; RP, Parana River; SAP, South American plate; SC, South Caribbean folded belt; SFR, São. Francisco River; SM, southeastern Brazilian Coastal Ranges; SU, Uatumã Supergroup, a set of magmatic rocks of 1900 ±200 Ma, (U.G. Cordani, personal communication, 1997); T, Tertiary; TU, Tucano, Reconcavo and Jatoba basins (mostly Lower Cretaceous); TRZ, Tranzamazonico domain. Modified from Lima (1999).

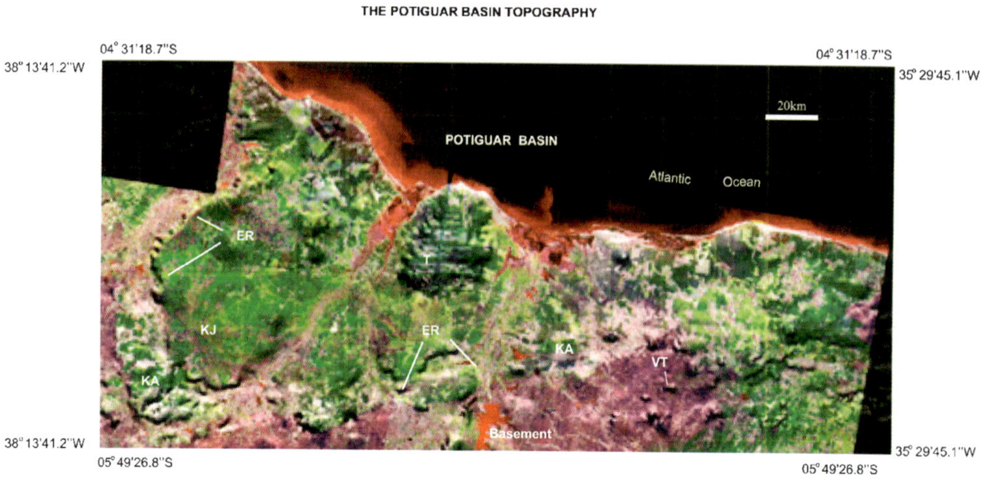

Fig. 7. Topographic expression of a basin undergoing incipient inversion. Overlay of a mosaic of Landsat TM Images and a shaded DTM covering the Potiguar basin (greenish hues) and its adjacent basement, which crops out at lower topographic levels (reddish hues). See the raised erosional edges of the basin, sustained by scarps (ER) that retreat towards its centre (T). These scarps (about 200–250 m high) are formed by Upper Cretaceous shallow marine limestones (KJ) conformably deposited over continental sandstones and shales (KA). Both units dip towards the centre of the basin, over which there is a local remnant topography sustained by retreating scarps formed by continental Tertiary sediments (T). VT is a Miocene volcanic neck whose top is about 500 m high.

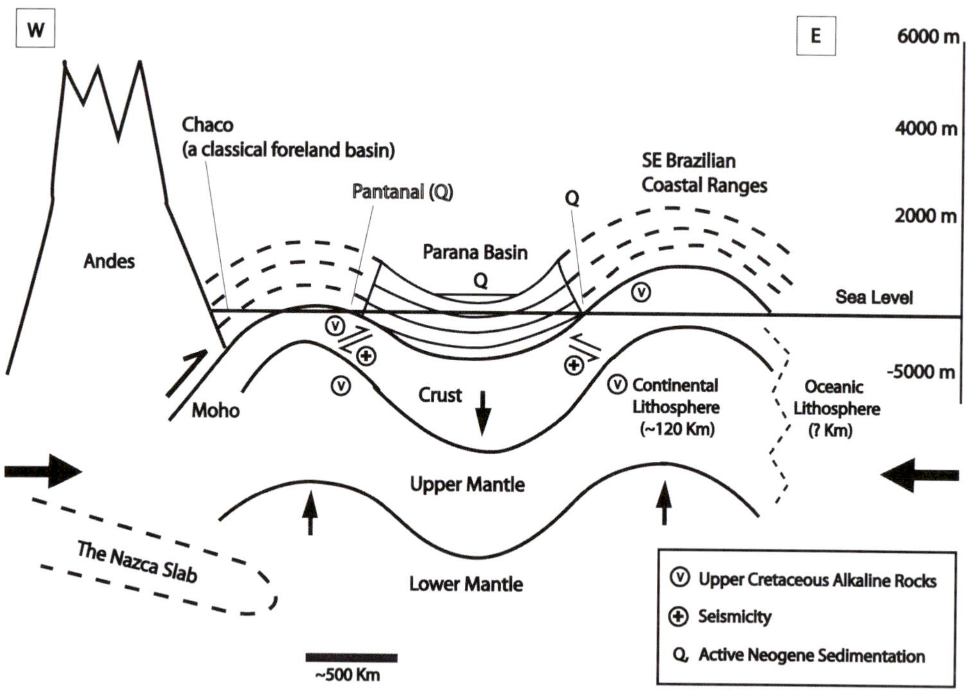

Fig. 8. Conceptual model for lithospheric deformation along a schematic transect between the Central Andes and the SE Brazilian Atlantic margin. Sketch showing the main features of the proposed conceptual model. Arrows indicate displacement. See text. Modified from (Lima 1999).

tectonic continuum implied in the conceptual model herein proposed.

In this model, the high topography of the SE Brazilian Coastal Range is supposed to be associated with a lithospheric antiform. Evidence of such a lithospheric structure could be the existence of low velocity zones in that area, which are possibly caused by a still-existent thermal anomaly below 100 km and extending to a depth of 600 km (Vandeccar *et al.* 1995). This antiform should be mostly explained by the compressional interaction of the continental lithosphere with the very strong oceanic lithosphere found eastwards. This is consistent with the observed compressional deformation found at the transitional limit between both lithospheres (Fig. 9; Bassetto 1997).

This conceptual model has been numerically tested and the preliminary results were successful in replicating the main aspects of the observed ongoing deformation of cratonic South America, namely the anticorrelation between lithospheric structure and topography (Fig. 10; Lima *et al.* 1999).

A schematic 3000 km long lithospheric section has been modelled using elasto-plastic rheologies (Fig. 10; Lima *et al.* 1999). The section incorporates an initial senoidal Moho topography (with a wavelength of about 2000 km) and a layered profile of lithospheric strengths. This profile takes into account a strong, 15–20 km thick, upper crust (including 5 km of sediments), overlying a 20 km thick, very weak lower crust, underlain by a 115 km thick, strong, upper mantle. The layered lithosphere overlies a 100 km thick, very weak asthenosphere, supposed to be attached to the lithosphere, as proposed by Vandecar *et al.* (1995) and Russo *et al.* (1994, 1996). Such a layered profile is supposed to roughly represent a lithosphere submitted to intermediate temperatures (Davy & Cobbold 1991).

In order to accumulate a shortening of about 0.25–0.30%, one end of the section was fixed and

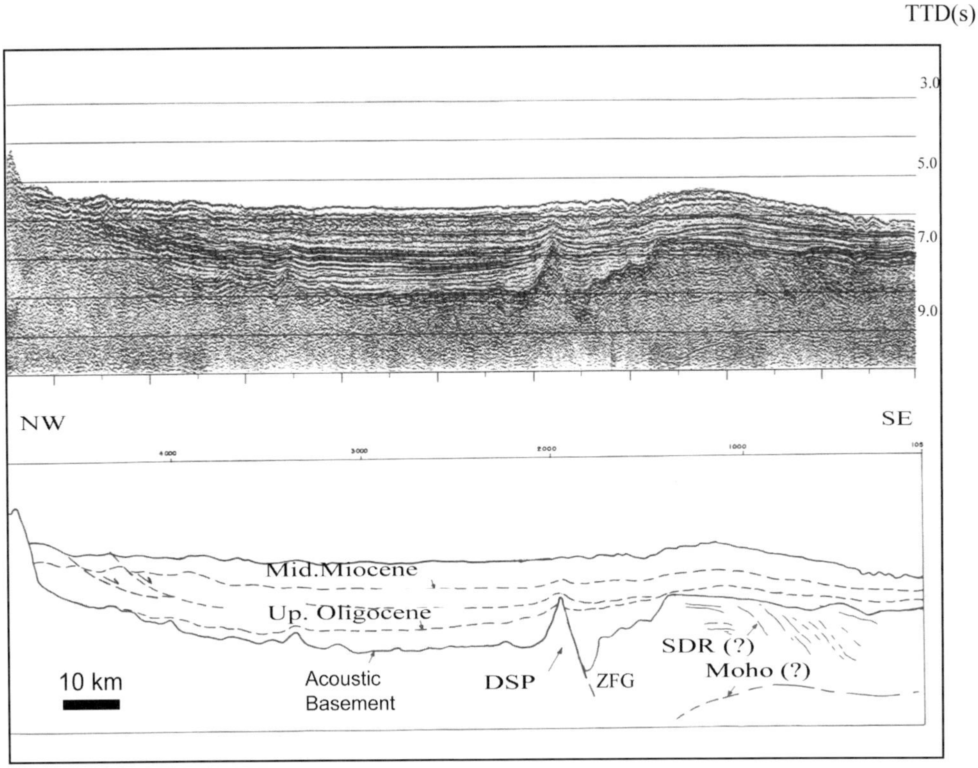

Fig. 9. Neogene compressional structures in the SE Brazilian Atlantic margin. Compressional structures affecting Neogene sediments and the sea bottom are found in several seismic lines, defining a 300 km long thrust front in the SE Brazilian Atlantic margin. Along the front, oceanic crust has been thrust upon thinned continental crust (Bassetto 1997). Consistently, all available focal mechanisms in the SE Brazilian margin indicate thrust regimes (Assumpção 1998). DSP, São Paulo dorsal; ZFG, Rio Grande fracture zone; SDR, seaward dipping reflectors. Modified from Bassetto (1997).

incremental displacements were applied to the other (Fig. 10a). This scenario is intended to simulate the action of Andean horizontal loads against the Brazilian craton (western edge of the section) pinned on a fixed and very strong oceanic lithosphere (eastern edge of the section). An average intraplate strain rate of about $6.97\times10^{-17}\,s^{-1}$ had been estimated from the shortening measured along the Cachoeria Paulista-Kourou base line of the DORIS space-based geodetic system (Lima 1999). Assuming this average intraplate strain rate could be linearly extrapolated, the shortening imposed by the models would have been accumulated over about 1–2 Ma.

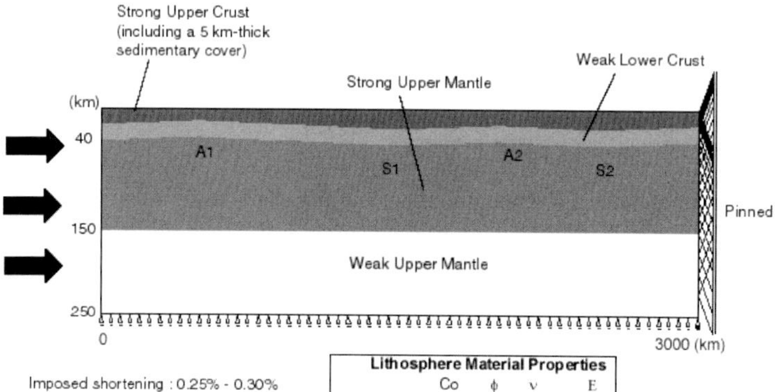

Lithosphere Material Properties

	Co (MPa)	ϕ	ν	E (MPa)
Sediments	20	27	0.33	20000
Upper Crust	60	32	0.33	60000
Lower Crust	20	15	0.49	1000
Upper Mantle	60	27	0.33	60000
Lower Mantle	20	0	0.49	1000

(a)

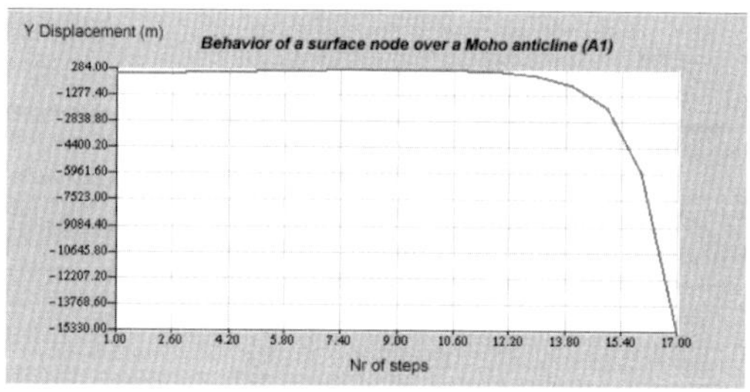

(b)

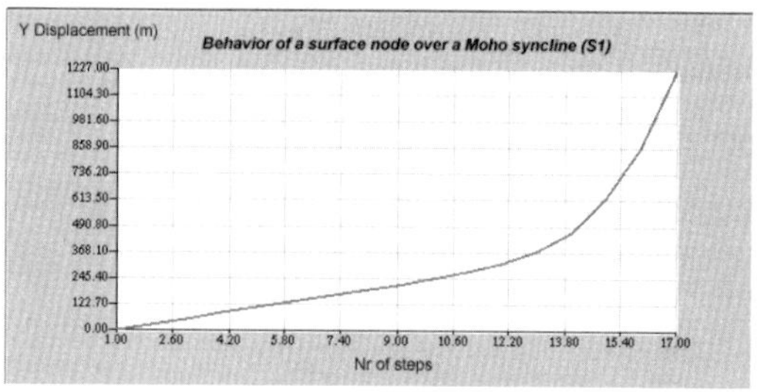

(b)

Figure 10. Numerical Model. (**a**) Initial configuration for a model of a 3000 km long schematic transect crossing a continental lithosphere containing an initial large wavelength senoidal Moho topography. A1 & A2, antiforms; S1 & S2, synforms; Co, cohesive strength; ϕ, internal friction angle; ν, Poisson's ratio; E, Young's modulus. Gravity and densities were taken into account. Arrows indicate displacement. (**b**) & (**c**) Modelling results after 0.25–0.30% of shortening, when the lithosphere was attached to the lower mantle. Modified from Lima *et al.* (1999).

Under the imposed boundary conditions, the deformed models show similarities to nature, qualitatively imitating the systematic anticorrelation between topography and structure observed across South America, where positive isostatic anomalies are mostly associated with flooded areas (Fig. 11). In the model, topographic depressions are formed over the original lithospheric antiforms, whereas high elevations are produced over the original lithospheric synforms, mainly on the limbs of the structure. The deformed models as a whole underwent uplift. The numerical modelling suggests that a dragging force pulls down the anticlines and pushes up the synclines. This is probably due to complex interactions between the rigid structures and the weak underlying material, the reaction forces which determine the movement being closely related to the shape of the rigid structures.

Discussion and conclusions

Assuming an elastic rheology, early studies of lithospheric flexure showed that for reasonable levels of compressional forces the induced vertical displacements of the lithosphere are negligible. This result, combined with the lack of evidence of natural examples until recently, led to withdrawal of attention from lithospheric folding (Cloetingh 1992).

Meanwhile, during the two past decades, several theoretical studies concerning numerical and analogue modelling based on other rheologies (viscous-plastic and elasto-plastic) have been made and natural examples have been recognized of lithospheric folding in both oceanic and in continental domains (Davy & Cobbold 1991; Martinod 1991; Martinod & Davy 1992, 1994; Beekmann 1994).

By combining tectonics with available plate-scale information on stresses, geodesy, seismology, topography, geology, gravity and geomorphology, Lima (1999) presented a convincing example for large-scale lithospheric folding and fracturing and incipient compressional reactivation of tensional basins. Numerical modelling based on elasto-plastic rheologies has been successful in qualitatively replicating the main observations, namely the anticorrelation between topography and lithospheric structure (Lima *et al.* 1999).

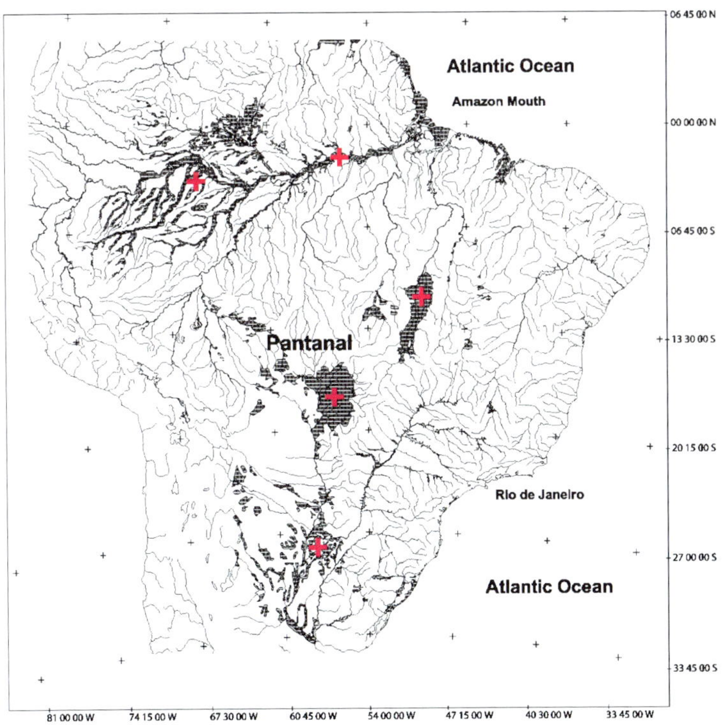

Fig. 11. Anticorrelation between lithospheric structure and topography in South American intraplate. Major flooded areas (dark shading) are mostly associated with positive isostatic anomalies (red crosses). One amongst them, the Pantanal, is the only known lithospheric antiform. Modified from Lima *et al.* (1999).

South American intraplate seismicity concentrates about the inflection points between the lithospheric antiforms and synforms. A similar relationship between lithospheric folding and fracturing has been found in nature and in numerical modelling of the intraplate deformation of Eurasia and Australia (Beekmann 1994), and in other numerical and analogue studies as well (Davy 1986; Davy & Cobbold 1991; Martinod & Davy 1992, 1994; Cobbold *et al.* 1993; Tondji Byio 1995).

The concepts that firstly, passive margins are "relaxed" (submitted only to minor extensional fault activity), and secondly, that the formation of structures have been terminated a long time ago, are deeply perceived within the oil industry. The perception that intraplate South American basins (and also probably basins of other intraplate areas) are in compression and actively deforming is opposed to these two concepts and opens up new perspectives for exploitation and exploration strategies.

In exploitation, the importance of taking into account the actual values of the current stress field in well design and secondary recovery has been theoretically recognized for many years (e.g. Bell 1990; Bol *et al.* 1992; Mody & Hale 1993; Santos 1997). Meanwhile, a gap probably still remains between theory and practice, at least for activities developed in 'passive' margin basins. Taking into account actual stress values, and so avoiding inaccurate assumptions such as the one that states that passive margins are 'relaxed', would improve well design and secondary recovery and probably help diminishing losses.

During the last few years, the role of *in situ* stresses in fracture reservoir characterization has been increasingly recognized and incorporated into reservoir modelling (e.g. Ameen 1999; Dholakia *et al.* 1999; Batchelor, 1999; Smart, 1999; Rogers & Evans 1999; Cacas & Daniel 1999). Nevertheless, as stated by Rogers & Evans (1999), the actual relationship between the flow of fluid through a fractured mass and the prevailing stress field remains poorly understood. For instance, conventional wisdom states that fracture flow is predominantly along fractures parallel to the maximum stress direction. However, the analysis of detailed fracture, stress field and flow data derived from a borehole drilled as part of the UK radioactive disposal programme has recently shown (Rogers & Evans 1999) that the most transmissive zones were those submitted to a very high shear/normal stress. Previous work has established that these critically stressed fractures have the potential to dilate and so are the most favourable pathways for fluid flow.

In a classic textbook of petroleum geology, F. K. North has remarked that, in spite of being a revolutionary stride, plate tectonics has not influenced the actual course of petroleum exploration in any significant way (North 1994, p. 23). His opinion was that, even though it can provide convincing explanations of exploratory outcomes after these become known, plate tectonics would be unlikely either to instigate or to deter their achievement. Lima (1999) pointed out that this statement has not significantly changed up to now, with plate tectonics still being used mostly in palaeogeographical reconstructions, which are designed to constrain the geological history of a given basin or a set of basins. For instance, over the past decade the World Stress Map Project has shown that most intraplate areas are in compression. Nevertheless, the impact of that information on exploration, if any, has probably been minimal. In summary, in exploration, the impact of plate tectonics has been minimal and neotectonics is usually simply disregarded, at least with regard to the exploratory activities developed in 'passive"' margin basins.

Nevertheless, we infer that neotectonics has an important role in the distribution and preservation of South American petroleum accumulations, since: (i) accumulations are ephemeral in a geological time scale, being strongly dependent on the seals's fine geometry and biodegredation; (ii) a strong positive correlation exists between permeability anisotropy and maximum horizontal stress (Heffer & Lean 1991; Bruno *et al.* 1991); (iii) the source rocks of the most important Subandean and intraplate petroleum systems are still in the oil generation window (Mello *et al.* 1994, 1997); and (iv) the disruption of 'kitchens' of generation is a positive factor to primary migration (Price 1994).

This inference seems to be confirmed, since the most important South American petroleum accumulations are found along the actively deforming border between the South American and Caribbean plates. Indeed, the most important accumulations in the so-called passive margin basins are found in the southeastern Brazilian margin, which has been deformed the most with respect to other margins during the Cenozoic, and continues to be the most seismically active.

I would like to thank Petrobras for the authorization to publish the present paper. In addition, I also thank to B. Müller and J. M. Jacques for the review and suggestions.

References

AMEEN, M. 1999. Characterisation of directional permeability zones for production and development purposes : a case study from a mature field, Saudi Arabia. *In*: *Fracture and In-situ Stress Characterization of Hydrocarbon Reservoirs*, Geological Society, London (Abstract).

ASSUMPÇÃO, M. 1992. The regional intraplate stress field in South America. *Journal of Geophysical Research*, **97**, 11889–11903.

ASSUMPÇÃO, M. 1998. Seismicity and stress in the Brazilian passive margin. *Bulletin of the Seismological Society of America*, **88**(1), 160–169.

ASSUMPÇÃO, M. & SUAREZ, G. 1988. Source mechanisms of moderate-size earthquakes and stress orientation in mid-plate South America. *Geophysical Journal*, **92**, 253–267.

ASSUMPÇÃO, M., JAMES, D. & SNOKE, A. 1997. Crustal thickness in SE Brazilian Shield with receiver function: isostatic compensation by density variations in the lithospheric mantle? *5th International Congress of the Brazilian Geophysical Society*. Expanded Abstracts, 895–897.

BASSETTO, M. 1997. *Análise morfo-estrutural do domínio oceânico da porção sul-sudeste da margem continental brasileira*. MSc. Thesis, Universidade Federal de Ouro Preto, Minas Gerais, Brazil.

BATCHELOR, T. 1999. *in situ* stresses: a key control on fracture permeability and reservoir pressure. *In*: *Fracture and In-situ Stress Characterization of Hydrocarbon Reservoirs*, Geological Society, London (Abstract).

BEEKMANN, F. 1994. *Tectonic modelling of thick-skinned compressional intraplate deformation*. PhD Thesis, Vrije Universiteit, The Netherlands.

BELL, J. S. 1990. Investigating stress regimes in sedimentary basins using information from oil industry wire line logs and drilling records. *In*: HURST, A., LOWELL, M. A & MORTON, A. C (eds) *Geological Applications of Wireline Logs*. Geological Society, London, Special Publications, **48**, 305–325.

BOL, G. M., WONG, S. W., DAVISON, C. J. & WOODLAND, D. C. 1992. Borehole Stability in Shales. *SPE European Petroleum Conference, Cannes, France, 16–18 November*. SPE Paper **24975**, 87–94

BRITO NEVES, B. B. 1991. Os dois Brasis geotectônicos. *Atas do XIV Simpósio de geologia do Nordeste*, 6–8.

BRUNO, M. S., BOVBERG, C. A. & NAKAGAWA, F. M. 1991. Anisotropic stress influence on the permeability of weakly cemented sandstones. *Rock Mechanics as a Multidisciplinary Science*. Balkema, Rotterdam, 375–383.

CACAS, M. C. & DANIEL, J. 1999. Discrete modeling of natural fracture networks : methods and software. *In*: *Fracture and In-situ Stress Characterization of Hydrocarbon Reservoirs*, Geological Society, London (Abstract).

COBBOLD, P. R., DAVY, P., *et al.* 1993. Sedimentary basins and crustal thickening. *Sedimentary Geology Journal*. **86**, 77–99.

COBBOLD, P. R., SZATMARI, P., LIMA, C. C. & ROSSELLO, E. A. 1996. Cenozoic deformation across South America: continent -wide data and analogue models. *Third ISAG, St Malo (France)*, 21–24.

COBLENTZ, D. D. & RICHARDSON, R. M. 1996. Analysis of the South American intraplate stress field. *Journal of Geophysical Research*, **101**, 8643–8657.

CLOETINGH, S. 1992. Intraplate stress and sedimentary evolution. *In*: BROWN, G. C., HAWKESWORTH, C. J. & WILSON, R. C. L. (eds) *Understanding the Earth: A New Synthesis*. Cambridge University Press, 415–434.

DAEMON, R. F., QUADROS, L. P. & SILVA, L. C. 1967. Devonian palynology and biostratigraphy of the Parana basin. *In:* BIGARELLA J. J. (ed.) *Problems in Brazilian Devonian Geology. Boletim Paranaense de Geociências*, **21/22**, 99–132.

DAVY, P. 1986. Modélisation thermo -mécanique de la collision continentale. Mémoires et Documents du Centre Armoricain d'Etude Structurale des Socles, Rennes, **8**.

DAVY, P. & COBBOLD, P. R. 1991. Experiments on shortening of a 4–layer model of the continental lithosphere. *Tectonophysics*, **188**, 1–25.

DHOLAKIA, S. K., ZOBACK, D. & BARTON, C. 1999. A case study of hydrocarbon transport along active faults and production-related stress changes in the Monterey formation, California. *In: Fracture and In-situ Stress Characterization of Hydrocarbon Reservoirs*, Geological Society, London (Abstract).

DNPM 1981. *Mapa geológico do Brasil e da área oceânica adjacente, incluindo depósitos minerais. Scale, 1:2.500.000*. Departamento Nacional de Produção Mineral, Brasília.

FERREIRA J. M., OLIVEIRA, R. T. *et al.* 1998. Superposition of local and regional stresses in Northeast Brazil: evidence from focal mechanisms around the Potiguar marginal basin. *Geophysical Journal International*, **134**, 341 – 355.

GEPHART, J. W. 1994. Topography and subduction geometry in the central Andes: clues to the mechanics of a noncollisional orogen. *Journal of Geophysical Research*, **99**, 12279–12288.

HEFFER, K. J. & LEAN, J. C. 1991. Earth stress orientation -a control on, and guide to flooding directionality in a majority of reservoirs. *In:* LINVILLE, B. (ed.) *International Reservoir Characterisation Technical Conference, Tulsa, Oklahoma*, vol. III, 800–822.

ILLIES, J. H. & GREINER, G. 1979. Holocene movements and state of stress in the Rhinegraben Rift System. *Tectonophysics*, **52**, 349–359.

ILLIES, J. H., PRODEHL, C., SCHINCKE, H. U. & SEMMEL, A. 1979. The Quaternary uplift of the Rhenish Shield in Germany. *Tectonophysics*, **61**, 197–225.

LANGE, F. W. & PETRI, S. 1967. The Devonian of the Parana basin. *In:* BIGARELLA, J. J (ed.) *Problems in Brazilian Devonian Geology. Boletim Paranaense de Geociências*, **21/22**, 5–55.

LIMA, C. 1999. *Expressions topographiques et structurales de l'état de compression généralisée au sein de la plaque sud-américaine*. PhD Thesis, Université de Rennes 1, France.

LIMA, C., NASCIMENTO, E. & ASSUMPÇÃO, M. 1997. Stress orientations in Brazilian sedimentary basins from breakout analysis: implications for force models in the South American plate. *Geophysical Journal International*, **130**, 112–124.

LIMA, C., AMARAL, C., MORAES, A. & DA COSTA, A. M. 1999. Are the large wave-length South American intraplate deformation and the incipient inversion of continental Brazilian basins manifestations of ongoing lithospheric/crustal folding? *6th International Congress of the Brazilian Geophysical Society, Rio de Janeiro*, Paper SBGf 305.

LIMA NETO, F. F. 1998. *Um exemplo da interferência de uma interface fraca na distribuição regional de esforços tectônicos: o campo atual de tensões da bacia Potiguar, Nordeste Brasileiro*. MSc Thesis, Universidade Federal de Ouro Preto (UFOP), Minas Gerais, Brazil.

LIMA NETO, F. F. 1999. O regime atual de tensões nas bacias sedimentares brasileiras. *VII Simpósio Nacional de Estudos Tectônicos*, Sessão 4, Lençóis, Brazil, 25–28.

LIMA NETO, F. F. & BENEDUZI, C. 1998. Using leakoff tests and acoustic logging to estimate *in situ* stresses at deep waters –Campos Basin. *AAPG International Conference & Exhibition, Rio de Janeiro, Brazil*, Extended Abstracts Volume, 224–225.

MARTINOD, J. 1991. *Instabilités périodiques de la lithosphère (flambage, boudinage) en compression et en extension*. Centre Armoricain d'Etude Structural des Socles, Memoire et Document, **44**.

MARTINOD, J. & DAVY, P. 1992. Periodic instabilities during compression or extension of the lithosphere: 1. Deformation modes from an analytical perturbation method. *Journal of Geophysical Research*, **97**, 1999 -2014.

MARTINOD, J. & DAVY, P. 1994. Periodic instabilities during compression of the lithosphere: 2. Analogue experiments *Journal of Geophysical Research*, **99**, 12057–12069.

MEIJER, P. TH. 1995. *Dynamics of active continental margins: the Andes and the Aegean region*. PhD Thesis, Utrecht University, The Netherlands.

MELLO, M. R., KOUTSOUKOS, E. A. M. *et al.* 1994. Selected Petroleum Systems in Brazil. *In:* MAGOON, L. B., & DOW, W. G. (eds) *The Petroleum System – From Source to Trap*. *AAPG*, Memoir **60**, 499–512.

MELLO, M. R., TRINDADE, L. A. F. *et al.* 1997. Geochemical Characterization of South American Subandean Petroleum Systems. *Simposio Bolivariano de Exploração Petrolera en las Cuencas Subandinas*, 324–336.

MILANI, E. J. 1997. *Evolução Tectono –Estratigráfica da Bacia do Paraná e seu Relacionamento com a Geodinâmica Fanerozoica do Gondwana Sul – Ocidental*. PhD Thesis, Universidade Federal do Rio Grande do Sul, Brazil.

MODY, K. & HALE, A. H. 1993. A borehole stability model to couple the mechanics and chemistry of drilling fluid to shale integration. *SPE/IADC Drilling Conference*, Amsterdam, 2/23–25/93. SPE Paper **25728**, 473–490.

NOBLET, C., LAVENU, A. & MAROCCO, R. 1996. Concept of a continuum as opposed to periodic tectonism in the Andes. *Tectonophysics*, **255**, 65–78.

NORABUENA, E., LEFFLER-GRIFFIN, L. *et al.* 1998. Space geodetic observations of Nazca –South America convergence across the Central Andes. *Science*, **279**, 358–362.

NORTH, F. K. 1994. *Petroleum Geology*. Chapman & Hall, London.

PETRI, S. 1948. Contribuição ao estudo do Devoniano paranaense. *Boletim da Divisão de Geologia e Mineralogia*, **129**, 1–125.

PRICE, L. C., 1994. Basin richness and source rock disruption –a fundamental relationship? *Journal of Petroleum Geology*, **17** (1): 5–38.

ROGERS, S. & EVANS, C. 1999. Critical stress and flow in fracture rock mass. *In*: *Fracture and In-situ Stress Characterization of Hydrocarbon Reservoirs*, Geological Society, London (Abstract).

RUSSO, R. M. & SILVER, P. G. 1994. Trench-parallel flow beneath the Nazca plate from seismic anisotropy. *Science*, **263**, 1105–1111.

RUSSO, R. M. & SILVER, P. G. 1996. Cordillera formation, mantle dynamics, and the Wilson cycle. *Geology*, **24**(6), 511–514.

SANTOS, H. M. R. 1997. *A new conceptual approach to shale stability*. PhD Thesis, University of Oaklahoma.

SHIRAIWA, S. & USSAMI, N. 1993. Flexura da litosfera continental sob os Andes centrais e a origem da bacia do Pantanal. *3° Congresso da Sociedade Brasileira de Geofisica*, 1096 – 1102.

SMART, B. 1999. Reservoir characterisation for stress sensitivity. *In*: *Fracture and In-situ Stress Characterization of Hydrocarbon Reservoirs*, Geological Society, London (Abstract).

SOUDARIN, L. 1995. *Positionement précis et tectonique des plaques à partir du système spatial DORIS*. PhD Thesis, Université Paul Sebatier, France.

TONDJI BYIO, J. J., 1995. *Chevauchements et bassins compressifs. Influence de l'érosion et de la sédimentation. Modélisation analogique et exemples naturels*. Mémoires et Documents du Centre Armoricain d'Etude Structurale des Socles, Rennes, **59**.

UNESCO, 1981. *Atlas géologique du monde* (sheets 4 and 5), 1:10.000.000.

USSAMI, N., SÁ, N. C. & MOLINA, E. C. 1993. Gravity map of Brazil: 2. Regional and residual isostatic anomalies and their correlation with major tectonic provinces. *Journal of Geophysical Research*, **98**, 2199–2208.

VANDECAR, J. C., JAMES, D. E. & ASSUMPÇÃO, M., 1995. Seismic evidence for a fossil mantle plume beneath South America and implications for plate driving forces. *Nature*, **378**(2), 25–31.

VIDOTTI, R. M. 1997. *Lithospheric structure beneath the Paraná and Parnaíba basins, Brazil, from regional gravity analyses*. PhD Thesis, The University of Leeds.

ZIEGLER, P. A. 1989. Geodynamic model for Alpine intraplate compressional deformation in Western and Central Europe. *In:* COOPER, M. A & WILLIAMS, G. D. (eds) *Inversion Tectonics*. Geological Society, London, Special Publication3, **44**, 63–85.

Fracture networks on the Belluno syncline, a fault-propagation fold in the footwall of the Belluno thrust, Venetian Alps, NE Italy

D. ZAMPIERI & P. GRANDESSO

CNR: Istituto di Geoscienze e Georisorse, Sezione di Padova/Dipartimento di Geologia, Paleontologia e Geofisica, University of Padova, Via Giotto 1, 35137 Padova, Italy (e-mail: dario.zampieri@unipd.it)

Abstract: The Belluno syncline is a structure located in the Venetian fold-and-thrust belt in the Italian Southern Alps. This SSE-vergent chain mainly developed during the Neogene as a back-thrust belt of the Alps deforming the northern margin of the Adria plate. The syncline is asymmetric in shape, with a steep backlimb, as the forelimb of the paired Monte Coppolo Anticline; it is the result of fault-propagation folding during the development of the underlying Belluno thrust. In plan view the Belluno thrust shows a prominent curvature interpreted as reactivation of inherited Mesozoic extensional structures. The hinge of the Belluno syncline presents a similar curvature. Rocks cropping out in the syncline core are Upper Chattian to Langhian fine-grained molassic sediments, mainly siltstones and marls.

Most fractures on the Belluno syncline have formed at high angles to bedding. The dominant fracture trends are subparallel and orthogonal to the fold hinge. Sometimes these fractures show plumose structures revealing extension (Mode I) fracture mechanism and therefore they are referred to as longitudinal (*bc*) joints and cross (*ac*) joints respectively. Other fracture trends oblique to the hinge line are found in measurement stations located on limbs distant from the fold hinge. They are referred to as oblique (shear) fractures. Throughout the western part of the Belluno syncline the fracture network maintains a roughly symmetrical distribution with respect to the fold hinge, suggesting a development in association with folding.

The Belluno syncline, located in the Venetian Alps fold-and-thrust belt, formed during the Neogene development of the Southern Alps, in NE Italy. The eastern Southern Alps are a SSE-vergent chain, which inverted the pre-existing Mesozoic passive margin of the Adria plate (e.g. Bertotti *et al.* 1993). The Tertiary sedimentary evolution of the Venetian Alps shows the interplay of two suborthogonal orogenic belts: the Dinarides and the Southern Alps (Fig.1). Therefore, two foreland basins of different ages and polarities overlap in time and space. Late Palaeocene to Middle Eocene external Dinaric flysch deposits accumulated within a NW-trending basin (Massari *et al.* 1986). Subsequent Upper Chattian to Langhian late-orogenic Dinaric molasse infilled a basin shifting southwards due to the rapid rising of the Alpine axial zone. A second stage of Alpine molasse accumulation began in the Serravallian and the basin axis assumed an ENE trend. From the Late Serravallian the basin was incorporated into the South Alpine thrust belt (Massari *et al.* 1986; Ghibaudo *et al.* 1996).

The Venetian Alps are arranged in an imbricate fan of thrust sheets involving the crystalline basement. In plan view the thrusts show several structural undulations controlled by inherited features, such as the NNE-trending normal faults (Doglioni 1992). The Belluno Line is a major thrust with a prominent ramp anticline (Monte Coppolo anticline) on the hanging wall and a syncline (Belluno syncline) on the footwall. It has a N75°E trend, but in the sector between Sospirolo and Cesiomaggiore it turns towards N45°E (Fig. 1), assuming an oblique ramp geometry with sinistral transpression (Costa *et al.* 1996). The near-vertical attitude of the forelimb of the hanging wall anticline and of the backlimb of the footwall syncline suggests fault-propagation folding (D'Alberto *et al.* 1995).

During folding the molasse sediments developed a prominent fracture network, which is of special interest to the applied geology of the area. The aims of this study are to present an example of natural fracture network on a syncline structure, and to compare the fracture network with other fold-related systems mentioned in the literature.

Stratigraphy of the Chattian–Langhian deposits of the Belluno syncline

The core of the Belluno syncline contains the Upper Chattian–Langhian siliciclastic molasse

From: AMEEN, M. (ed.) *Fracture and In-Situ Stress Characterization of Hydrocarbon Reservoirs*. Geological Society, London, Special Publications, **209**, 101–106. 0305-8719/03/$15.00
© The Geological Society of London 2003.

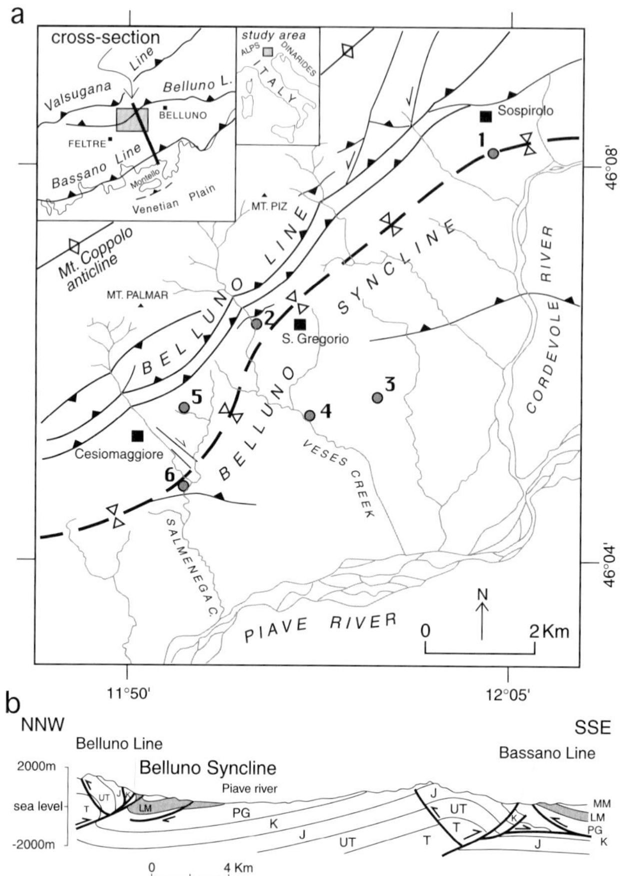

Fig. 1. (**a**) Structural map of the western part of the Belluno syncline (after Costa *et al.* 1996) with the six fracture measurement stations: (1) Rorei, (2) San Vittore, (3) Rive di Ignan, (4) Velos, (5) Boschi di Villa, (6) Cavallea. (**b**) Cross-section normal to the thrust belt. T, Lower-Middle Triassic; UT, Upper Triassic; J, Jurassic; K, Cretaceous; PG, Palaeogene; LM, Lower Miocene; MM, Middle Miocene. The investigated stratigraphic interval is shaded.

deposits that unconformably overlie the Eocene Belluno Flysch. The lithostratigraphic logs of representative sections with a tentative correlation of stage boundaries of the molasse deposits cropping out in the study area are shown in Figure 2. They comprise a number of arenaceous and muddy lithostratigraphic units respectively referred to, from base to top, as Belluno Glauconitic Sandstone, Bastia Siltstone, Orzes Sandstone, Casoni Siltstone, Libano Sandstone, Bolago Marl, S. Gregorio Sandstone, Monfumo Marl and M. Baldo Formation. These units constitute a clastic wedge in the area between Belluno and Feltre thinning out progressively westwards along the Dinaric foreland ramp.

According to Massari *et al.* (1986) and Ghibaudo *et al.* (1996), in the Upper Chattian–Lower Langhian local succession, four third-order depositional sequences, bounded by regional-scale unconformity surfaces often marked by glauconitic key beds, can be recognized.

The lower sequence (Upper Chattian–Lower Aquitanian in age) is represented by a basal transgressive glauconitic fossiliferous sand sheet (Belluno Glauconitic Sandstone) overlain by offshore muddy deposits (Bastia Siltstone). The second sequence (Upper Aquitanian–Lower Burdigalian in age) comprises three units: Orzes Sandstone, Casoni Siltstone and Libano Sandstone. The Orzes Sandstone is a discontinuous siliciclastic arenitic body with an erosional basal surface, interpreted as an estuary mouth valley-fill deposit. The Casoni Siltstone and Libano Sandstone represent a coarsening- and shallowing-upward succession of pro-delta to delta-front deposits respectively.

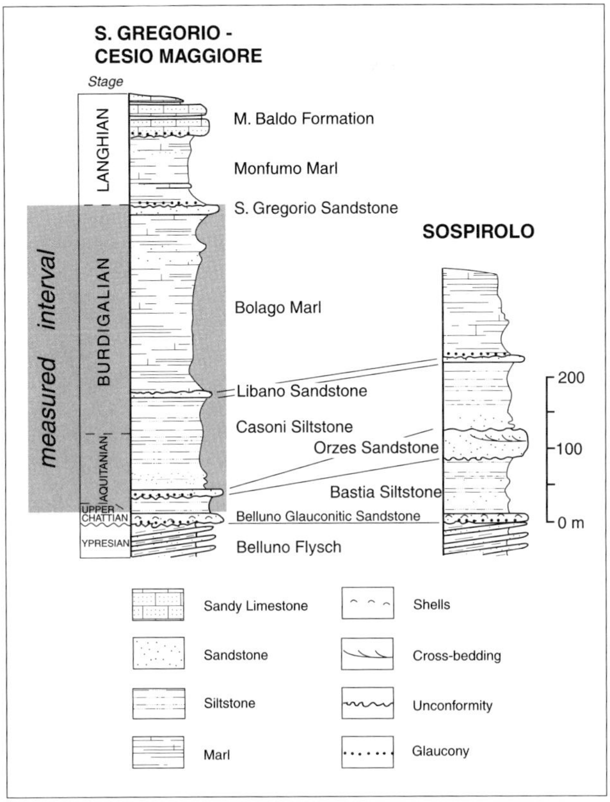

Fig. 2. Lithostratigraphic columns of the studied succession.

The third sequence (Upper Burdigalian in age) also records an episode of deltaic progradation. It is represented by the Bolago Marl (pro-delta silty marls) and the S. Gregorio Sandstone (delta-front arenites).

The upper sequence (Lower Langhian in age) is made up of outer shelf to epibathial silty marls of the Monfumo Marl unit, that represents the maximum deepening of the basin. The sequence is bounded at the top by the erosional base of the overlying M. Baldo Formation.

Fracture network

Data

Due to the low elevation of the study area and widespread presence of glacial deposits, rock outcrops are rare and localized only along stream sections. Six measurement stations were established at different structural positions of the Belluno syncline between Sospirolo and Cesiomaggiore: 1 and 6 are hinge zone stations, 2 and 5 are backlimb stations, 3 and 4 are forelimb stations (Fig. 1). Fracture data were collected on bedding surfaces and on bedding-perpendicular exposures. The lithostratigraphic units investigated are the Bastia Siltstone (stations 6, Cavallea), the Casoni Siltstone (lower part at station 6 and upper part at station 4 – Velos), the Libano Sandstone (station 4), the Bolago Marl (lower part at stations 3 – Rive di Ignan, 4 and 5 – Boschi di Villa; middle part at station 1 – Rorei; upper part at station 2 – S. Vittore), S. Gregorio Sandstone (station 2).

Most fractures have formed at high angles to bedding (Fig. 3). In each station fractures are well grouped in a few distinct sets (Fig. 4): stations 2 and 6 show only one fracture set, stations 1 and 3 show two sub-orthogonal fracture sets, station 5 shows two main fracture sets making an acute angle of *c*. 70°, while station 4 shows the most complex fracture system, composed of four main fracture sets. The fractures show a roughly planar geometry with a trace

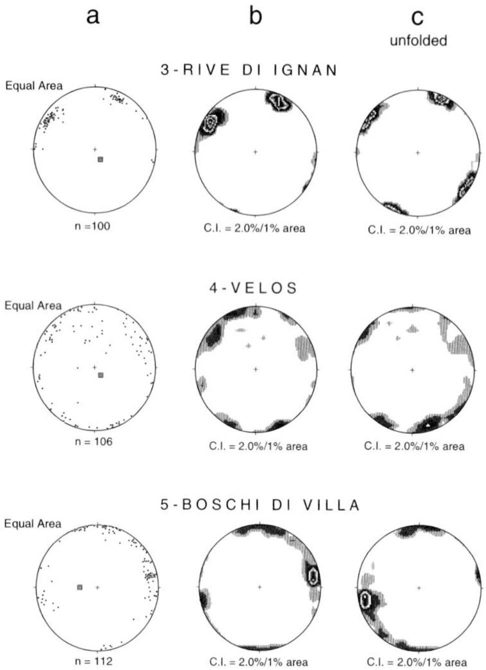

Fig. 3. Examples of lower hemisphere projections for fractures measured in fold limbs. (**a**) Poles to fractures (dots) and to bedding (square); (**b**) 1% area contour of poles to fractures; (**c**) 1% area contour of poles to fractures after bedding rotation to the horizontal.

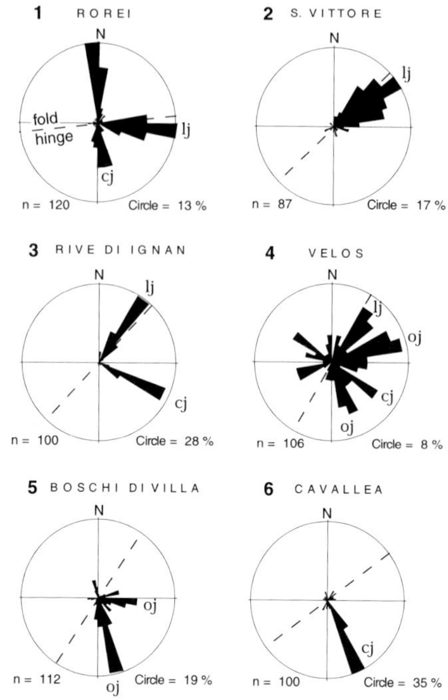

Fig. 4. Rose diagrams of fracture orientation as measured in the field, with interpretation of different sets in a layer and fold-control system: cj, cross joints; lj, longitudinal joints; oj, oblique joints. The dashed line is the local trend of the fold hinge.

length from one to several metres. The fractures are closed, without any measurable slip or dilation. In some cases (sets of stations 1 and 6) typical hackle plume ornaments (Hodgson 1961; Roberts 1961) were found on fracture faces (Fig. 5), providing critical information to interpret their origin. These fractographic features point to a brittle failure during Mode I loading at the crack tip (e.g. Pollard & Aydin 1988; Kulander *et al.* 1979). In contrast, on fracture faces of the sets oblique to the local trend of the fold hinge plumose structures have never been found supporting their shear fracture interpretation.

Fracture orientation versus fold geometry

Stereographic rotations to the horizontal around the layering strike were performed in order to investigate the relationship between fractures and fold . The maximum dip angle of the beds in measured stations is 30°. After rotation the trend of all fracture sets does not change significantly (Fig. 3). This is a consequence of the symmetrical arrangement of fractures with respect to the fold hinge. Given such symmetry, the orientation of similar sets should change from locality to locality according to variations in hinge orientation. For example, in our interpretation the N20–30°W set of station 6, which is normal to the local fold hinge, should not correspond to the N20–30°W set of station 4, which is oblique to the local fold hinge.

Discussion and conclusions

The geometric arrangement of the fracture system with respect to the Belluno syncline hinge strongly suggests a close relationship between fractures and fold. The plumose structures found on fracture sets parallel and orthogonal to the fold hinge points to an extension fracture mechanism (Mode I) for these fractures. With reference to the commonly used notation for fold fractures (e.g. Price 1967) they are inferred to be *bc* and *ac* respectively, where *bc* is a plane normal to bedding and parallel to the fold hinge, whereas *ac* is a plane normal to bedding and the fold hinge. The other fracture sets, normal to bedding but oblique to the fold hinge, are inferred to be shear or hybrid fractures (e.g. Hancock 1985)

Fig. 5. Near-vertical and normal-to-bedding joint face of N30°W set of station 6 showing metre-scale plumose structure. Such ornaments suggest an extension fracture mechanism ('Mode I') for these joints. Because tey are perpendicular to the fold hinge these fractures are inferred to be 'cross joints'.

and then they are labelled oblique fractures enclosing an acute angle about the *a* axis (Fig. 6). In some stations (1 and 4) some conjugate fractures enclosing an acute angle about *c* are also observed (Fig. 3).

From the Chattian to the Langhian the Venetian basin behaved as a Dinaric foreland basin. Due to low rates of thrust propagation, the basin evolved under weak tectonic control. From the Serravallian onwards an imbricate stack of overthrusts advanced rapidly towards the SSE shedding large volumes of clastic material into the Venetian basin (Massari *et al.* 1986; Ghibaudo *et al.* 1996). The maximum principal stress axis of the Neoalpine contractional phase in the Venetian Alps is recognized to trend N20°W (Doglioni 1992). Nevertheless, a fracture set parallel to the maximum stress axis is completely absent in some measured stations (1, 2 and 3), along with the lack of a system of conjugate fractures enclosing an acute angle about the same stress axis. Therefore, a post-folding origin of the fractures may be excluded, at least for most of them. Pre-folding fracturing may also be excluded, due to the very weak tectonic control of the basin during the Langhian interval, when a change from Dinaric to Alpine control occurred. This points to a fracture development coeval with folding.

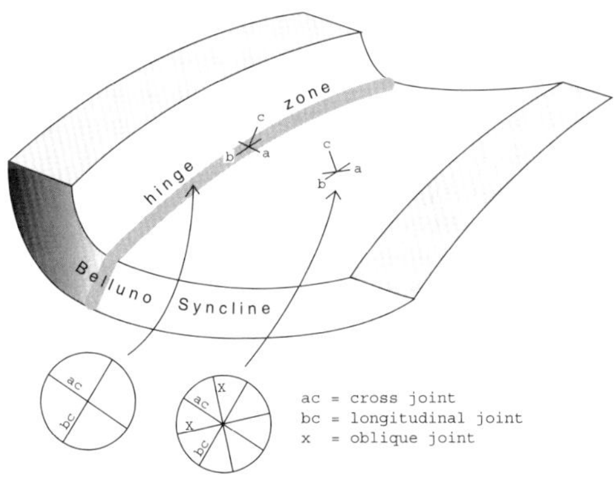

Fig. 6. Structure of the Belluno syncline with *abc*-coordinate reference system.

The analysis of the fracture network of the western part of the Belluno syncline shows that fractures formed in association with folding and that there is distinctly more variability in the fracture orientation patterns in the limbs than in the hinge zone. This result is consistent with those obtained on other anticline (Norris 1967; Winsor 1979; Hancock *et al.* 1983; Dunne 1986; Cooper 1992; Ameen 1995; Jamison 1997) and syncline (e.g. Hancock 1985) structures. Studying the natural fracture systems of folds, particular attention has been focused on anticline structures, probably due to their importance as hydrocarbon traps. We think our study may increase the limited number of examples of natural fracture systems on syncline structures and contribute to a better understanding of the fracture characterization in the mountains belts.

This work was supported by CNR (Centro di Studio per la Geodinamica Alpina, Padova) and MIUR (Cofin 2002). The authors thank Danilo Giordano for field assistance, S. Nardon, G. Minelli and J. W. Cosgrove for their helpful comments and review of the manuscript.

References

AMEEN, M. S. 1995. Regional joint pattern in the Taurus-Zagros Range, northern Iraq. *In*: MYER, L. R., COOK, N. G. W., GOODMAN, R. E. & TSANG, C.-F. (eds) *Fractured and Jointed Rock Masses.* Balkema, Rotterdam, 133– 143.

BERTOTTI, G., PICOTTI, V., BERNOULLI, D. & CASTELLARIN, A. 1993. From rifting to drifting: tectonic evolution in the South-Alpine upper crust from the Triassic to the Early Cretaceous. *Sedimentary Geology*, **86**, 53–76.

COOPER, M. 1992. The analysis of fracture systems in subsurface thrust structures from the Foothills of the Canadian Rockies. *In*: MCCLAY, K.R. (ed.) *Thrust Tectonics.* Chapman & Hall, London, 391–405.

COSTA, V., DOGLIONI, C., GRANDESSO, P., MASETTI, D., PELLEGRINI, G. B. & TRACANELLA, E. 1996. *Note illustrative del F° 063 Belluno della Carta Geologica d'Italia alla scala 1:50.000.* Istituto Poligrafico e Zecca dello Stato, Roma.

D'ALBERTO, L., BOZ, A. & DOGLIONI, C. 1995. Structure of the Vette Feltrine (Eastern Southern Alps). *Memorie di Scienze Geologiche*, **47**, 189–199.

DOGLIONI, C. 1992. The Venetian Alps Thrust Belt. *In*: MCCLAY, K.R. (ed.) *Thrust Tectonics.* Chapman & Hall, London, 319–324.

DUNNE, W. M. 1986. Mesostructural development in detached folds: an example from west Virginia. *Journal of Geology*, **94**, 473–488.

GHIBAUDO, G., GRANDESSO, P., MASSARI, F. & UCHMAN, A. 1996. Use of trace fossils in delineating sequence stratigraphic surfaces (Tertiary Venetian Basin, northeastern Italy). *Palaeogeography, Palaeoclimatology, Palaeoecology*, **120**, 261–279.

HANCOCK, P. L. 1985. Brittle microtectonics: principles and practice. *Journal of Structural Geology.* **7**, 437–457.

HANCOCK, P. L., DUNNE, W. M. & TRINGHAM, M. E. 1983. Variscan deformation in south-west Wales. *In:* HANCOCK P. L. (ed.) *The Variscan Fold Belt in the British Isles.* Adam Hilger, Bristol, 47–73.

HODGSON, R. A. 1961. Classification of structures on joint surfaces. *American Journal of Science*, **259**, 493–502.

JAMISON, W. R. 1997. Quantitative evaluation of fracture on Monkshood Anticline, a detachment fold in the Foothills of Western Canada. *The AAPG Bulletin*, **81**, 1110–1132.

KULANDER, B. R., BARTON, C. C. & DEAN, S. L. 1979. *The application of fractography to core and outcrop fracture investigations.* Morgantown Energy Technology Center, United States Department of Energy, METC/SP-79/3.

MASSARI, F., GRANDESSO, P., STEFANI, C. & JOBSTRAIBIZER, P. G. 1986. A small polyhistory foreland basin evolving in a context of oblique convergence: the Venetian basin (Chattian to recent, Southern Alps, Italy). *In:* ALLEN, P. A. & HOMEWOOD, P. (eds) *Foreland Basins.* International Association of Sedimentologists Special Publications, **8**, 141–168.

NORRIS, D. K. 1967. Structural analysis of the Queensway folds, Ottawa, Canada. *Canadian Journal of Earth Sciences*, **4**, 299–321.

POLLARD, D. D. & AYDIN, A. 1988. Progress in understanding jointing over the past century. *Geological Society of America Bulletin*, **100**, 1181–1204.

PRICE, R. A. 1967. The tectonic significance of mesoscopic subfabrics in the southern Rocky Mountains of Alberta and British Columbia. *Canadian Journal of Earth Sciences*, **4**, 39–70.

ROBERTS, J. C. 1961. Feather fractures and the mechanics of rock jointing. *American Journal of Science*, **259**, 481–492.

WINSOR, C. N. 1979. The correlation of fracture directions with sediment anisotropy in folded rocks of the Delamerian fold belt at port Germein gorge, South Australia. *Journal of Structural Geology*, **1**, 245–254.

A new method for smoothing orientated data and its application to stress data

B. MÜLLER[1], V. WEHRLE[2], S. HETTEL[1], B. SPERNER[1] & K. FUCHS[1]

[1] *Geophysical Institute, University of Karlsruhe, Hertzstrasse 16, 76187 Karlsruhe, Germany (e-mail: birgit.mueller@gpi.uni-karlsruhe.de)*

[2] *SAP, 69189 Walldorf, Germany*

Abstract: Smoothing algorithms provide a means of identifying significant patterns in sets of orientated data, eliminating local perturbations within the observations and predicting patterns of orientated data in places which lack observations. Here we present the smoothing of orientation data with a distance-related method of data weighting as an alternative to previous weighting algorithms. The data weighting and smoothing method presented here in theory and practice is developed on the basis of a statistical smoothing algorithm. The method can be applied to orientation data of 180° periodicity such as maximum horizontal tectonic stresses (S_H) as compiled in the World Stress Map database. Our smoothing algorithm enables discrimination between local (<250 km of lateral extent) and regional (*c.* 250–5000 km of lateral extent) stress fields, and allows comparison of S_H with other directional data such as fault trends or strain data. We present smoothed stress maps for northeastern America, the Himalayas and western Europe. By varying the scale and smoothing parameters we illustrate their influence on the accuracy and smoothness. We give recommendations for the appropriate choice of these parameters.

Orientation data, such as the azimuth of the maximum horizontal stress (S_H), are often heterogeneous due to differing data densities, locally significant scattering, varying qualities or sporadic erroneous values. In these cases, as well as identifying major trends in stress patterns, interpolation or smoothing algorithms are necessary. In this paper we discuss arguments for the smoothing of orientation data, present a distance-based method of weighting, thus modifying the smoothing algorithm of Hansen & Mount (1990), and apply it to three test areas with stress observations compiled in the World Stress Map (WSM).

The WSM project compiles data about the contemporary tectonic stress field in the Earth's crust (Zoback 1992*a*; http://www.world-stress-map.org). Data from more than 10 000 locations are now available and are frequently displayed in the form of stress maps, in which the orientation of the maximum horizontal stress S_H is plotted. Based on the 1997 WSM release we investigated the distribution of stress orientations with depth for three different tectonic settings (namely NE America, the Himalayas, and western Europe) with good data coverage from the surface to 35 km depth. Northeastern America (Fig. 1) represents a Precambrian craton, the Himalayas (Fig. 2) an orogenic belt and western Europe (Fig. 3) a conglomerate of tectonic settings including rift zones (Rhinegraben), plume areas (Massif Central), and sedimentary basins (northern Germany). The three areas of application are characterised by different stress patterns, as described in Richardson *et al.* (1979), Gowd *et al.* (1992), Müller *et al.* (1992), Zhonghuai *et al.* (1992), Grünthal & Stromeyer (1986, 1992), Zoback & Zoback (1981), and Zoback (1992*b*). Table 1 shows that data from different depth intervals give comparable results in mean stress orientation confirming statements of earlier publications (Zoback *et al.* 1989, Zoback 1992*a*; Wehrle 1998). Thus, the joint usage of data from different depths for the creation of smoothed stress maps is legitimate. In other places stress reorientations with depth might occur and cause scattering of the stress data. This has to be checked in the individual cases.

The stress patterns can be summarized as follows. In northeastern America (Fig. 1) the dominance of a NE-SW S_H orientation is clearly visible. Deviations from this trend, such as in the coastal and off-shore areas between 36° N and 45° N, appear subordinate. In the Himalayas (Fig. 2) S_H varies with the tectonic position: at the plate boundary near the thrust front S_H is orientated perpendicular to the mountain belt and a thrust regime is prevailing. In elevated areas of the orogen normal faulting occurs with S_H perpendicular to the mountain belt. Towards

From: AMEEN, M. (ed.) *Fracture and In-Situ Stress Characterization of Hydrocarbon Reservoirs.* Geological Society, London, Special Publications, **209**, 107–126. 0305–8719/03/$15.00
© The Geological Society of London 2003.

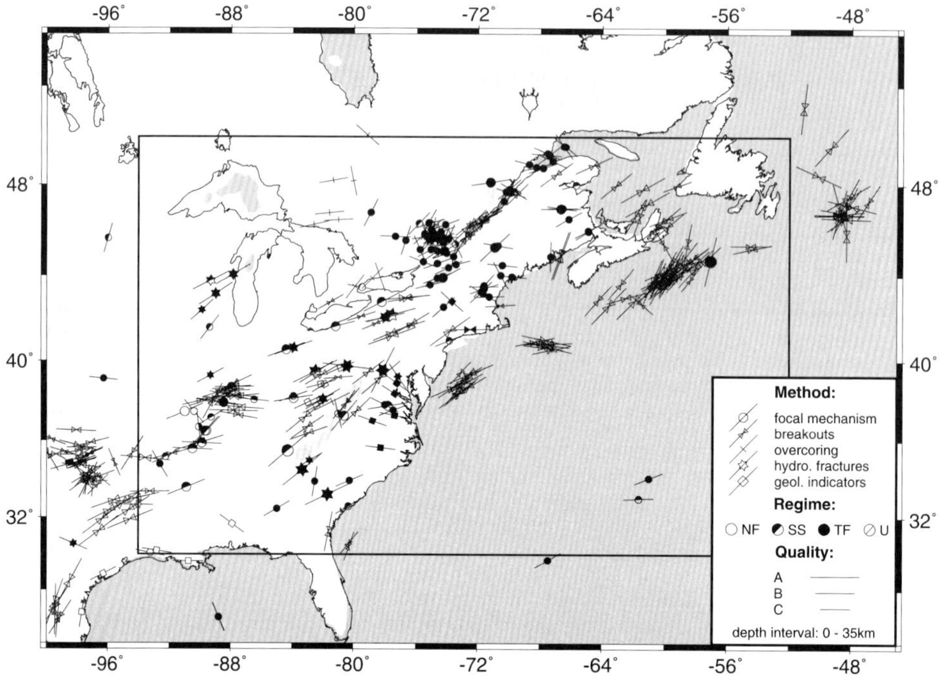

Fig. 1. Map of maximum horizontal stress orientations S_H for Northeastern America. Different symbols stand for different stress indicators (e.g. borehole breakouts, earthquake focal mechanisms; Zoback *et al.* 1989; Müller *et al.* 1992, 1997). Symbol length indicates data quality, and symbol filling provides information about the tectonic regime (NF, normal faulting, SS, strike-slip TF, thrust faulting, U, unknown stress regime) (Anderson 1942). The rectangle marks the area for which the smoothed stress pattern in Figure 10 is computed.

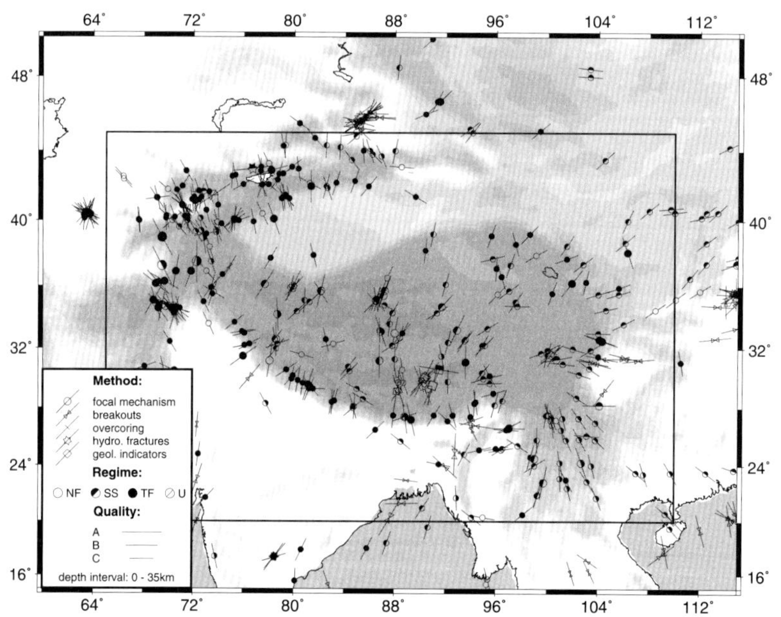

Fig. 2. Map of maximum horizontal stress orientations S_H for the Himalayan region. The rectangle marks the area for which the smoothed stress pattern is computed (Fig. 11). For explanation of symbols see Figure 1 caption.

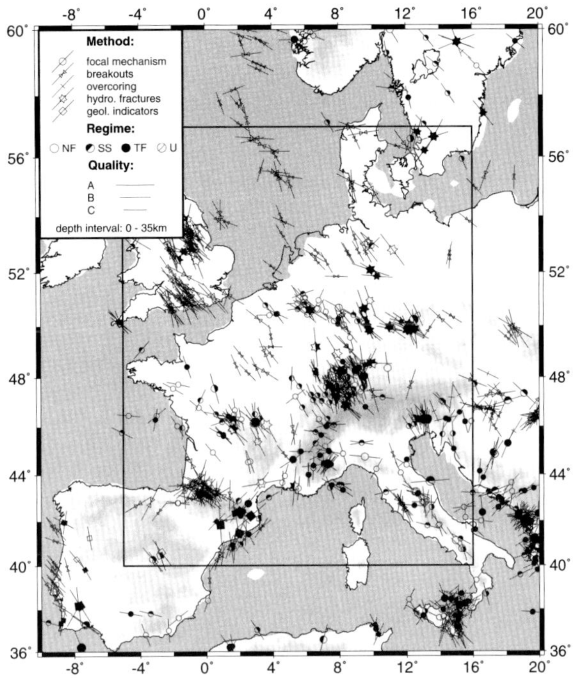

Fig. 3. Map of maximum horizontal stress orientations S_H for western Europe. The rectangle marks the area for which the smoothed stress pattern is computed (Fig. 12). For explanation of symbols see Figure 1 caption.

Table 1. Average S_H azimuth and standard deviation (SD) in different depth intervals illustrating the stability of stress orientation with depth for 3 selected areas. Average orientations are calculated using circular statistics (Mardia 1972).

Depth interval (km)	NE America		Himalaya		Western Europe	
	S_H±SD	No.	S_H±SD.	No.	S_H±SD	No.
0–5	N 65° E±27°	289	N 47° E±2°	177	N 143° E±41°	516
5–10	N 52° E±49°	25	N 2° E±41°	58	N 161° E±49°	150
10–15	N 81° E±44°	27	N 172° E±46°	106	N 151° E±44°	112
15–20	N 62° E±53°	23	N 173° E±48°	45	N 162° E±36°	33
20–25	N 75° E±33°	2	N 4° E±46°	36	N 155° E±47°	35
25–30	N 47° E±40°	3	N 3° E±53°	28	N 164° E±22°	11
30–35	–	0	N 3° E±44°	55	N 95° E±60°	13
0–35	N 63° E±32°	369	N 9° E±53°	506	N 143° E±43°	870

the east, the stress regime alters to strike-slip with a radial pattern of S_H orientations. In western Europe (Fig. 3) NW–SE orientations dominate but with marked deviations. In the western Alpine area (approx. long. 5° E, lat. 44° N) a radial stress pattern can be detected. A NNE trend along the NE coast of Spain (approx. long. 3° E, lat. 42° N) and a strong scatter, for example in the western Pyrenees (approx. long. 0° E, lat. 44° N), are obvious.

Unwanted features in orientation data such as scatter of data and variation in qualities can be partly eliminated by interpolating or smoothing the data. The interpolation method uses measured values at the observation points and interpolates between them. In contrast, smoothing algorithms first place a regular grid over the study area and then calculate orientation data at each grid-point by averaging data from a certain area around the grid-point.

Investigations into interpolation and smoothing of orientation data have been made by Watson (1985), Rebai *et al.* (1992), Lee & Angelier (1994), Coblentz & Richardson (1995) and Bird & Li (1996). Rebai *et al.* (1992) obtained average orientations using a linear interpolation method by means of a grid of triangular elements. Grid nodes were either placed in the centres of data clusters or, in places with a paucity of data, at the locations of isolated data points. Average orientations were computed for each grid node, taking into account the relative data quality, and the stress field inside the triangular elements was computed by linear interpolation (for details see Rebai *et al.* 1992). Lee & Angelier (1994) used quality-weighted data to determine the averaged stress field. In contrast to these interpolation approaches, Bird & Li (1996) and Coblentz & Richardson (1995) used different statistical methods to average the stress field. Hansen & Mount (1990) suggested a smoothing procedure based on the statistical smoothing algorithm of Watson (1985). Our approach resembles the smoothing procedure of Hansen & Mount (1990), but uses a different distance weighting procedure (Hüsges 1995). Whereas Hansen & Mount (1990) consider a fixed number of measurements in the vicinity of the point where the smoothed stress should be determined, we use the measured data within a fixed radius around this point. We first explain the smoothing method, then compare and discuss different weighting methods. By means of artificial stress data, we will test the reliability of the two weighting methods and apply them to data sets from the WSM to illustrate more realistic examples of their usage.

Applications of stress data smoothing

There are a number of reasons for the smoothing of stress data. In the following we briefly summarize the major aspects.

Data distribution

Stress data are distributed irregularly on regional scales as well as on global scales. On a global scale, there are areas with tens of measurements within a few square kilometres such as in Western Europe, parts of Asia, and northern America, in which stress orientation smoothing enables main trends to be better identified. This is particularly useful when the observations are clustered and exhibit azimuthal variations. On the contrary, there are large regions with an almost complete lack of data, for example eastern Europe and northwestern Asia or the oceans. In these regions stress orientations can be predicted by using the results of nearby observations.

Scattering stress orientations

In places with clusters of observations, especially when the observed S_H data scatter strongly, smoothing is a tool to identify the dominating stress orientation.

Distinction of regional from local stress field

According to Zoback (1992*a*) stress orientations are often homogeneous over large areas, thus showing regional trends as for example in western Europe, the Andes, or northern America. These areas extend laterally over distances of 250–5000 km; such distances exceed several times the lithospheric thickness. Within these large-scale stress provinces, local stress fields occur extending laterally over areas with dimensions of up to 100–250 km, which is equivalent to several times the average crustal thickness of 35 km. The distinction of regional and local stress fields is usually obtained visually (Zoback 1992*a*, *b*). However, the comparison of stress orientations of different regions by visual analysis is always subjective and might be unreliable. Furthermore, the visual interpretation of stress distributions has the disadvantage that the boundaries of the stress provinces can only be identified in areas of dense data distribution. The prediction of the lateral extent of a stress province into neighbouring regions with only scattered data or lack of data is difficult. The identification of the local stress field is not only of scientific interest, but also has economic relevance, for example in defining boundary conditions for reservoir stress state management or in identifying the slip tendency of faults in seismic risk assessment.

The distinction of regional and local stress fields is also significant for the numerical modelling of stress and deformation. Numerical modelling commonly uses either displacement or stress boundary conditions. If stress boundary conditions are selected, it is the regional, not the local trend of stress orientation which has to be applied to the model boundaries.

Estimate of stress magnitudes from the superposition of stress fields

Local stress fields may originate from (1) isolated loading, for example, glaciers or volcanic islands (Stein & Pelayo 1991), (2) lateral density contrasts at continental margins or in sedimentary basins, (3) flexural bending (Turcotte & Schubert

1982) or (4) weak faults (Zoback 1991). The superposition of regional and local stress fields poses the chance to estimate the regional stress magnitude, if the magnitude and the orientation of the local field are known or can be estimated (Sonder 1990; Zoback 1992*a*). This approach requires a quantitative separation of regional and local stress fields.

Comparison of stress data with other orientation data

Regional tectonic stress is ascribed to originate from plate boundary forces (slab pull, ridge push, trench suction), which are also responsible for the plate motions. Thus, the general qualitative coincidence between plate motion trajectories (DeMets *et al.* 1990) and stress orientations is unsurprising (Zoback *et al.* 1989; Zoback & Zoback 1991; Müller *et al.* 1992; Richardson 1992). A measure of this coincidence was obtained by averaging the angular difference between the plate motion trajectories and the stress observations at the individual measuring locations (e.g. Zoback *et al.* 1989; Müller *et al.* 1992). However, because the distribution of stress observations is varying in space and because stress orientations show irregularities due to the superposition of local effects this approach permits only a limited quantitative comparison of the parallelism between stress orientations and plate motions. Smoothing of the stress data may help to reduce these effects.

Method

Watson (1985) developed a smoothing method for observations of 180° periodicity which we summarize here. Suppose that n orientation data with azimuth Θ_i ($i=1,\ldots,n$) and unit length have been observed at locations $\underline{x}_i=(x_i^{(1)},x_i^{(2)})$. The observed orientation can be represented by a vector $\underline{y}_i=(y_i^{(1)},y_i^{(2)})$ for which $\tan\Theta_i=y_i^{(2)}/y_i^{(1)}$. It is the objective of the smoothing procedure to find a function $\underline{F}(\underline{x})$ for which the residuals $\{\underline{F}(\underline{x}_i)-\underline{y}_i,\ i=1,\ldots n\}$ are small; $\underline{F}(\underline{x})$ is an approximation or prediction of the 'true', but unknown stress field function. If $\underline{F}(\underline{x}_i)=\underline{y}_i$ for all i, we have obtained a maximum accuracy in terms of replicating the observed stress data exactly. In contrast, maximum stress field smoothness is obtained when $\underline{F}(\underline{x}_i)$ is a constant, but this corresponds, in general, to high residual values at each observation point. Thus, $\underline{F}(\underline{x})$ should be chosen so as to obtain an optimal compromise between smoothness and accuracy. For the determination of $\underline{F}(\underline{x})$, Watson (1985) suggested a statistical approach whereby a function $\underline{f}(\underline{x})$, defined at the locations $\underline{x}_i$, maps any point from the plane $\varkappa^2$ to the points within a unit circle. To calculate a smoothed stress field for the locations $\underline{x}_i$, $\underline{f}(\underline{x})$ is chosen to maximize the objective function $K(\underline{f})$, where

$$K(\underline{f})=\sum_{i=1}^{n}\left(\begin{array}{l}\left[\underline{f}(\underline{x}_i)^T\,\underline{y}_i\right]^2 d(\underline{x}_i,\underline{x}_i)+\\ \lambda\displaystyle\sum_{j=1,j\neq i}\left[\underline{f}(\underline{x}_i)^T\,\underline{f}(\underline{x}_j)\right]^2 d(\underline{x}_i,\underline{x}_j)\end{array}\right) \tag{1}$$

The first term in the overall summation controls the accuracy with which the observations are reproduced, since $\sum_{i=1}^{n}\{\underline{f}(\underline{x}_i)^T\underline{y}_i\}^2$ is large when $\underline{f}(\underline{x}_i)$ is parallel or anti-parallel to $\underline{y}_i$. The second term controls the smoothness of $\underline{f}(\underline{x})$: $\underline{f}$ is smoother when the term is close to *1* for $\underline{x}_j$ in the vicinity of $\underline{x}_i$. The scalar l controls the influence of both parts, accuracy and smoothing. Large values of l result in a smooth $\underline{f}(\underline{x})$, at the expense of accuracy at the observation points, whereas for small l the function $\underline{f}(\underline{x})$ is less smooth, but in good agreement with the observations (Hansen & Mount 1990; Fig. 4). The function $d(\underline{x}_i,\underline{x}_j)$ is a weight term for quality and distance (see below) which has to be a smooth function (here smooth in the sense that it is continuous together with its first derivative), which decreases steadily with increasing distance between $\underline{x}_i$ and $\underline{x}_j$, and which is limited for $\|\underline{x}_i-\underline{x}_j\|\rightarrow 0$. To determine the maximum of $K(\underline{f})$ Equation 1 is rearranged:

$$K(\underline{f})=\sum_{i=1}^{n}\underline{f}(\underline{x}_i)^T\left(\begin{array}{l}\left[\underline{y}_i\underline{y}_i^T\right]d(\underline{x}_i,\underline{x}_i)+\\ \lambda\displaystyle\sum_{j=1,j\neq i}\left[\underline{f}(\underline{x}_j)\underline{f}(\underline{x}_j)^T\right]^2 d(\underline{x}_i,\underline{x}_j)\end{array}\right)\underline{f}(\underline{x}_i) \tag{2}$$

$K(\underline{f})$ is maximized when, for all i, $\underline{f}(\underline{x}_i)$ is the eigenvector corresponding to the maximum eigenvalue of the matrix:

$$M=\left[\underline{y}_i\underline{y}_i^T\right]d(\underline{x}_i,\underline{x}_i)+\lambda\sum_{j=1,j\neq i}\left[\underline{f}(\underline{x}_j)\underline{f}(\underline{x}_j)^T\right]^2 d(\underline{x}_i,\underline{x}_j) \tag{3}$$

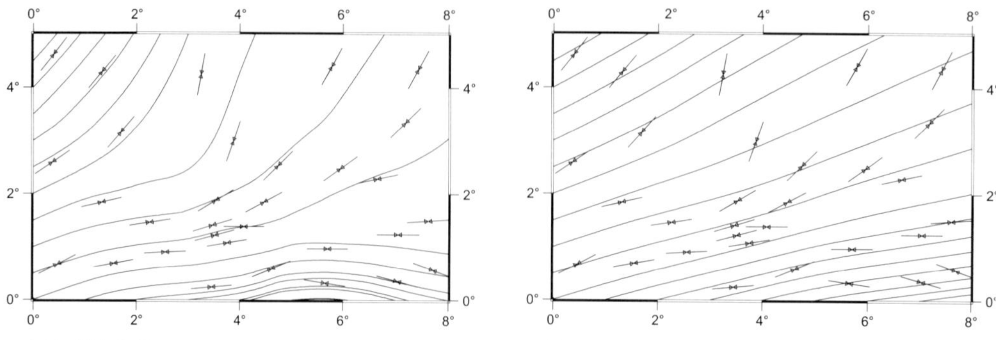

Fig. 4. Stress trajectories calculated for an artificial data set using the FNN method: (**a**) balanced for high data accuracy by using $\lambda=0.2$ and the five nearest neighbours (R) considered, and (**b**) balanced for greater smoothness by using $\lambda=5$ and $R=10$. The symbols represent the observed maximum horizontal stress orientations S_H.

Watson (1985) suggested an iterative method to determine $\underline{f}$ starting from $\underline{f}^{(0)}(\underline{x}_i)=\underline{y}_i$, thus beginning the iteration using the observed data. The iteration stops at the Nth step when $K(\underline{f}^{(N-1)})-K(\underline{f}^{(N)})$ is less than a specified amount. The iteratively computed n vectors $\{\underline{f}^{(N)}(\underline{x}_i)\}$ are estimates for the smoothed stress field at the n locations where stress observations are available. The smoothing predictor $\underline{F}(\underline{x})$ at any location is determined from the eigenvector of the largest eigenvalue:

$$\underline{F}(\underline{x}) = \text{eigenvector of} \sum_{i=1}^{n}\left[\underline{f}^{(N)}(\underline{x}_i)\underline{f}^{(N)}(\underline{x}_i)^T\right] d(\underline{x},\underline{x}_i) \quad (4)$$

Weight

In general, the weight term $d(\underline{x}, \underline{x}_i)$ consists of two factors $d(\underline{x}, \underline{x}_i)= q(i) \bullet w(\underline{x}, \underline{x}_i)$, where $q(i)$ represents the quality weight and $w(\underline{x}, \underline{x}_i)$ represents the distance weight.

Quality weight. The quality weight, $q(i)$ corresponds to the robustness weight of Hansen & Mount (1990). $q(i)$ controls the effect of each observation's quality in the smoothing procedure. Stress data in the WSM are ranked according to their reliability as indicators for the contemporary tectonic stress orientation. In the WSM database the qualities range from A to E, with A being the best quality. Data of qualities A, B and C are considered to be reliable stress indicators, and are used in the stress maps. In the smoothing procedure, we use different weight factors $q(i)$ for these three quality categories, namely 1 for A, 0.75 for B, and 0.5 for C quality.

Distance weight. The distance weight controls the influence of the individual observation $\underline{y}_i$ at location $\underline{x}_i$ on the computation of the smoothed stress field at another location $\underline{x}$. Hansen & Mount (1990) suggested a distance weight in which a fixed number of nearest neighbours is taken into account (FNN method). Here, we present a distance weight with a fixed search radius (FSR method). The FNN method uses local weights with a varying area of coverage which is determined by the data density, whereas the FSR method uses a global weight independent of the data distribution and dependent only on the Euclidean distance $||\underline{x}-\underline{x}_j||$.

In the FNN method, a fixed number R of nearest neighbours is used for the computation of the smoothed stress field. The distance weight $w(\underline{x}, \underline{x}_i)$ is zero, if $\underline{x}_i$ does not belong to the R nearest neighbours of $\underline{x}$. The weight function $w(\underline{x}, \underline{x}_i)$ is chosen such that it decreases from $w(\underline{x}, \underline{x}_i)=1$ for $\underline{x}=\underline{x}_i$ to $w(\underline{x}, \underline{x}_{R+1})=0$ for the $(R+1)^{\text{th}}$ neighbour of $\underline{x}$. Hansen & Mount (1990) followed Cleveland (1979) and used the tricube weight function

$$w(\underline{x},\underline{x}_i)=\begin{cases}\left[1-\left(\dfrac{\|\underline{x}-\underline{x}_i\|}{h(\underline{x})}\right)^3\right]^3 & \text{for } \|\underline{x}-\underline{x}_i\|<h(\underline{x})\\ 0 & \text{else}\end{cases}$$

with $h(\underline{x})$ being the distance between $\underline{x}$ and the $(R+1)^{\text{th}}$ neighbour. Figure 5 shows an example of stress trajectories computed with the FNN method for an artificially produced, random set of stress observations. It clearly demonstrates the big differences in size of the smoothing areas which arises due to variable data density.

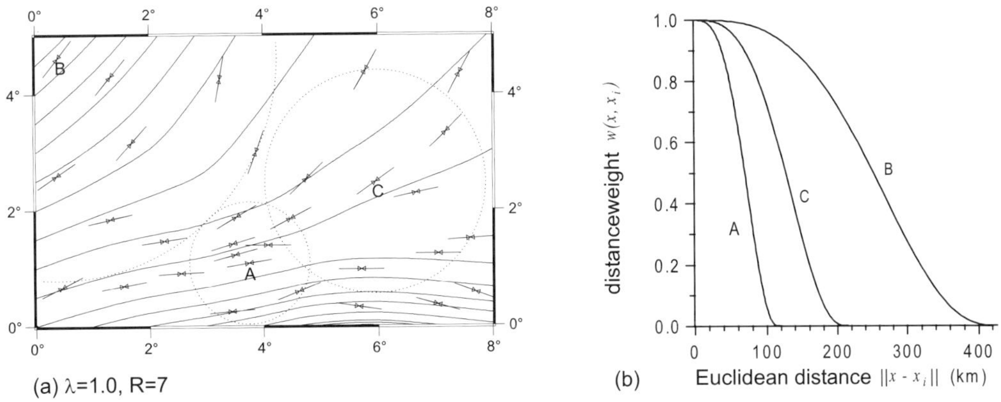

Fig. 5. Stress data smoothed with the FNN method. (**a**) Original observations and trajectories of the smoothed stress field. The circles around the points A, B and C show the distance to the $(R+1)$th nearest neighbour. Note the big difference in the size of the search radii. (**b**) Plot of the tricube weight function at points A, B and C. The smoothed stress value at A is calculated over a smaller area than the values at B and C.

Hansen & Mount (1990) recommended rescaling the distance weight to preserve the balance between the smoothing and accuracy terms over the entire area (see, for example, fig. 3 in Hansen & Mount 1990), in the case that all R nearest neighbours of $\underline{x}_i$ are at a similar distance to $\underline{x}_i$. In this particular case, the distance weight is small for all nearest neighbours and the first term of Equation 3, the accuracy term, dominates. The smoothing term is not significant and thus the stress field is only weakly smoothed. This might be critical for the case that an anomalous orientation is surrounded by homogeneous orientations at equal distance. To obtain a balance between smoothing and accuracy which is not biased by the data distribution Hansen & Mount (1990) suggested:

$$\tilde{w}(\underline{x}_i, \underline{x}_j) = \frac{w(\underline{x}_i, \underline{x}_j)}{G(\underline{x}_i)} \quad \text{for} \quad i \neq j$$

$$\tilde{w}(\underline{x}_i, \underline{x}_i) = w(\underline{x}_i, \underline{x}_i) = 1 \quad \text{else}$$

with

$$G(\underline{x}_i) = \sum_{j \neq i}^{n} w(\underline{x}_i, \underline{x}_j)$$

Then, the sum of weight terms of all nearest neighbours is equal to the weight term of the data point itself. In the case of rescaling with $\lambda = 1.0$, there is a balance between smoothing and accuracy; for $\lambda > 1.0$ smoothness prevails against accuracy, and for $\lambda < 1.0$ accuracy prevails. Rescaling is not necessary for the computation of the prediction $\underline{F}$ (Equation 4) because Equation 4 only uses the second term of Equation 3.

The FSR method is a global weight function, constant over the study area, because it uses a fixed search radius A to define the nearest neighbours. In evaluating the smoothed stress field at $\underline{x}_i$, only those observations for which the Euclidean distance to $\underline{x}_i$ is equal to or less than A are considered:

$$w(\underline{x}, \underline{x}_i) = \begin{cases} a\|\underline{x} - \underline{x}_i\| & \text{for} \quad \|\underline{x} - \underline{x}_i\| < A \\ 0 & \text{else} \end{cases}$$

We choose the distance weight $a\|\underline{x} - \underline{x}_i\|$ to be the tricube weight function:

$$a\|\underline{x} - \underline{x}_i\| = \left[1 - \left(\frac{\|\underline{x} - \underline{x}_i\|}{A}\right)^3\right]^3$$

following Hansen & Mount's (1990) suggestion, because $0 \leq a(\underline{x}, \underline{x}_i) \leq 1$, and $a(\underline{x}_i, \underline{x}_i) = 1$ and $a(A) = 0$. That is, the distance weight depends on the Euclidean distance and decreases to zero at the distance A. Figure 6 shows the data illustrated in Figure 5 smoothed with the FSR method. Because the value of the smoothing term depends on the amount of data considered, the significance of the smoothing term varies over the area of investigation. To obtain a balance

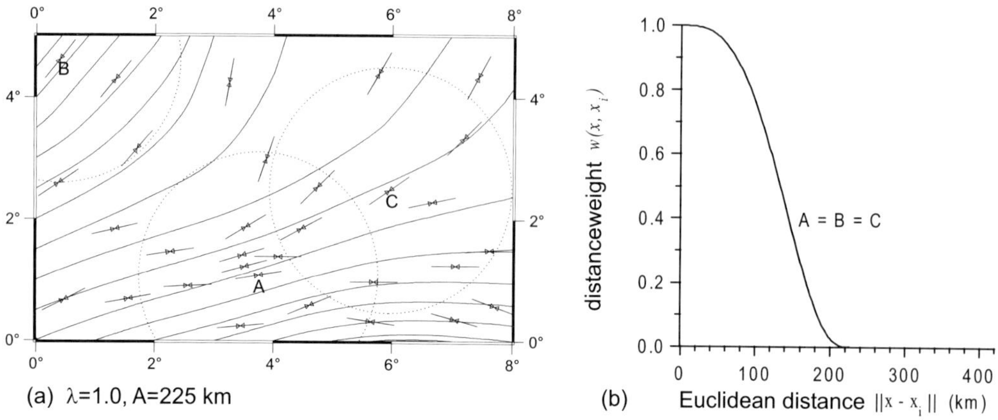

Fig. 6. Stress data smoothed with the FSR method. (**a**) Original observations and trajectories of the smoothed stress field. Search radii for A, B and C are constant. (**b**) Plot of the tricube weight function: it is the same for the locations A, B and C, but the number of data used in the computation varies.

between accuracy and smoothing terms, Wehrle (1998) suggested a normalization of the weight function analogous to that of Hansen & Mount (1990) by the number of nearest neighbours n_n within the search radius A:

$$\tilde{w}(\underline{x}_i,\underline{x}_j) = \frac{w(\underline{x}_i,\underline{x}_j)}{n_n} \qquad for \quad i \neq j$$

$$\tilde{w}(\underline{x}_i,\underline{x}_i) = w(\underline{x}_i,\underline{x}_i) = 1$$

The number of nearest neighbours which are used for the smoothing varies from point to point. To ensure, that in cases of low data density or poor quality, a sufficiently high number of data are taken into account, Wehrle (1998) proposed to use one of the following threshold criteria:

(1) a minimum number n_{min} of nearest neighbours, e.g. $n_n = n_{min} = 3$;
(2) a minimum value sq for the sum of the quality weight factors of the nearest neighbours, e.g. $sq = \sum_{i=1}^{n_{min}} q(i) \geq 1.0$;
(3) a minimum value sw for the sum of the weight terms of the nearest neighbours, e.g. $sw = \sum_{i=1}^{n_{min}} d(\underline{x},\underline{x}_i) = \sum_{i=1}^{n_{min}} w(\underline{x},\underline{x}_i)q(i) \geq 0.4$.

The smoothed stress field is displayed only at locations where the chosen criterion is met. This leads to gaps in the smoothed map indicating regions where the data are of insufficient quality or too sparse.

Reliability of smoothed orientations

A measure of the reliability of a smoothed stress orientation is the average weighted angular difference between the smoothed orientation and the orientations at neighbouring points:

$$\bar{\alpha} = \frac{1}{nso}\sum_{i=1}^{r} \alpha_i d(\underline{x},\underline{x}_i) = \frac{1}{r}\sum_{i=1}^{r} \alpha_i w(\underline{x},\underline{x}_i)q(i)$$

with nso being the number of stress orientations $\underline{y}_i$, which were used to compute the smoothed stress field $\underline{F}(x)$ at location $\underline{x}$. α_i is the angular difference between $\underline{F}(\underline{x})$ and $\underline{y}_i$. The average angular difference is large when nearby observations scatter strongly and small when they show a homogeneous trend. The angular difference can be shown in maps using for example a colour code for the symbols of the smoothed stress field.

Comparison of FNN and FSR distance weight methods

We have compared the FNN and FSR smoothing methods by means of a chess-board-type stress pattern with a varying density of observations (Wehrle 1998). The stress orientations vary between N 0°E and N 90°E (Fig. 7a). From this pattern of stress domains, with side lengths of 225 km each, 'observations' have been selected randomly, but with a decreasing observation density from left to right (Fig. 7b).

The FNN method uses a fixed number R of nearest neighbours. Therefore, the distance to the Rth nearest neighbour is smaller in an area of high data density than in an area of low data density, resulting in varying search radii at different locations, which is a severe disadvantage for

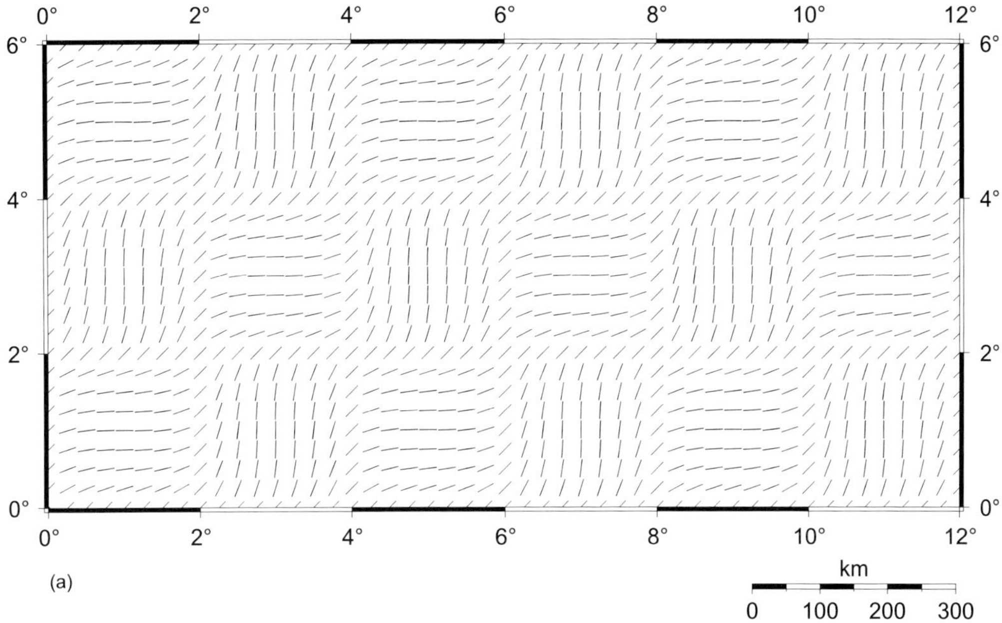

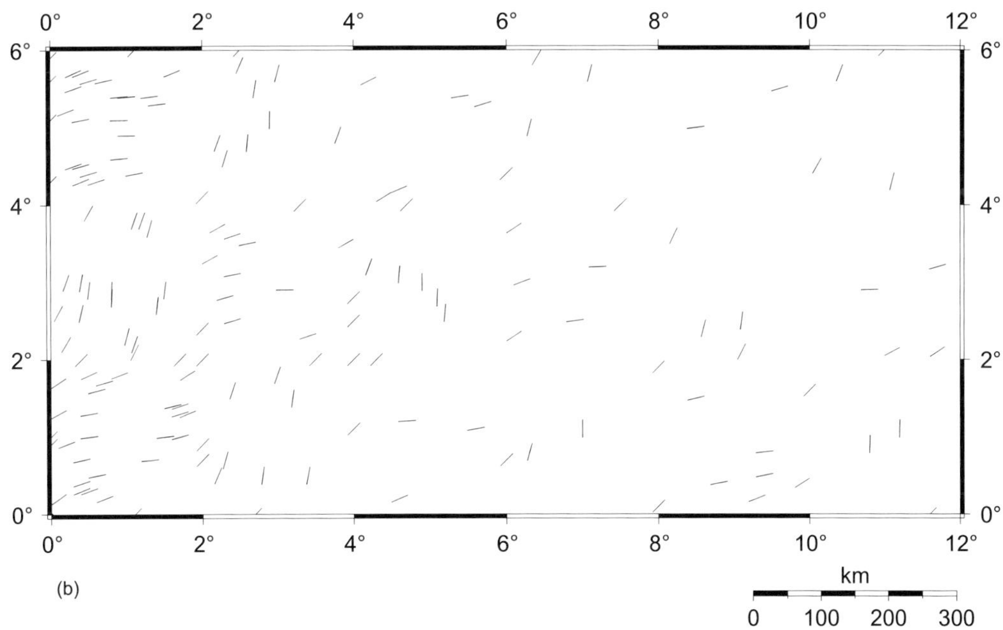

Fig. 7. (**a**) Artificially produced set of chess-board-type stress domains. (**b**) Randomly selected stress 'observations' of the stress field in (a) with an increase in data density from right to left.

regions with low or heterogeneous data density. The resolution of stress domains is only possible for places with a high density of observations (left sides in Fig. 8a & b.). For regions with low data density, local trends are smeared out (right sides in Fig. 8a & b). For a higher number of nearest neighbours (Fig. 8b) this 'smearing zone' is much larger.

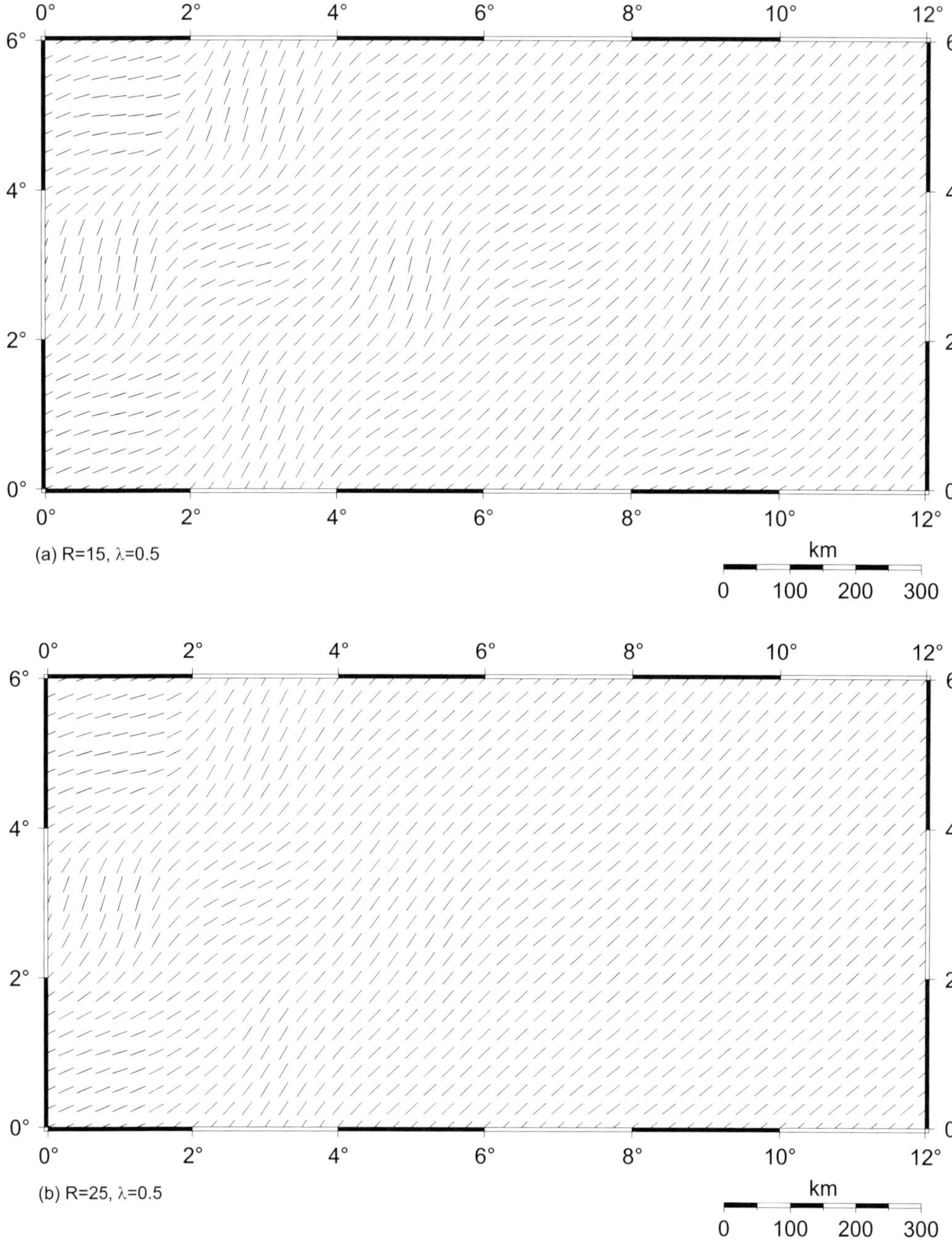

Fig. 8. Smoothed stress field computed with the FNN method based on the data of Figure 7b. Computations are done with (**a**) $R=15$ and (**b**) $R=25$ nearest neighbours. Stress domains are only visible on the left side, where the data density is high (see Fig. 7b).

With a fixed search radius (FSR method) it is possible to focus either on the large-scale trend (regional stress field) or on small domains (local stress field). If the goal is to resolve small scale stress patterns, the search radius should be small. For a search radius in the order of the side length of the stress domain or less, the chess-board-type pattern is resolved throughout its extent (Fig.

9a). If one aims at the regional trend a large search radius A has to be selected (Fig. 9b). The result is independent from the density of observations.

Application of the FSR method

To obtain a map with smoothed stress data for a specific area it is important to extend the original area of investigation by at least one search radius

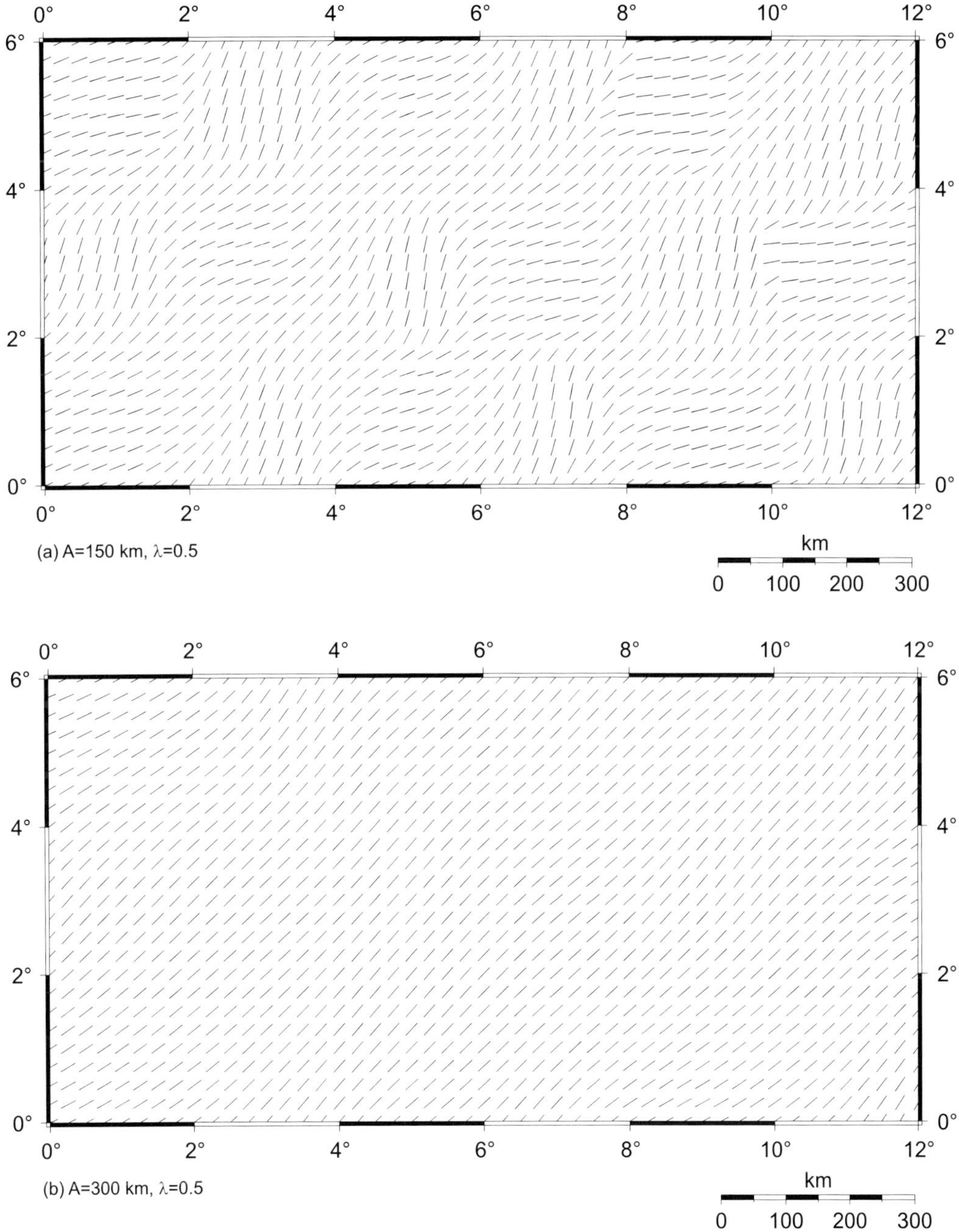

Fig. 9. Smoothed stress field computed with the FSR method based on the data of Figure 7b. (**a**) Computation with a small search radius A: the local trend of stress domains is clearly visible over the whole area. (**b**) Computation with a large search radius A: the regional trend is obtained for the whole study area.

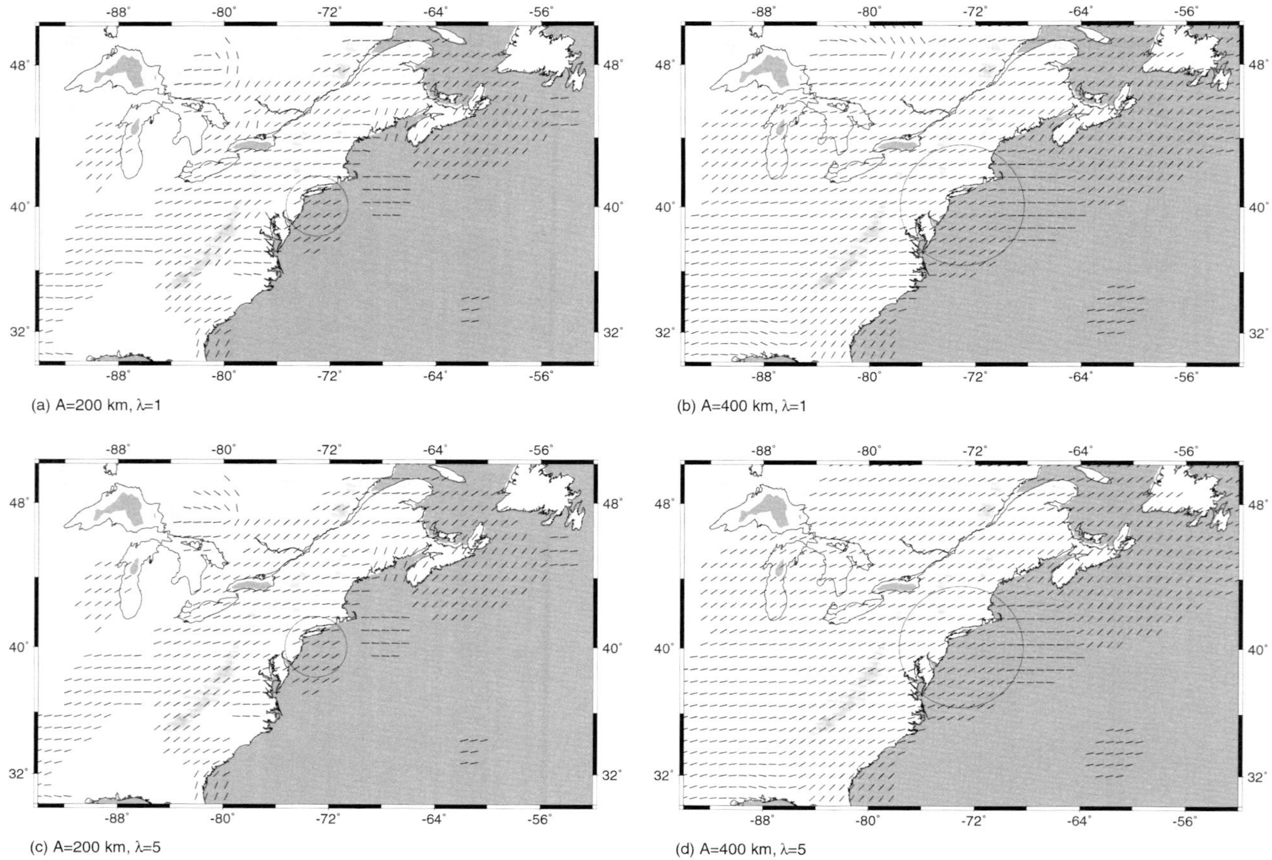

Fig. 10. Smoothed stress map of northeastern America. The circle represents the search radius A used. Observed stress data are displayed on Figure 1. (**a**) Accentuation of local stress patterns due to the choice of a small search radius A and a small value for the smoothing parameter λ. Local variations in the continental on-shore stress patterns are resolved. (**b**) Accentuation of regional stress patterns due to the choice of a larger search radius A. (**c**) Same as (a) with a higher value for the smoothing parameter λ. Differences are only small. (**d**) Same as (b) with a higher value for the smoothing parameter λ resulting in more gradual transitions. The regional NE–SW pattern extends from Florida to Canada.

A to get reliable data along the margins of the area of interest. The stress data for our test areas (Figs 1, 2 & 3) are smoothed by the FSR method and are displayed as gridded maps (Figs 10, 11 & 12) with a 0.75°×0.75° grid. The smoothed S_H orientations are plotted as short lines, centred at the appropriate grid point.

Smoothing of the stress data in each individual case was performed using threshold values for the sum of the weight terms, *sw,* of 0.2

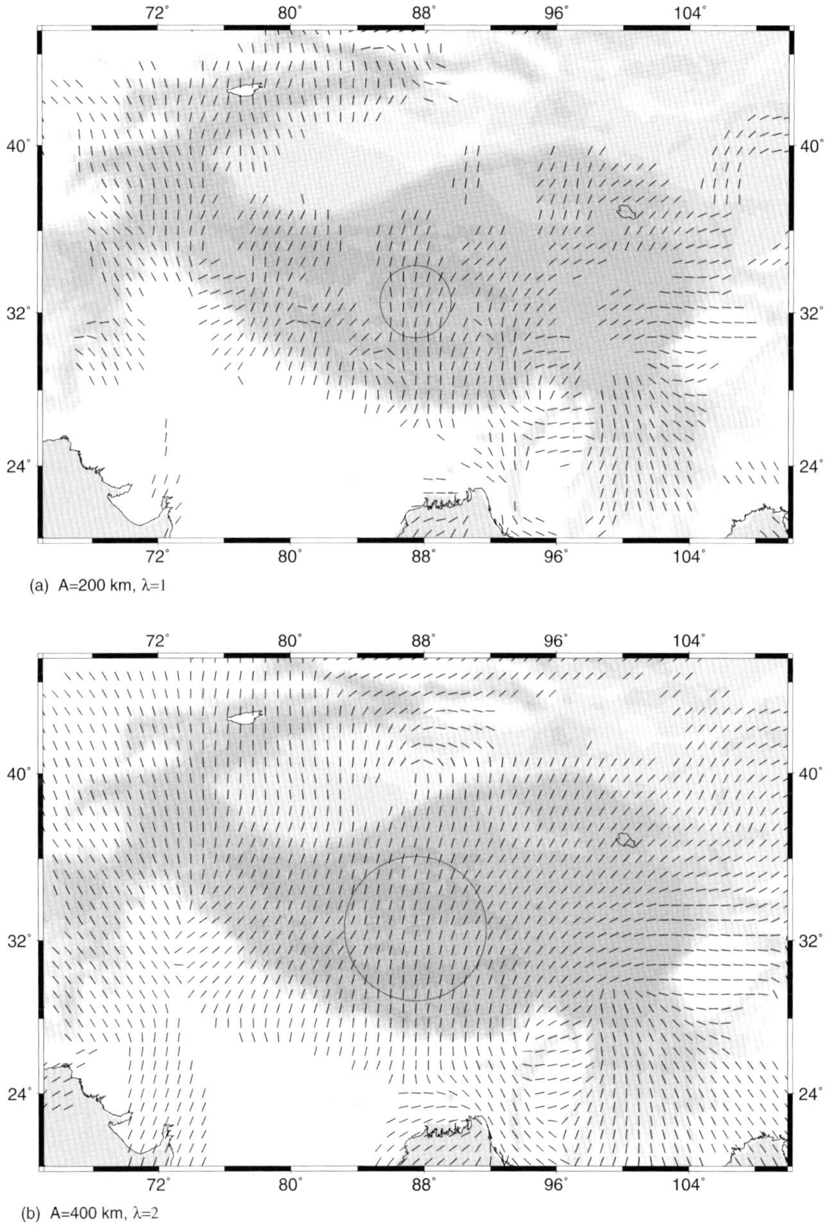

Fig. 11. Smoothed stress map of the Himalayan region. The circle represents the used search radius A. Observed stress data are displayed on Figure 2. (**a**) Accentuation of the local stress patterns which show significant variations even in the central part of the Himalayas. (**b**) Accentuation of the regional stress patterns illustrating a general north–south trend in the central region and a radial pattern in the east.

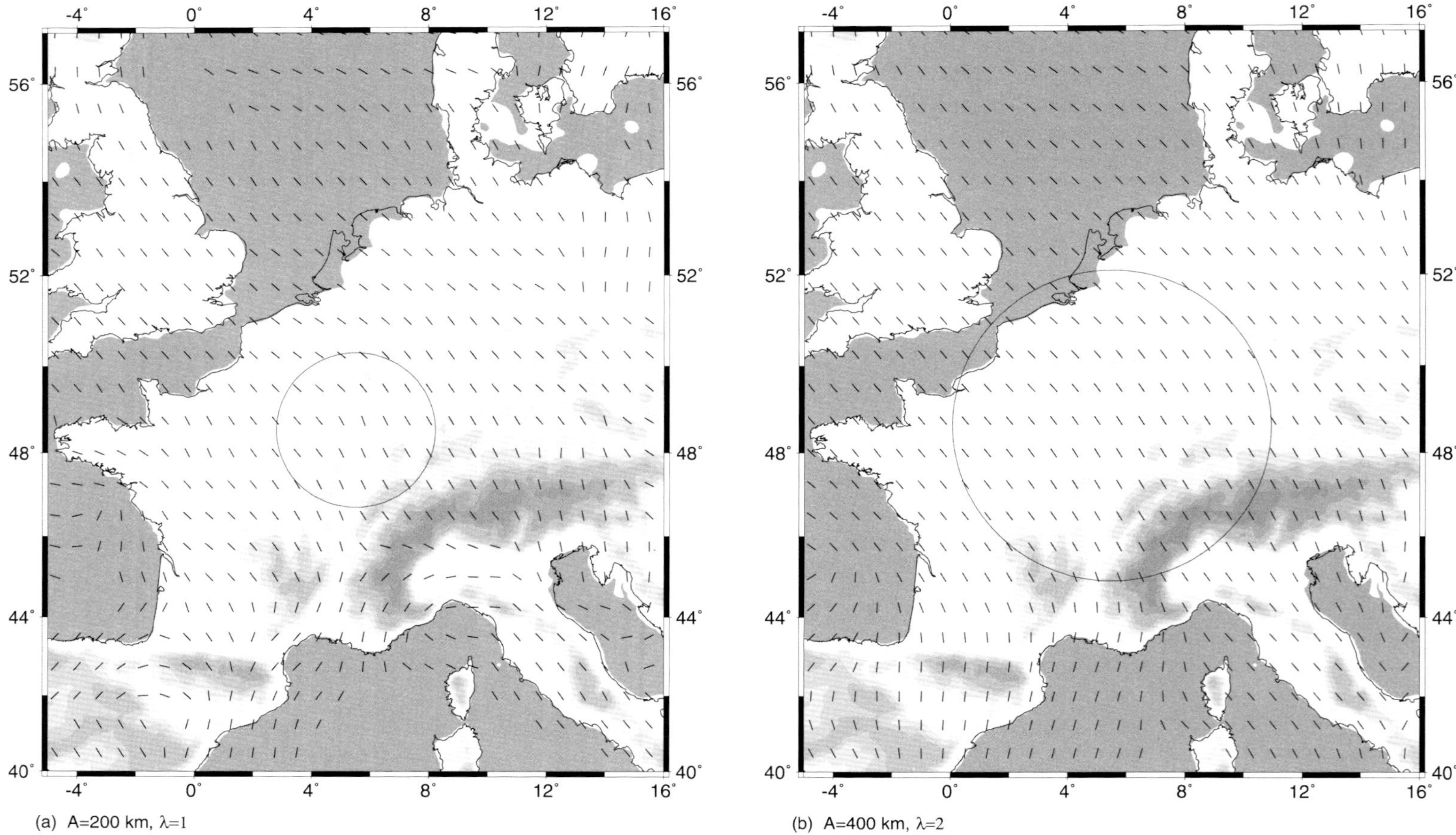

Fig. 12. Smoothed stress map of western Europe. The circle represents the used search radius A. Observed stress data are displayed on Figure 3. (**a**) Accentuation of the local stress patterns. The bending of stress orientations in the Western Alps is clearly visible. (**b**) Accentuation of the regional stress patterns with north–south orientation in the SW and the NE and a NW–SE trend elsewhere.

to 0.8. The search radius A varies from 100 km to 600 km, and the smoothing parameter λ is either 1, 2, 3, 4 or 5. The quality weights are the same for each computation, namely 1 for A, 0.75 for B and 0.5 for C quality data. Smoothing results for the three test areas are shown in Figs 10, 11 & 12. For these plots sw is 0.4.

We show examples of the influence of variations in λ and A for the test area of northeastern America (Fig. 10a-d). Smoothing the stress data with a search radius of $A=200$ km and a smoothing parameter $\lambda=1$ results in a gridded map where the dominating NE–SW trend shows lateral variations on the continent as well as in off-shore areas (Fig. 10a). An increase in the search radius to $A=400$ km reduces the gaps in the gridded maps (Fig. 10b). A change to $\lambda=5$ for $A=200$ km does not show much difference. In general, the transitions between the domains are more gradual, visible, for example, in the area NE of the Great Lakes (Fig. 10c). For $A=400$ km the change from $\lambda=1$ to $\lambda=5$ extends the areas with uniform S_H orientation, e.g. the domain of NE-orientated S_H in the region NE of the Great Lakes (Fig. 10d).

Smoothing the data of the Himalayas to accentuate local trends resolves a radial pattern of stress orientations at the southern and eastern borders of the Himalayas while the internal parts show stress orientations which vary from north–south to NE–SW (Fig. 11a). The local trends in the internal parts are less distinct when a larger search radius and an increased smoothing parameter are chosen (Fig. 11b). Nevertheless, the search radius is small enough to preserve large-scale trends. The local stress pattern of western Europe as displayed in Figure 12a shows a western trend of S_H in the western Pyrenees, NNE orientations in the Mediterranean Sea offshore NE–Spain, and a radial pattern in the western Alps. The other regions are characterized by a NW–SE stress orientation. Increasing the search radius and the smoothing parameter results in NW–SE stress orientations everywhere except Spain and its off-shore surroundings (Fig. 12b). The latter show north–south orientations.

Discussion

Application and comparison of FNN and FSR distance weight methods

Figure 13 shows the World Stress Map with the A–C quality stress data (stress maps can be obtained from http://www.world-stress-map.org). It displays areas of highest data density in northern America and Europe as well as in parts of Asia and Indo-Australia. The map demonstrates the limitation of stress maps on this scale: it is hard to detect even first-order trends.

Figures 14 and 15 show the results of smoothing with the FNN -method and the FSR method, respectively, for a 2°×2° grid. For maps of this global scale we choose a high value for the smoothing parameter ($\lambda=4$) to identify the major patterns. The FNN method results in smoothed orientations even in regions with a complete lack of data, such as the oceans or wide areas of Africa, eastern Europe and northern Asia. In these regions the FNN smoothing results in a map pretending homogeneous stress orientations (Fig. 14). For example, in western Africa where no data are available (Fig. 13), the FNN method predicts the existence of different stress domains with sharp boundaries in regions with sparse data (Fig. 14). In contrast, Figure 15 displays a world map of smoothed stress orientations calculated with the FSR method. The regional trends are resolved over scales of *c*. 600 km. The density of tick marks plotted on any map must be such that the user can immediately obtain an impression of the distribution and orientation of the stress observations. Clearly, the density of data plotted need not be the same as that of the underlying data. From experience using the algorithm, we recommend that at least three grid-points are always within the search radius.

The main advantages of the FSR method in comparison with the FNN method as deduced from the analysis of the example data of chessboard type and from application to real data are therefore:

(1) With the FSR method the choice of the search radius A and smoothing parameter λ enables us to focus on local or regional stress orientations without dependence on the distribution of observations;
(2) threshold values prevent computation of smoothed stress fields in areas of weak data distribution or quality.

Rules of thumb for smoothing stress orientations with the FSR-method

The choice of suitable parameters is strongly dependent on the database and on the purpose of the analysis. For local investigations, the search radius A and smoothing parameter l must both be small, while they have to be large for regional investigations (Table 2). However, the search radius has to be adapted to the data density and the map scale. We can make the following general recommendations for the FSR method.

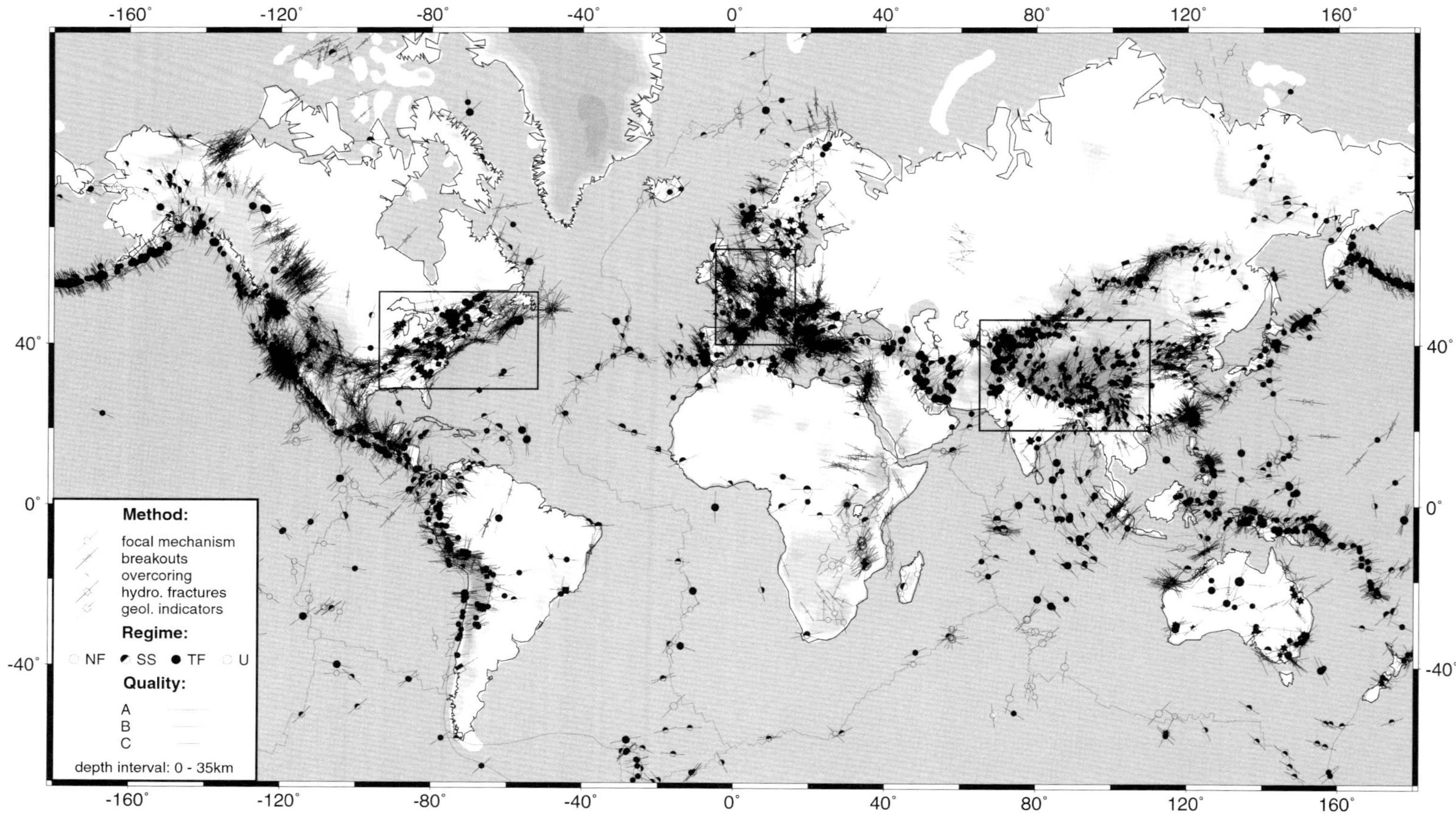

Fig. 13. World Stress Map with data of A–C quality. The data distribution is inhomogeneous and the stress domains and their borders are hard to detect in areas of high data density. For explanation of symbols see Figure 1 caption.

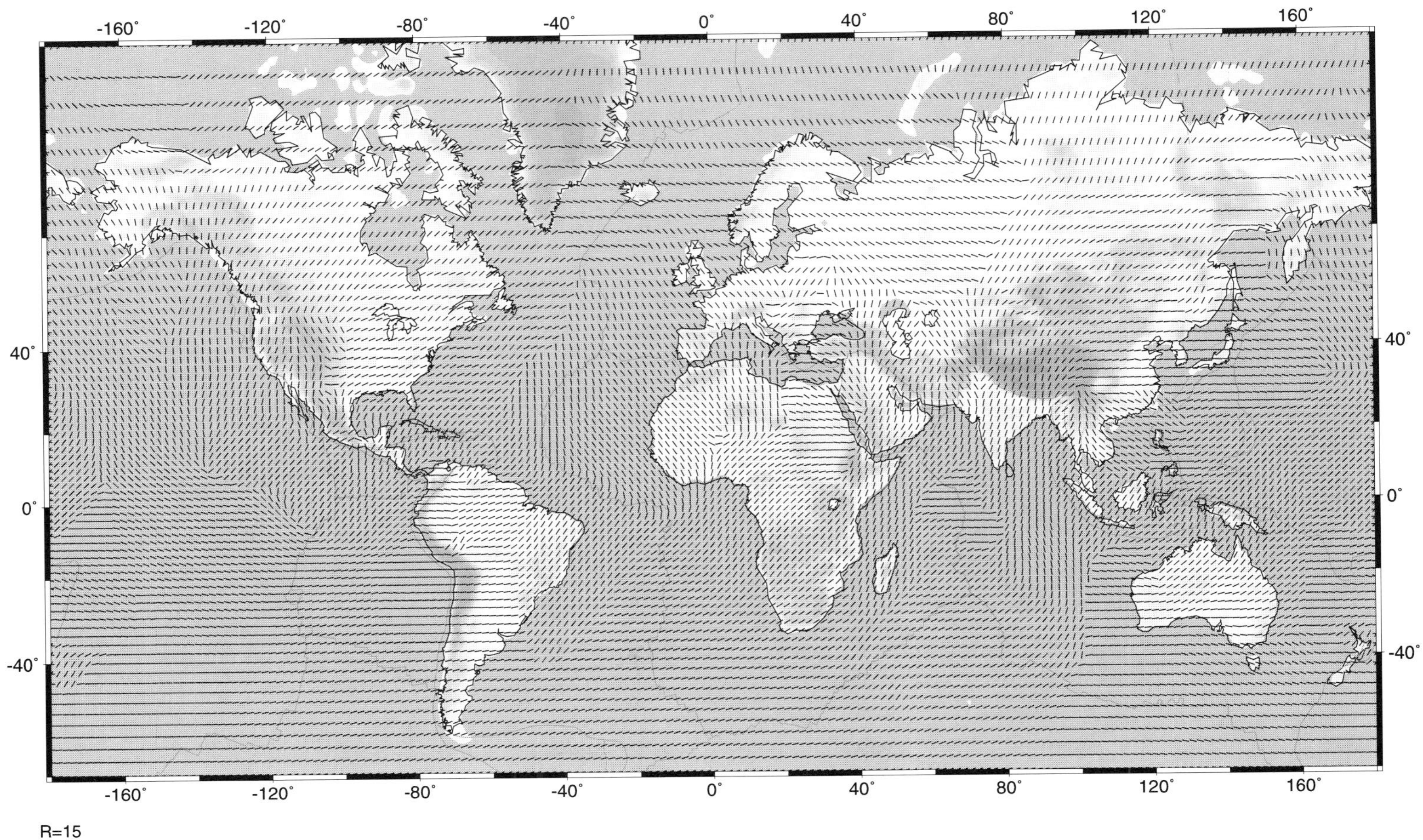

Fig. 14. Data of Figure13 smoothed with the FNN method, which provides smoothed orientations even at large distances from next data points.

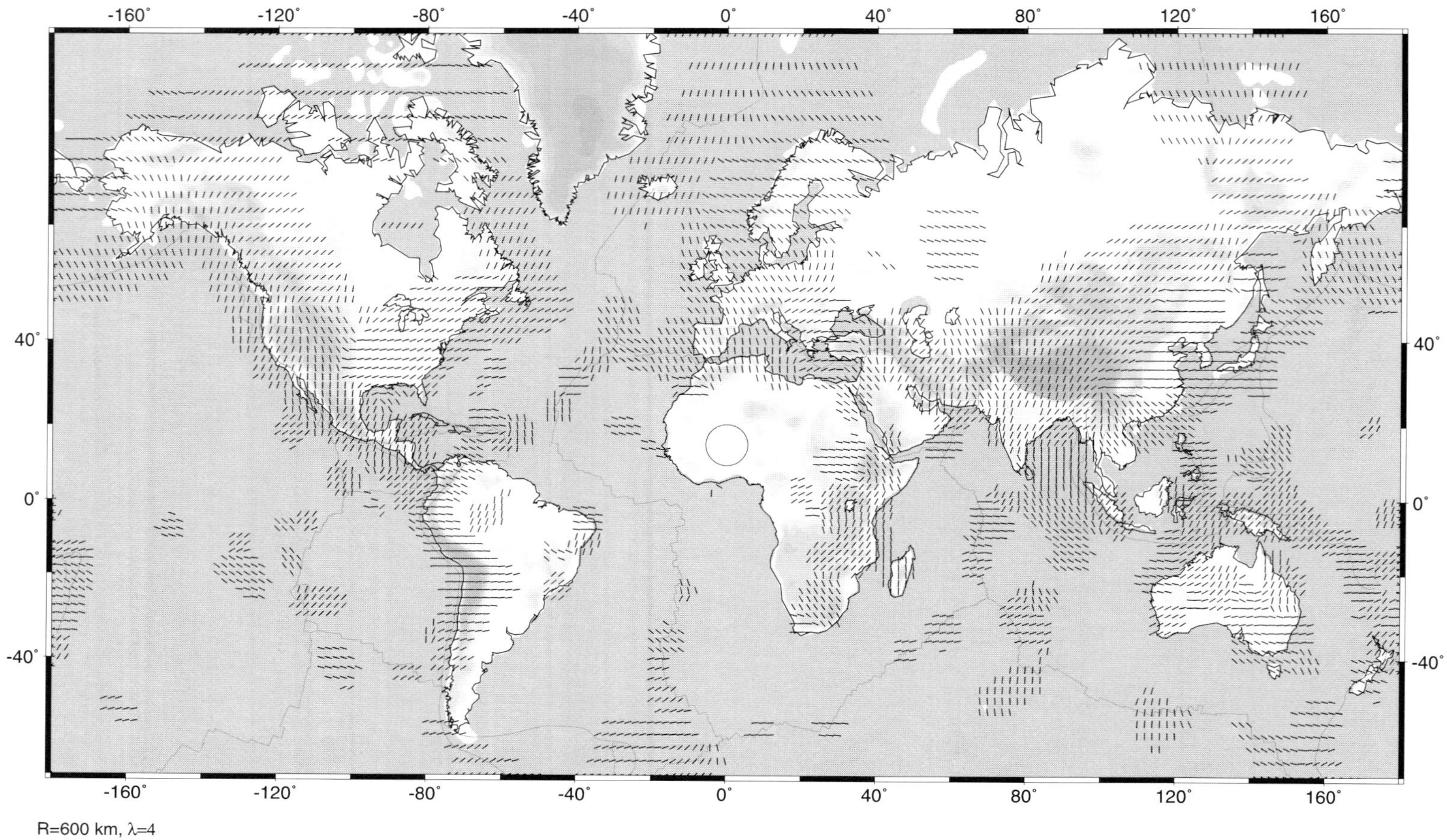

Fig. 15. Data of Figure 13 smoothed with the FSR method. The circle represents the search radius *A* used. In comparison with the FNN method (Fig. 14), the FSR-method is more conservative and leaves gaps in areas where data are sparse.

Table 2. *Recommendations for the choice of the search radius and the smoothing parameter for the FSR method.*

	Local stress field investigations	Regional stress field investigations
Search radius A (km)	60–200	>400
Smoothing parameter λ	0.5–1.0	>2.0

- Only reliable data should be used for smoothing (A–C qualities for WSM data).
- Distance weights have to be normalized.
- Smoothing is rather independent of the exact distance weight function used as long as it is a continuous function, with continuous derivative, and decreases with distance. Possible distance weight functions are for example the tricube weight as suggested in Hansen & Mount (1990).
- The smoothing area should be larger than the study area to avoid boundary effects. We recommend enlarging the area of interest on each side by at least one search radius.
- The smoothed stress data should be displayed in combination with the original data or in such a way as to indicate the angular difference between the smoothed and observed data.
- Use a threshold value for n_{min}, *sw* or *sq*.
- Select a reasonable grid for the smoothing in relation to the size of the study area, the search radius *A*, and the variation of the stress field. The grid itself should be chosen such that a minimum of three grid points is within the search radius.

Conclusions

In comparison with raw orientation data, a smoothed orientation field has a number of advantages. It enables an estimation of orientation data in areas where data are sparse and helps to identify the main trends at various scales by getting rid of local fluctuations. In this study, we improved a smoothing algorithm suggested by Hansen & Mount (1990) by using a fixed search radius (FSR method). The FSR method is an efficient tool for discriminating between local and regional trends in the distribution of orientation data. We have described a number of guidelines for handling the smoothing algorithm.

The FSR method provides a practical smoothing procedure which extends the possibilities of stress data analysis. The derivation of the local stress field is important for applications in the industrial sector such as in stabilization of underground openings or the increase of hydrocarbon recovery through optimum oil- or gas-field design. It is equally essential in the assessment of seismic risk from identification of potential slip on preexisting fault planes.

The smoothing algorithm can also be applied to other orientation data, such as lineaments, fault strikes, velocity or strain data. It helps to quantify the analysis of orientated data and therefore is a potential means for the comparison and combination of stress data with other directional observations, such as spatial orientation of faults or GPS observations.

This work constitutes part of the World Stress Map Project of the Heidelberg Academy of Sciences and Humanities (http://www.world-stress-map.org and http://www.haw.baden-wuerttemberg.de/). For assistance during the preparation of this paper we especially thank H. Bäßler. We thank the reviewers N. Koutsabeloulis and especially J. Townend for detailed and helpful suggestions to improve the paper.

References

ANDERSON, E. M. 1942. *The dynamics of faulting.* Oliver and Boyd, Edinburgh.

BIRD, P. & LI, Y. 1996. Interpolation of principal stress directions by nonparametric statistics: Global maps with confidence limits. *Journal of Geophysical Research*, **101**(3), 5435–5443.

CLEVELAND, W. S. 1979. Robust locally weighted regression and smoothing scatterplots. *Journal of the American Stat. Association*, **74**, 829–836.

COBLENTZ, D. D. & RICHARSON, R. M. 1995. Statistical trends in the intraplate stress field. *Journal of Geophysical Research*, **100**, 20245–20255.

DEMETS, C., GORDON, R. G., ARGUS, D. F. & STEIN, S. 1990. Current plate motions. *Geophysical Journal International*, **101**, 425–478.

GOWD, T. N., SRIRAMA RAO, S.V. & GAUR, V. K. 1992. Tectonic stress field in the Indian subcontinent. *Journal of Geophysical Research*, **97**(B8), 11879–11888.

GRÜNTHAL, G. & STROMEYER, D. 1986. Stress pattern in Central Europe and adjacent areas. *Gerlands Beiträge zur Geophysik*, **95**, 443–452.

GRÜNTHAL, G. & STROMEYER, D. 1992. The recent crustal stress field in central Europe –trajectories and finite element modelling. *Journal of Geophysical Research*, **97**(B8), 11805–11820.

HANSEN, K. M. & MOUNT, V. S. 1990. Smoothing and extrapolation of crustal stress orientation measurements. *Journal of Geophysical Research*, **95**(B2), 1155–1165.

HÜSGES, S. 1995. *Modellierung der tektonischen Spannungen der Eurasischen Platte*. Diploma thesis, University of Karlsruhe.

LEE, J.-C. & ANGELIER, J. 1994. Paleostress trajectory

maps based on the results of local determinations: the Lissage program. *Computers & Geosciences*, **20**, 161–191.

MARDIA, K. V. 1972. *Statistics of directional data: Probability and mathematical statistics*. Academic Press, San Diego, Calif., 357 pp.

MÜLLER, B., ZOBACK, M.L., FUCHS, K., MASTIN, L., GREGERSEN, S., PAVONI, N., STEPHANSSON, O. & LJUNGGREN, C. 1992. Regional patterns of tectonic stress in Europe. *Journal of Geophysical Research*, **97**(B8), 11783–11803.

MÜLLER, B., WEHRLE, V. ZEYEN, H. & FUCHS, K. 1997. Short scale variations of tectonic regimes in the western European stress province north of the Alps and Pyrenees. *Tectonophysics*, **275**(1–3), 199–220.

REBAI, S., PHILIP, H. & TABOADA, A. 1992. Modern tectonic stress field in the Mediterranean region: Evidence for variation in stress directions at different scales. *Geophysical Journal International*, **110**, 106–140.

RICHARDSON, R.M. 1992. Ridge forces, absolute plate motions, and the intraplate stress field. *Journal of Geophysical Research*, **97**(B8), 11738–11748.

RICHARDSON, R. M., SOLOMON, S. C. & SLEEP, N. H. 1979. Tectonic stress in the plates. *Review of Geophysics and Space Physics*, **17**(5), 981–1019.

SONDER, L. J. 1990. Effects of density contrasts on the orientation of stresses in the lithosphere: Relation to principal stress directions in the Transverse ranges, California. *Tectonics*, **9**, 761–771.

STEIN, S. & PELAYO, A. 1991. Seismological constraints on stress in the oceanic lithosphere. *In*: WHITMARSH, R. B., BOTT, M. H. P., FAIRHEAD, J. D., KUSZNIR, N. J. (eds), *Tectonic Stress in the Lithosphere*. Royal Society of London, London.

TURCOTTE, D. L. & SCHUBERT, G. 1982. *Geodynamics*. John Wiley & Sons, New York.

WATSON, G. S. 1985. Interpolation and smoothing of directed and undirected line data. *In*: KRISHNAJAH, P. R. (ed.), *Multivariate Analysis-VI*. Elsevier Science, New York, pp. 613–625.

WEHRLE, V., 1998. *Analytische Untersuchung intralithosphärischer Deformationen und Numerische Bestimmung krustaler Spannungsdomänen*, TH Karlsruhe, Karlsruhe, 167 pp. http://www.ubka.uni-karlsruhe.de/cgi-bin/psview?document=1998/physik/3.

ZHONGHUAI, X., SUYUN, W., YURUI, H. & AJIA, G. 1992. Tectonic stress field of China inferred from a large number of small earthquakes. *Journal of Geophysical Research*, **97**(B8), 11867–11877.

ZOBACK, M. D. & ZOBACK, M. L. 1991. Tectonic stress field of North America and relative plate motions. *In*: D.B. Slemmons, E.R. Engdahl, M.D. Zoback & M.D. Blackwell (Eds), *Neotectonics of North America*. Geological Society of America, Boulder, Colorado, pp. 339–366.

ZOBACK, M. L. 1992a. First and second order patterns of stress in the lithosphere: The World Stress Map Project. *Journal of Geophysical Research*, **97**(B8), 11703–11728.

ZOBACK, M. L. 1992b. Stress field constraints on intraplate seismicity in eastern North America. *Journal of Geophysical Research*, **97**(B8), 11761–11782.

ZOBACK, M. L. & ZOBACK, M. D. 1981. State of stress in the conterminous United States. *Journal of Geophysical Research*, **85**, 6113–6156.

ZOBACK, M. D. 1991. State of stress and crustal deformation along weak transform faults. *Philosophical Transactions of the Royal Society of London*, **337**, 141–150.

ZOBACK, M. L., ZOBACK, M. D., ADAMS, J. *et al.* 1989. Global patterns of tectonic stress. *Nature*, **341**, 291–298.

The role of spectral curvature mapping in characterizing subsurface strain distributions

T. J. WYNN[1] & S. A. STEWART[2]

[1]*TRACS International, Falcon House, Union Grove Lane, Aberdeen, AB10 6XU, UK (e-mail: tim@tracsint.com)*

[2]*BP, Chertsey Road, Sunbury on Thames, Middlesex, TW16 7LN, UK*

Abstract: The curvature of structured geological surfaces can be used to assess the degree of strain they have undergone. In many hydrocarbon reservoirs, this strain is expressed as brittle fracturing that may significantly impact reservoir performance. Here we describe the development of an algorithm for measuring the curvature of gridded surfaces derived from seismic data. For any grid node, the algorithm calculates the magnitude and orientations of the two principal curvatures, K_1 and K_2, from which other curvature measurements can be derived, such as Gaussian curvature and summed absolute curvature (K_1+K_2). The algorithm has also been used to generate plots of summed absolute curvature as a function of grid node separation (k versus λ). These 'spectral' or $k\lambda$ plots can be generated for each grid node and allow the definition of short-wavelength, high-amplitude noise cut-off lengths. They also deliver intermediate wavelength features such as fault drag or buckle folding and the identification of long-wavelength (basin-scale) curvatures. Portions of these data can be collapsed into single values by calculating the integral of the $k\lambda$ curve. Further filters designed to screen the effects of background tectonic, or non-tectonic, curvatures can be applied to the $k\lambda$ integral.

This algorithm has been tested using data from several North Sea chalk fields. A range of alternative types of curvature and curvature spectra are compared with other approaches to curvature calculation and other factors relevant to the calibration of such techniques in terms of the distribution of brittle fractures in sedimentary rocks.

The $k\lambda$ integral provides a relatively simple approach to calculating the degree of multi-wavelength strain present at a particular grid node. Freeing algorithms from the restriction of the 'arbitrarily' selected minimum grid node spacing is a key step towards calibrating measured curvature against strain mechanisms. However, care must be taken to separate intrinsic and tectonic curvatures when generating and interpreting $k\lambda$ plots and their integrals.

The presence of small-scale faults and fractures in a hydrocarbon reservoir can have a significant impact on the permeability characteristics of the reservoir. Small-scale fractures can be defined as those not explicitly defined by 'remote' mapping methods such as 3D seismic data or aerial photography. It has long been recognized that fold-related curvature may result in fracturing important for oilfield permeability. Murray (1968) related the radius of curvature of a competent unit to fracture porosity for an oilfield in North Dakota. Stearns (1964) and Stearns & Friedman (1972) provided models of fracture orientations and distributions for non-cylindrical folds which have proved an enduring reference point for fold-related fractures in hydrocarbon reservoirs (Cooke-Yarborough 1994). A number of studies have related the degree of curvature of a bed or sequence of beds and the fracture densities mapped within the bed. Relatively recent examples include Narr (1991) who related increased fracture density to the location of a doubly plunging fold hinge in the Point Arguello oilfield, California and Ericsson *et al.* (1998) who defined a strong relationship between fracture density and curvature for the Fateh Field, offshore Oman.

The curvature of a cylindrical fold structure can be defined as the reciprocal of the radius of curvature (fold radius) measured in a plane perpendicular to the fold axis (the profile plane, see Fig. 1a). In a more complex fold structure, such as those commonly mapped from 3-D seismic surveys, the folding may not be adequately described by one component of curvature. Lisle & Robinson (1995) detailed a more sophisticated method for measuring the curvature of geological surfaces, which is represented by two orthogonal principal curvatures, the maximum (K_1) and the minimum (K_2) principal curvatures. The product of these two curvatures gives the Gaussian curvature, which is a measure of the amount of double curvature of a fold structure (see Fig. 1b).

From: AMEEN, M. (ed.) *Fracture and In-Situ Stress Characterization of Hydrocarbon Reservoirs*. Geological Society, London, Special Publications, **209**, 127–143. 0305-8719/03/$15.00
© The Geological Society of London 2003.

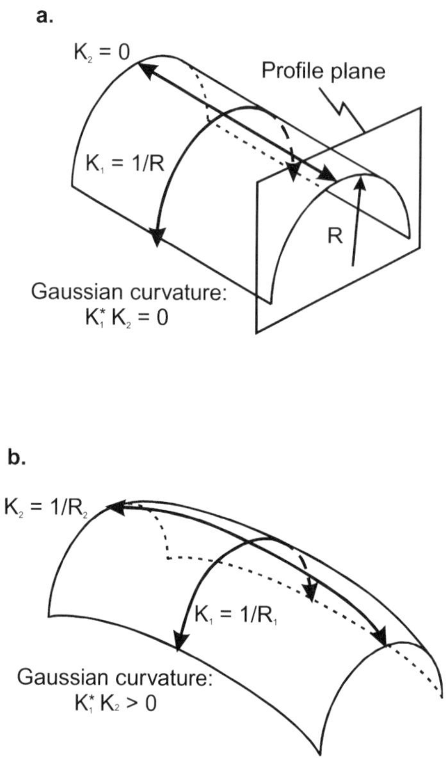

Fig. 1. (**a**) Model of a cylindrical fold with the principal curvatures and the profile plane shown. The curvature is measured within the profile plane.; (**b**) Model of a doubly curved fold surface. The product of the two curvatures is not equal to zero and is termed the Gaussian curvature.

There are several methods for measuring curvature (Stewart & Podolski 1998); mapped surface dip or azimuth change (Ericsson *et al.* 1998; Steen *et al.* 1998); position with respect to the crest of an anticline or pericline (McQuillan 1973; Agarwal *et al.* 1997; Hanks *et al.* 1997; Ericsson *et al.* 1998) or Gaussian curvature (Lisle 1994; Agarwal *et al.* 1997). Whilst these methods are sufficient in many cases, they have little or no potential to accurately measure the principal curvatures of a surface in most situations. The purpose of this paper is to outline a practical implementation of the methods for measuring curvature outlined in Lisle & Robinson (1995) coupled with a method for representing and analysing curvature variations across several spatial wavelengths. The algorithm used is described and several examples of spectral curvature estimates are presented from North Sea oilfield reservoir surfaces. These examples are compared and discussed with respect to models of curvature versus strain and other factors influencing the development of fractures in deformed strata.

Algorithm for surface curvature

The principles of surface curvature outlined in Calladine (1986), Nutbourne & Martin (1988) and Lisle & Robinson (1995) allow the description of various parameters of curvature on a surface. These papers define the characteristics of the space curve and surface curve reference frames that allow the magnitudes and orientations of the principal curvatures to be calculated. It is beyond the scope of this paper to reiterate the principles and methodologies outlined in the above papers. However, the algorithm used to calculate the curvature parameters discussed in this paper draws heavily on these principles. The reader is referred to those works to gain a more detailed knowledge of curvature analysis.

A computer algorithm has been developed based on the methodology outlined in Lisle & Robinson (1995). The algorithm calculates the principal curvature magnitudes and orientations from an orthogonal grid of points on a surface, such as those obtained from 3D seismic mapping packages. Figure 2 shows a summary of the process for calculating the principal curvatures for a given grid node. The input and output formats of the algorithm are grids of XYZ data with orthogonal X and Y directions. To provide meaningful curvature values, the XYZ values must all be in the same units; in the real examples used here the units are metres. Processing time is reduced if the grid boundaries are parallel to the grid directions, but the algorithm can handle input and output grids with irregular boundaries. The preferred grid geometry is for points with equal spacing in the X and Y directions (square grids). Grids with unequal X and Y direction spacings (rectangular grids) will result in bias of the curvature measurements. This is shown in Figure 3a where three curvature estimates are clustered spatially on a highly anisotropic rectangular grid. The algorithm also makes corrections for diagonal direction curvature calculations to compensate for the extra length (equal to a factor of root 2 in orthogonal, equispaced grids). This can only be accomplished at grid offsets greater than the minimum grid spacing. Figure 3b shows the general methodology for calculating the curvature variables using grid offsets of $2s$ and $8s$ (where s is the grid spacing). The following variables are output by the algorithm: K_1, K_2, Absolute K_1+Absolute K_2, Gaussian

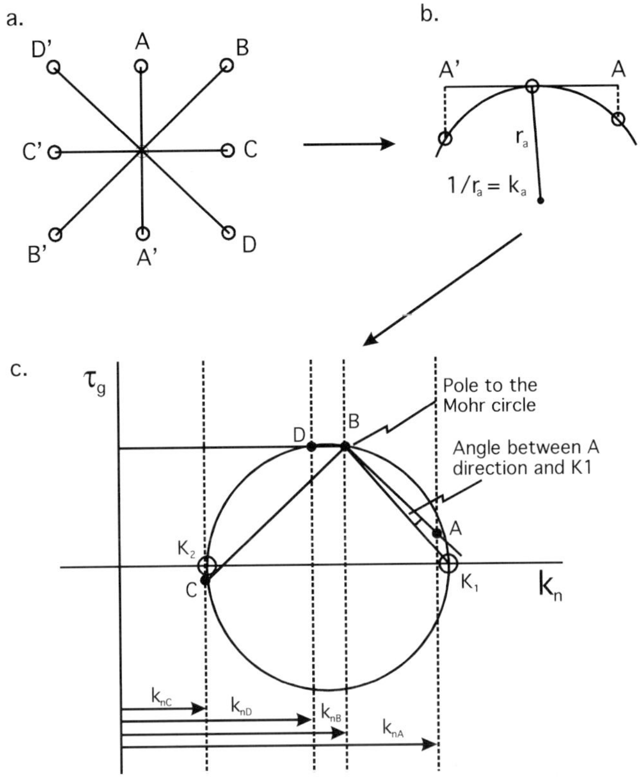

Fig. 2. Features of the algorithm for calculating curvature from gridded data points. (**a**) Plan view of arrangement of a node with its eight nearest neighbours. Curvature is measured on four vertical planes through the central node and opposite pairs. The planes are labelled A, B, C and D. (**b**) Section view of three nodes. The curvature is calculated by fitting an arc of a circle through the three points, finding the coordinates of the centre and calculating the radius of the circle. Curvature is the reciprocal of the radius and is calculated for each set of A–A', B–B', C–C' and D–D'. The curvatures are then plotted on a Mohr circle diagram. (**c**) The Mohr circle construction allows plotting of the data points in normal curvature (k_n) versus surface torsion (t_g) space without prior knowledge of the position of the k_n axis. The procedure for calculating normal curvatures and surface torsion is described in Lisle & Robinson (1995). The principal curvatures lie in directions of no surface torsion. A least squares best fit circle is fitted through the four points and the centre used to position the k_n axis. The principal curvatures K_1 and K_2 are calculated from the intersection of the circle with the k_n axis.

curvature, theta and the root mean square variation of the best fit circle to the data on the Mohr circle construction. Theta is the angle between the K_1 direction and the grid Y direction and ranges from 0° to 180° clockwise from the Y direction. Figure 3b also illustrates the reduction in length of the diagonal section planes at larger grid offsets. A key feature of the algorithm is the ability to calculate the curvature variables across the grid at consecutive grid offsets or at multiples of the consecutive grid offsets. Therefore, for each grid node it is possible to calculate the six output variables at many different sampling offsets. This allows the construction of the spectral curvature plots, which are discussed below.

Absolute K_1+Absolute K_2 (referred to as K_1+K_2) is used in this study as it is a good indicator of the total curvature affecting a point (Stewart & Podolski 1998). Considerable strain may be associated with cylindrical or near-cylindrical fold structures; however, Gaussian curvature values for these structures will show low or zero values because K_2 will be low or zero. It is unlikely that any features will be exactly cylindrical, but the very low values of K_2 will correspondingly reduce the Gaussian curvature values. The absolute values are used in the K_1+K_2 measure because just using the sum of the principal curvatures can lead to low or zero values if one curvature is positive and the other

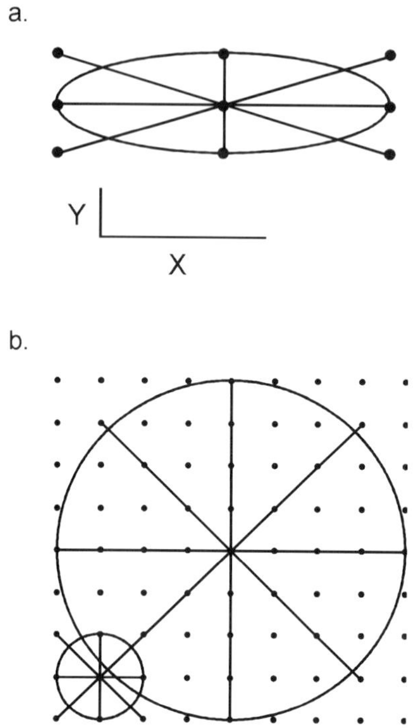

Fig. 3. (**a**) Plan view of an example of a highly anisotropic grid with X spacing approximately three times the Y spacing. It can be seen that the curvature is undersampled in the X direction compared with the Y direction. (**b**) Curvature can be calculated at a range of grid offsets. The small circle shows the data points used for an offset of one grid spacing either side of the chosen node. The larger circle shows the arrangement of points for an offset of four grid spacings. Note that the diagonal measurements on the larger circle are at an offset of three to compensate for the extra distance between the nodes in the diagonal direction. The optimum diagonal offset can be calculated for all offsets greater than one.

negative, as in saddle structures. The curvature sign convention used in this paper is positive for convex-up curvatures. In this paper, K_1, K_2 and K_1+K_2 are expressed in units of m^{-1} and Gaussian curvature is expressed in units of m^{-2}.

Relationship of curvature to geological structures

A variety of different structural features can result in changes of curvature of a geological surface such as a bed boundary. Obvious examples are buckle folds, fault related folds, basement-controlled structures (e.g. inversion anticlines in post-rift cover) and salt domes.

Buckle folds

The two mechanisms of folding relevant to buckle folds are tangential longitudinal strain and flexural slip (Ramsay & Huber 1987; Price & Cosgrove 1990). If the curvature of a surface can be shown or inferred to have resulted from tectonic activity, then an additional problem is how to relate the curvature to brittle strain within a unit or group of units. The strains in a cylindrical fold can be defined by the following relationship (from Price & Cosgrove 1990)

$$\varepsilon = kt/2 \qquad (1)$$

where ε is the fibre strain in the outer arc of the fold (assuming tangential longitudinal strain), k is maximum curvature and t the thickness of the folded unit. However, it is generally not clear where the neutral surface of a lithological unit lies with respect to the gridded surface. As seismic reflectors are produced at acoustic impedance contrasts, bed boundary reflectors will dominate them. If the top of a competent lithomechanical unit has been imaged, then the extensional strains will be at a maximum over areas of positive curvature. Core data will help to constrain the lithostratigraphy and if combined with field analogue data can provide an estimate of the mechanical stratigraphy. In addition, several outcrop studies of large-scale folds dominated by the flexural slip folding mechanism indicate that fracturing distributions can be unrelated to curvature or even concentrated in the relatively planar fold limbs rather than the more curved fold hinges (Jamison 1997; Hank *et al.* 1997; Couples *et al.* 1998). Clearly an assessment of the appropriate folding mechanism is important when trying to relate mapped surface curvature to strain.

Fault-related curvature

Extensional fault offsets of a bedding surface produce a number of issues with respect to curvature mapping. Faults with large throws (several hundred metres plus) in a surface with an area of a few tens of square kilometres will dominate the curvature spectra, especially at points adjacent to the fault. However, this effect is primarily a result of the offset of the beds and not necessarily related to the curvature of the bedding surface itself. In these situations it is recommended that the curvature mapping is subdivided into subareas defined by these large-scale faults (Stewart & Podolski 1998). This will allow the mapping of curvature on rollover anticlines (Gibbs 1984), fault drag folding or microfaulting immediately adjacent to the fault (Steen *et al.* 1998) without interference from the large-scale fault offset.

Basement-controlled features in cover rocks

A number of features may form in sedimentary rocks deposited unconformably over a pre-existing structure in the underlying basement rocks. On a large scale, these features could be monoclinal flexures of the cover sequence that either formed over an existing topographic feature or were formed by reactivation of the basement structure. Another common manifestation of basement reactivation is the propagation of faults and fractures into the cover sequence. Such features have been described from many oilfields such as the Asmari Formation reservoirs, SW Iran (McQuillan 1985). Clearly, it is important to separate flexures produced from the drape of cover over topography from flexures produced by reactivation of the basement features. The former are unlikely to contain faults or fractures systematically related to the fold geometry. To differentiate between these two mechanisms some form of geological history needs to be established for the reservoir. This could be built up from identifying the facies types and structures from well data and assessing their relative timings. An input to this could be basin development modelling incorporating strain restoration techniques.

Salt-cored features

Salt-cored features are another common mechanism for forming structural closures with the potential for trapping hydrocarbons. Salt domes are rarely perfectly symmetrical and often form periclinal features (Jackson *et al.* 1994). Fault structures associated with salt domes may be radial such as those found on the Kyle Field, North Sea, or they may be aligned to one or both of the periclinal axes. Withjack & Scheiner (1982) have shown from experimental modelling that extensional faults form radial patterns on the flanks of circular and periclinal domes in the absence of an applied regional strain. With an applied extensional strain, the extensional faults become aligned perpendicular to the extension direction. Clearly the fault geometries on such structures are not determinable from the contemporary curvature alone. If the faulting on such a structure cannot be directly resolved from remote mapping methods (e.g. 3D seismic), then an appreciation of the strain history of a structure may be required in conjunction with the curvature analysis of a particular surface.

Spectral curvature analysis

Variable versus offset (v–λ) plots

The various types of data output by the algorithm outlined above can be plotted on a variable (v) versus grid offset (λ) plot (cf. Stewart & Podolski 1998). These plots show the variations of curvature with increasing offset from a single grid node and are a novel way of analysing the spectrum of curvature data measured at a single point. They represent a complementary approach to using multiple colour-shaded or contour maps of curvature at different grid offsets. The sections below describe the main features of the v–λ plots for the Top Ekofisk chalk surface of the Fife Field, North Sea. $K_1+K_2-\lambda$ plots from the Top Tor surface of the Valhall Field are also discussed. The surfaces from both fields were depth-converted prior to calculating the v-λ spectra.

Characteristics of spectral curvature plots

The curvature parameters plotted are K_1, K_2, K_1+K_2 and Gaussian curvature. All the plots show a general trend of extreme values at smaller offsets that change in a non-linear way to less extreme values at larger offsets. Figure 4 shows the main characteristics of typical K_1+K_2 spectral plots. Figure 4a shows the effects on K_1+K_2 of superimposing random noise on a plane. This was achieved by adding a random number to each grid node in the range $\pm0.1\%$ of the grid spacing. This plot was included to allow a comparison to the average of the K_1+K_2 spectra of the Top Ekofisk surface from the Fife Field shown in Figure 4b. The plots have very similar shapes and highlight the importance of short-wavelength features on the characteristics of the curvature spectra at sampling wavelengths greater than the noise wavelength. The similarity between the two plots in Figure 4 is due to the effect of aliasing from a sampling wavelength that is greater than the wavelength being measured. The aliased wavelengths are larger than the actual wavelengths causing the slope on the spectral plots to decrease over an interval rather than at a single point. To effectively sample a wavelength without aliasing, the sampling wavelength must be less than or equal to half the smallest wavelength of interest. This minimum wavelength is termed the Nyquist wavelength and in the practical situations described here, is determined by the minimum grid spacing (Stewart & Podolski 1998).

From the spectral plots of random noise on a planar surface shown in Figure 4a it can be seen that artefacts can occur at intermediate offsets that are not related to any genuine curvatures. Therefore, care must be taken when interpreting any specific features from individual spectral plots. However, a characteristic feature of the spectra from the noise superimposed on the

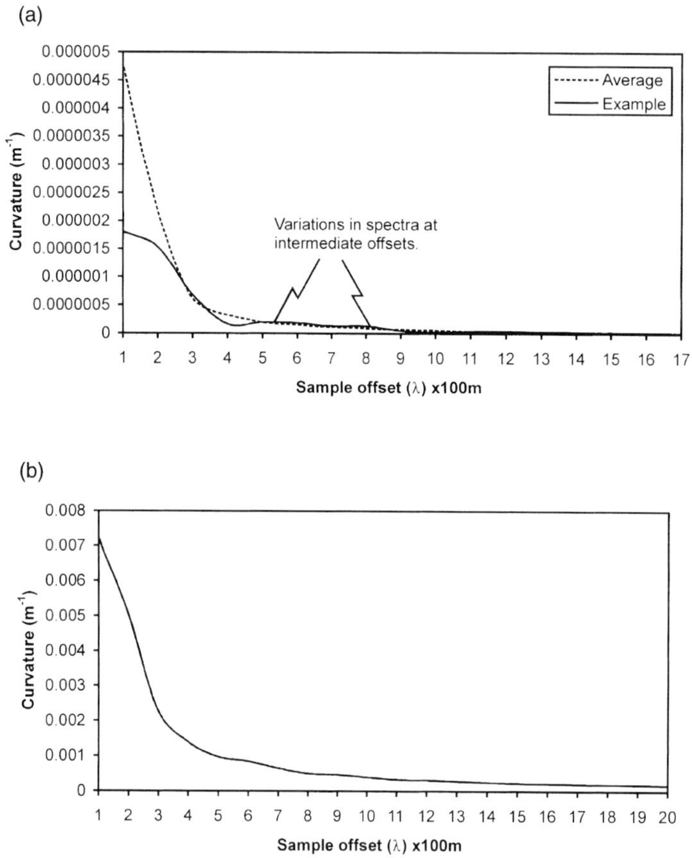

Fig. 4. (**a**) Average spectra and an example from a grid of points on a plane with random noise added to each grid node at ±0.1% of the grid spacing. Note the similarity of the spectra to (**b**) which is the average K_1+K_2 spectrum for the Fife Field Top Ekofisk surface.

planar surface is that the slope of the curve at smaller offsets is much steeper than that for the average of spectra of the Fife K_1+K_2 dataset. This can be quantified from the fitting of power law trend lines that appear to give the best fit to the average data from both the noise on the plane plot and data from the North Sea fields. Two North Sea fields were used, the Fife Field and the Valhall Field, both situated in the East Central Graben. The locations of the spectra discussed below are shown on Figure 5 for the Fife Field and Figure 6 for the Valhall Field.

On both plots shown in Figure 4, the trend line R^2 values are greater than 0.99. The power law exponent for the trendline fitted to the average K_1+K_2 spectra of noise superimposed on a plane is −2.0266. The trendline power law exponents for the average spectra of the real datasets are −1.3427 for the Fife K_1+K_2 data and −1.1355 for the Valhall K_1+K_2 data. The averages of the spectra from both the Fife and Valhall datasets were taken from spectra calculated on subgrids across the surfaces. A total of 132 spectra were used for the Fife average and 106 spectra for the Valhall average. The difference in power law exponent between the plane with noise and the real datasets may allow differentiation between-noise dominated data and natural curvature features. However, a detailed examination of the power law (fractal) characteristics of surfaces and their relationship to spectral curvature plots is beyond the scope of this work.

Spectral curvature plots from North Sea chalk fields

Figure 5 shows the topography of the Top Ekofisk surface from the Fife Field used in this study. The data are in a grid with 50 m spacing

Fig. 5. Contour map of the Top Ekofisk surface for the Fife Field, East Central Graben. The surface has been depth converted and regridded to 50 m. No smoothing has been applied to the data. All values in metres.

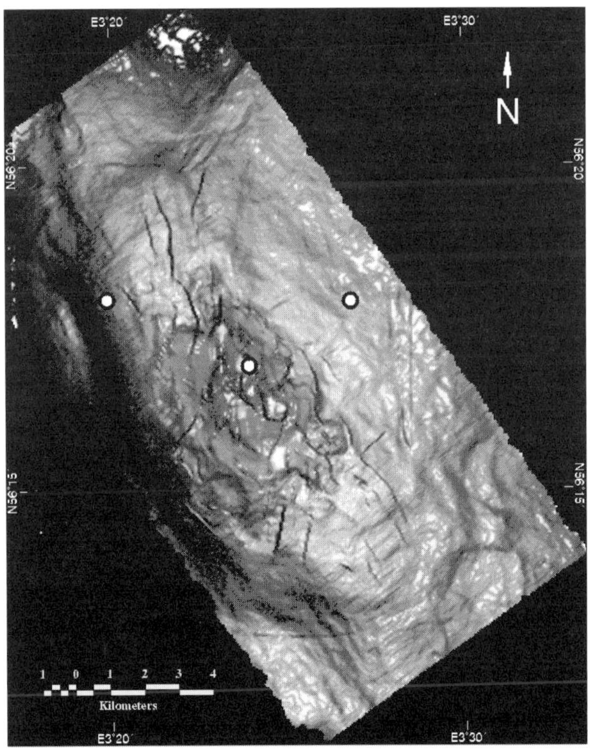

Fig. 6. Shaded relief map of the Top Tor surface for the Valhall Field, East Central Graben. The surface has been depth converted and no smoothing was applied to the data. The grid spacing is 50 m.

that was regridded from a 25 m spacing. The regridding was performed to reduce the dataset to a manageable size, although this will result in aliasing of the shorter (*c*. 50 m) wavelength features. It can be seen that in the centre of the field is a broad, flat-topped dome. The northerly closure is steeper than the east and west closures and represents topography of the chalk over a basement feature. This is probably a fault reactivated in the Late Cretaceous to Palaeogene (Gowers *et al.* 1993). There is also part of a smaller, steep-sided closure visible west of the main dome (E509200, N6206000). The north–south to NNW–SSE trending feature on the west side of the dome may represent reactivated faulting affecting the overlying chalk surfaces. It can also be seen that a number of 500 m to 1000 m scale domes and basins occur on the surface which will affect any curvature measurements. It is possible to smooth a surface prior to calculating the curvature to remove the effects of noise (Stewart & Podolski 1998). However, gentle smoothing is unlikely to remove all the noise, and aggressive smoothing is likely to remove some or all of the natural short-wavelength features. For this reason, it was decided to use an unsmoothed surface in this study.

The main characteristics of the K_1+K_2 spectra from the Top Ekofisk surface allow a description of the spectra in terms of the combination of two main components. The first is a short-wavelength, high-amplitude curvature that may represent random 'noise' associated with the acquisition and processing of the seismic data (Stewart & Podolski 1998; Brown 1991). The second component represents long-wavelength, low-amplitude curvature that probably represents the overall 'background' curvature of the basin. It must be noted that not all of the short-wavelength, high-amplitude curvature will be noise: some will represent the effects of real features such as parasitic folds or faults.

Superimposed on the general negative slope of the plots can be a variety of local changes in the rate of decrease of slope or even local increases in slope. These variations are discussed below where a variety of $k\lambda$ plots with departures from the general form are shown. For the unsmoothed Top Ekofisk surface of the Fife Field data, these departures can occur at any sampling wavelength but are most common at intermediate wavelengths of about 400 m to 1300m.

Fife Field K_1 and K_2 plots

Figure 7 shows typical spectra for the K_1 and K_2 variations at the crest of the main dome on the Top Ekofisk surface and the north–south trending ridge on the west side of the main dome (see Fig. 5 for locations). These spectra are highly variable at smaller offsets and become less variable at larger offsets; the values vary from positive to negative for both K_1 and K_2 but generally seem to converge on very low (<0.0003) positive curvatures at offsets greater than about 1700 m. The exception is the west flank K_1 curve that displays negative values at all sample offsets. There also appear to be changes in the slopes of all the spectra in Figure 7 at sample offsets of 700–900 m. This may reflect a characteristic wavelength of the Fife Field although it is not

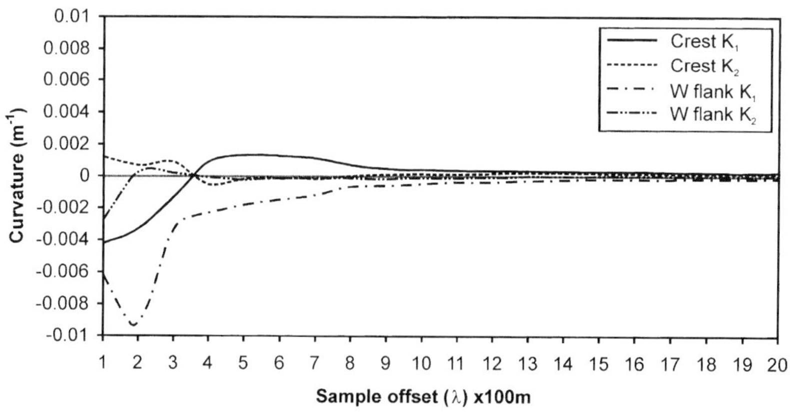

Fig.7. K_1 and K_2 versus sample offset spectra from two grid nodes on the crest and the west flank of the Fife Field Top Ekofisk surface.

clear if the origins of features at this wavelength are tectonic or sedimentary.

Fife Field K_1+K_2 plots

Figure 8 shows the basic features of a typical K_1+K_2 versus l plot. It can be seen that in common with the $k\lambda$ plots for K_1 and K_2, higher curvature values are measured at smaller offsets that fall off to a background level of lower curvatures at higher offsets.

The spectra taken from the crest show relatively low values at sample offsets of 100 to 400 m when compared to the spectra from the west flank of the dome. However, at sample offsets of 700 m to 1200 m the curves converge and at sample offsets of 1200 m to 2000 m the spectra from the crest show higher values. These characteristics are to be expected for these locations as the crest of the dome shows little topography on the 100 m to 500 m scale; conversely the west flank of the dome is dominated by the north–south trending ridge that produces the high values at sample offsets of 100m to 400m.

Valhall Field K_1+K_2 plots

Figure 6 shows the Top Tor surface from the Valhall Field, East Central Graben. The locations of the spectra shown in Figure 9 are also marked. The origins of this field are similar to

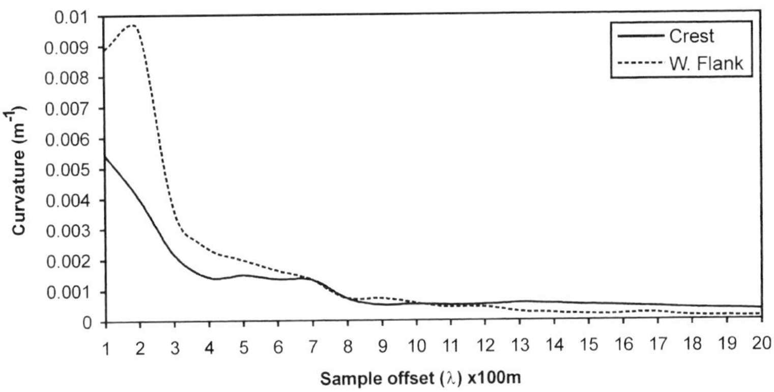

Fig. 8. Absolute K_1+Absolute K_2 (K_1+K_2) versus sample offset spectra from two grid nodes on the crest and the west flank of the Fife Field Top Ekofisk surface. Note the localized increase in curvature at 400 m sample offset for the crestal node.

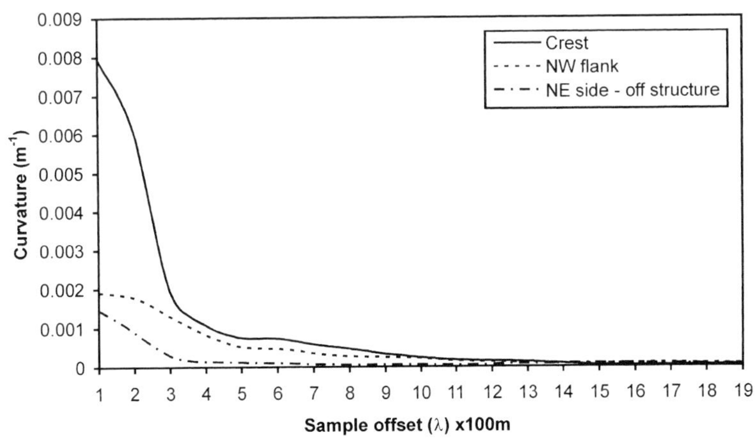

Fig. 9. Absolute K_1+Absolute K_2 (K_1+K_2) versus grid offset spectra from three grid nodes on the Valhall Field Top Tor surface. The nodes are situated on the crest of the main pericline, the NW flank of the pericline and the NE side of the dataset (off structure).

those of the Fife Field in that the main structure formed during late Cretaceous to Palaeogene inversion of the Feda Graben (Gowers *et al.* 1993). It can be seen that there are numerous faults concentrated around the crestal part of the main NW–SE trending pericline. The grid spacing for this dataset is also 50 m but the overall field size is much larger than Fife. The spectra show different characteristics, which appear to be related to their locations on the Top Tor surface. The spectrum from the crest shows very rapid decay from high values (0.006–0.008) at sample offsets of 200 m to 300 m. The curve decreases to low values (<0.0002) at offsets greater than 1100 m. This may be related to the fact that the crestal part of the pericline is relatively flat with the topography dominated by the faulting. The spectrum from the NW flank of the main pericline shows much lower values at sample offsets up to 300 m. This is expected as there is little or no faulting in this part of the structure. At sample offsets greater than 300 m, the NW flank spectrum is similar in shape to the crestal spectrum. This indicates that both spectra are controlled by the main periclinal structure at sample offsets greater than 300 m. The spectrum from the NE side of the dataset shows consistently lower values than the other two spectra up to sample offsets of 1400 m. This can be related to its location on the relatively flat NE flank of the structure with little or no faulting and minimal influence from the main periclinal structure.

The general shapes of the spectra are similar to those observed for the Fife Field but the slopes of the Valhall curves decay to lower values (<0.0002) at sample offsets of 1000 m to 1100 m for the spectra associated with the pericline. For the spectra associated with the flank this value is reached at a sample offset of 300 m. The K_1+K_2 spectra for the Fife Field tend to decay to these values at sample offsets of 1300 m to 1400 m. This difference can be related to the topography of the two surfaces. The Valhall Top Tor surface appears to be a relatively simple, large-scale pericline with 1–3 km long faults clustered around the crest. Therefore, the faulting and the large-scale pericline influence the curvature which occurs at two different scales. However, the Fife Top Ekofisk surface shows a greater degree of variability in topography at a wider range of scales than the Valhall Field. This spread of scales for the features in the Fife dataset has probably contributed to 'holding up' the decay of the short-wavelength curvature from short to long sample offsets.

Fife Field Gaussian curvature plots

Figure 10 shows the variations in Gaussian curvature with sample offset. The features of these plots are similar to those for the K_1 and K_2 plots. However, in general the variability at intermediate offsets (400 m to 1300 m) is less than for either the K_1, K_2 or the K_1+K_2 plots. All the curvatures measured at the scale of the Fife Field are less than unity. Therefore, when the product of K_1 and K_2 is calculated, the magnitude of the result is often considerably less than either of the two components. Consequently, Gaussian curvature from the Fife Field is more usefully plotted on maps to show the distribution

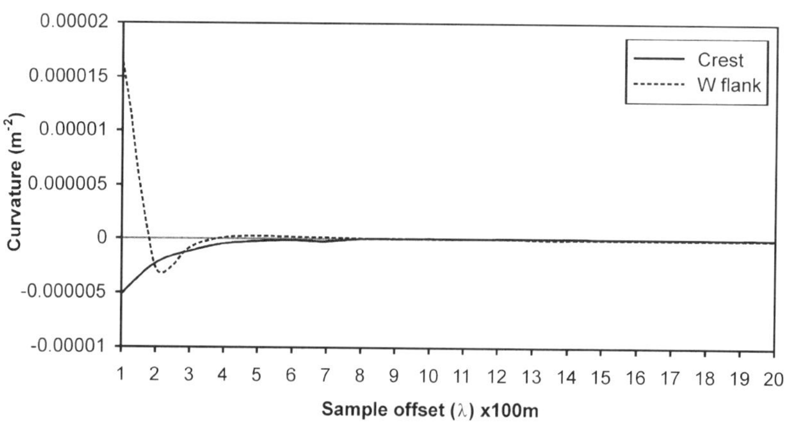

Fig. 10. Gaussian curvature versus grid offset spectra from two grid nodes on the crest and the west flank of the Fife Field Top Ekofisk surface. The curvature values for these spectra fall to very low values at sample offsets of approximately 400 m to 500 m.

of non-cylindrical curvature across the field. Alternatively, showing only the curves above a certain cut-off value may better represent these plots. This would allow an examination of the more subtle variations.

Fife Field integrals of K_1+K_2 plots

The spectral plots for the variables discussed above provide a useful insight into the wavelengths and magnitudes of curvatures affecting each grid node. However, for each grid node, six different spectral plots can be generated. In large gridded datasets it is not uncommon to have in excess of 500 nodes in each grid direction resulting in 250 000 nodes or more. Clearly, the time required for the analysis of individual plots becomes prohibitive for anything but the smallest subareas of a dataset. A method is required to extract the relevant information from one or more of these plots and present it as a map of values. In this study, it was decided to use the integral of the K_1+K_2 plots to collapse the multi-wavelength information into a single value for each node. The K_1+K_2 plots are amenable to this analysis because they are always plotted in the same Cartesian quadrant and an integral will allow a qualitative comparison of collapsed curvatures when plotted on a map. Using the integrals of the curve also allows the removal of components of curvature associated with noise at shorter wavelengths and basin scale curvature at longer wavelengths.

The integral is calculated by using the trapezium rule, which is adequate for data of this type. Figure 11 shows the main features of the method for calculating the integral and removing the noise- and basin-scale-related integrals. The noise-related curvatures are removed by applying a predetermined lower cut-off for the data, in this case 300 m, prior to calculating the integral. It is realized that this is a crude approach and because of the gradual decay of short-wavelength features over a range of wavelengths, some short-wavelength-related curvature will remain. The basin-scale curvatures are identified from the minimum measured curvature at a particular node with the assumption being that the smallest curvatures occur at the largest offsets. Removal of this basin-scale curvature has the effect of subtracting a constant value from the integral. The presence of uncharacteristically low curvatures at small or intermediate offsets will reduce the effectiveness of this approach. However, a visual inspection of the spectra from the Fife Field Top Ekofisk dataset revealed, for this dataset, that this is unlikely to occur. It may also be possible to filter the noise and basin-scale curvatures by integrating between the raw spectra and a normalized power law curve representing the contribution of noise- and basin-scale curvatures in the dataset. The input parameters for such a 'filter spectrum' may be determined by examining selections of spectra or the average of the spectra for the surface being considered.

After removal of the noise- and basin-scale-related integrals, the remaining area may give some measure of the different wavelengths of tectonically induced curvatures affecting the surface at any one point. The advantage of using this approach is that it is simple to calculate and

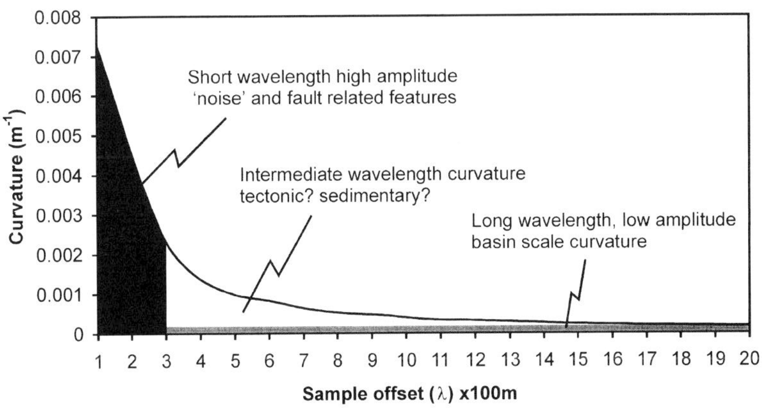

Fig. 11. Schematic diagram of how the integrals are distributed under a K_1+K_2 curve. The noise and basin curvature integrals are calculated and then removed from the raw integral to leave the area associated with intermediate sample offset curvatures. The integrals are calculated using the trapezium rule.

does not require a priori knowledge of the offsets of a tectonic wavelength or wavelengths affecting the surface. However, the influence of tectonic curvatures at the same wavelengths as any short-wavelength noise will be lost in the filtering procedure. It may be more fruitful to calculate the integrals over shorter sections than has been done here. A subset of the spectra or the average spectra of a dataset could be examined to identify regions of the curves that have broadly similar characteristics. For example, based on the average spectra for the Fife Field used in Figure 11, the spectra could be subdivided into domains at 100–300 m, 300–1000 m and 1000–2000 m sample offsets.

Maps of calculated variables

The various spectra discussed above can also be used to identify an offset or groups of offsets with features of interest that can be plotted on a map or contoured surface to assess their areal distributions. For the Fife dataset, the variables plotted as part of this study were: K_1+K_2, Gaussian curvature, and the integrals of the K_1+K_2 spectra both with and without filters applied for noise and basin-scale curvature. For the Valhall dataset the variable plotted was K_1+K_2.

K_1+K_2 maps from the Fife Field

Figure 12 shows the variations in K_1+K_2 for a subset of offsets for the Top Ekofisk surface of the Fife Field. Figure 12a shows the curvature variations for sample offsets of 100 m. The main features visible are broadly north–south trending linear zones of higher curvature in the west part of the dataset. These ridges of curvature are probably related to geological features that represent the effect on the cover rocks of the reactivation of faults in the deeper strata. Also visible throughout the dataset are aligned 'nodes' of low and high curvature values. It is possible that the isolated nodes are an artefact of the contouring process; alternatively these nodes could represent the areal distribution of the noise identified from the spectral plots at offsets equal to or less than 300 m.

The 1000 m offset curvatures are shown in Figure 12b where it can be seen that the isolated nodes visible on Figure 12a are less prominent and the higher values of curvature are distributed in broader 'patches'. These may be related to the 500 m to 1000 m scale features visible on the topographic map. The most obvious feature at this sample offset is the small NE–SW trending closure in the west part of the dataset.

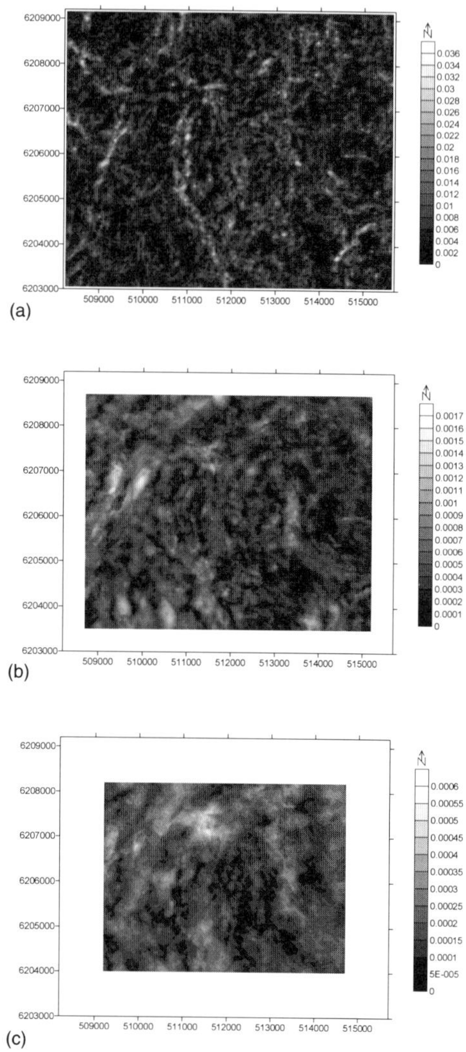

Fig. 12. Maps of K_1+K_2 for the Fife Field Top Ekofisk surface at different offsets: (**a**) 100m sample offset; (**b**) 1000m sample offset; (**c**) 2000 m sample offset (XY scale in m; curvature values in m^{-1}).

The 2000 m offset curvatures seen in Figure 12c show an even wider distribution of patches of curvature than those visible in Figure 12b, although the north–south trending ridge of curvature visible on the west side of the dome in Figure 12a is also visible on this plot. The most obvious feature is the patch of higher curvature values over the crest of the dome.

It can clearly be seen that in general, the areal distributions of curvature generally become more broadly distributed with increasing sample wavelength. This can be interpreted to be a result of the removal of short-wavelength features at increasing sample wavelengths. This has the effect of revealing more subtle features at larger sample offsets. If used in conjunction with the spectral curvature plots, these maps can help determine the range of scales over which a given feature is influencing curvature.

K_1+K_2 *maps from the Valhall Field*

Figure 13 shows the K_1+K_2 variations from the Top Tor surface of the Valhall Field. The 100 m sample offset curvatures shown in Figure 13a clearly isolate the faulting on the crest of the main pericline. Very little curvature is seen between the faults on the crest or on the flanks of the dome indicating that the curvature at short wavelengths is predominantly related to faulting rather than more dispersed features such as acquisition-related noise. However, there are some linear zones of curvature parallel to grid lines E529000 and N6230000. These are almost certainly related to the acquisition or processing or interpretation of the data rather than to any geological features.

Figure 13b shows the 1000 m sample offset curvatures. The fault-related curvatures are still obvious but affect larger areas around the faults. This could be related to fault related drag or rollover; however, in this case it is more likely to be a result of aliasing of the fault-related curvatures (throws) immediately adjacent to the faults. Also visible are some patches of curvature associated with the margins of the larger-scale periclinal feature (e.g. E520000, N6239000). The map in Figure 13c shows the 1900 m sample offset curvatures. This map is dominated by the larger distributions of curvature associated with the main periclinal structure. However, even at these offsets there are obvious fault related curvatures in discrete zones within the crestal region although they are less marked than in the other maps.

Fault-related, high curvature values are visible at E522500, N6235500 on all three K_1+K_2 maps. The persistence of this feature over a large range of sample offsets highlights the need for subdividing the curvature map into subdomains to remove the effects of faults with throws large enough to be seen across many sample offsets. However, for the smaller throw faults it can also be seen that, as the sample offset is increased, their influence on curvature is decreased.

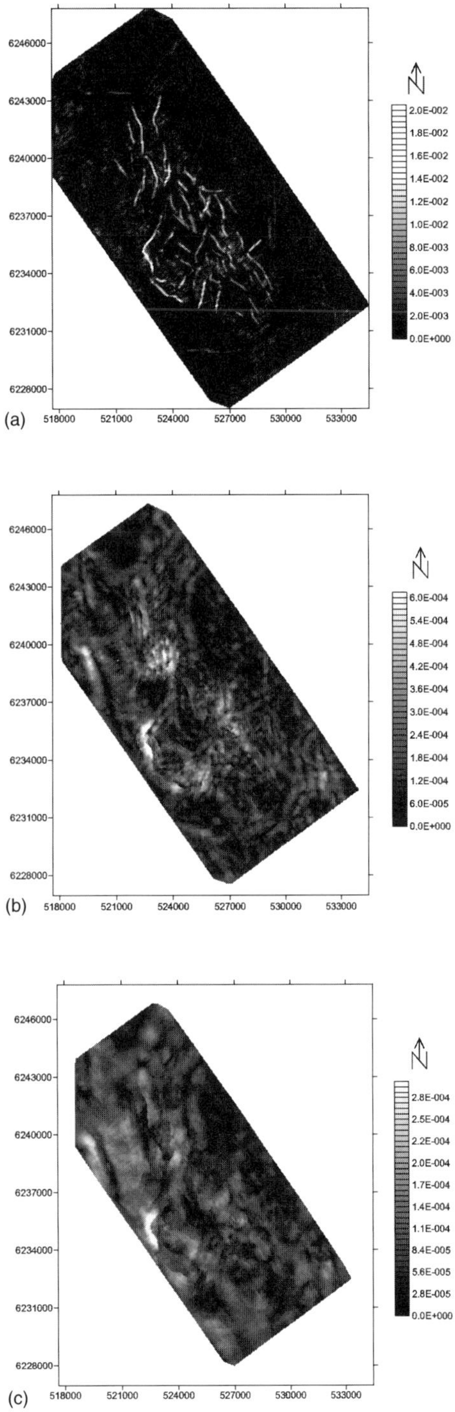

Fig. 13. Maps of K_1+K_2 for the Valhall Field Top Tor surface at different offsets: (**a**) 100 m sample offset; (**b**) 1000 m sample offset; (**c**) 1900 m sample offset (XY scale in m; curvature values in m^{-1}).

Fife Field Gaussian curvature maps

Figure 14 shows the variations in Gaussian curvature with increasing offset for the Top Ekofisk

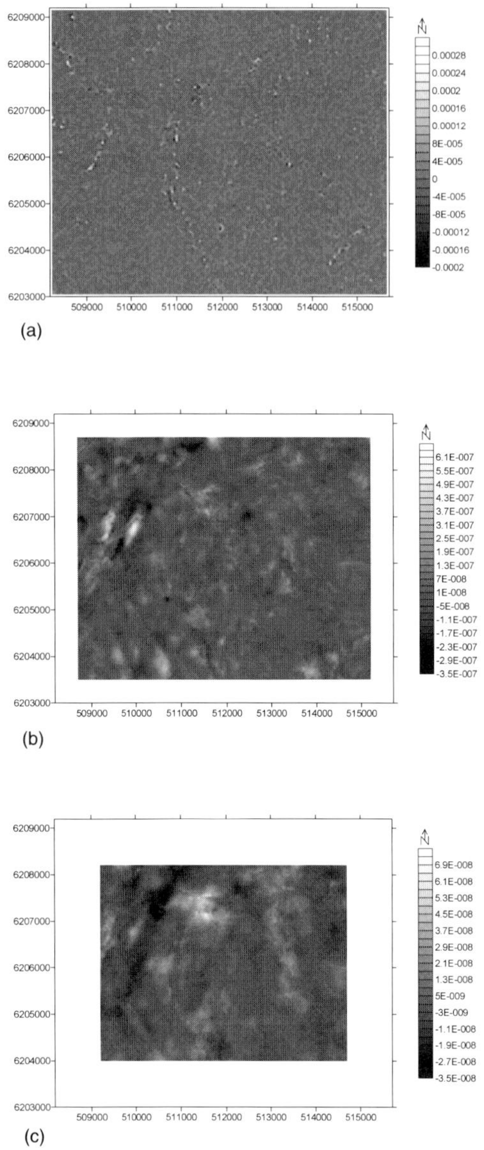

Fig. 14. Maps of Gaussian curvature for the Fife Field Top Ekofisk surface at different offsets: (**a**) 100 m sample offset; (**b**) 1000 m sample offset; (**c**) 2000 m sample offset (XY scale in m; curvature values in m^{-2}).

surface of the Fife Field. The spatial variability at offsets of 100 m (Fig. 14a) is markedly less than for the K_1+K_2 maps although some small, aligned 'hot spots' are visible on the west side of the main closure and on the smaller closure to the west. At intermediate sample offsets of 1000 m (Fig. 14b) the most obvious feature is a NNE trending dome in the west part of the dataset. This feature is not very clear at the 100 m and 2000 m sample offsets. No alignment of features is visible to the east of the main dome at sample offsets of 100 m and 1000 m. However, at sample offsets of 2000 m (Fig. 14c) a 1 km wide, NNW trending zone of curvature is visible in the east part of the dataset.

In general, the Gaussian curvature maps appear to provide a clearer distribution of the main features on the Fife surface when compared to the K_1+K_2 maps. This is perhaps due to the suppression of grid nodes where the K_2 curvatures are very low, irrespective of the value of the K_1 curvatures. On the K_1+K_2 maps these nodes will show up if the K_1 values are high.

K_1+K_2 spectral integral maps

Figure 15 shows maps of the integrals of the spectra of K_1+K_2 for each grid node of the Top Ekofisk surface of the Fife Field. Figure 15a shows the raw integral and Figure 15b shows the map of the integral filtered for noise-related curvatures at offsets below 300 m and basin-scale curvatures at offsets above 1000 m. It can be seen that there is very little difference between Figure 15a and Figure 15b although the filtered map appears to show a smoother distribution of curvature. Any differences are probably due to the removal of the short-wavelength noise-related integrals. The effects of the removal of longer-wavelength curvatures are probably not visible in Figure 15b. This is because the values of curvature are consistently very small above offsets of about 1000 m so the larger offset curvatures will have a negligible input to the integral. It can also be seen on both plots that in general the variability in curvature across the surface is low but that there are a few 'hot spots' distributed throughout the field.

The maps demonstrate two features of using the integral of the spectra. Firstly, the effects of noise can be reduced so that the regularly spaced 'nodes' visible on the 100 m sample offset map of K_1+K_2 in Figure 12a are not present. Secondly, the integral map may allow the assessment of the distributions of the most important wavelength or wavelengths of curvature without a priori knowledge of those wavelengths.

Discussion

Spectral curvature analysis

From the various plots and maps presented above, it can be seen that curvature mapped at different grid offsets can have markedly different distributions across the mapped surface. The K_1+K_2 spectra derived from a synthetic plane

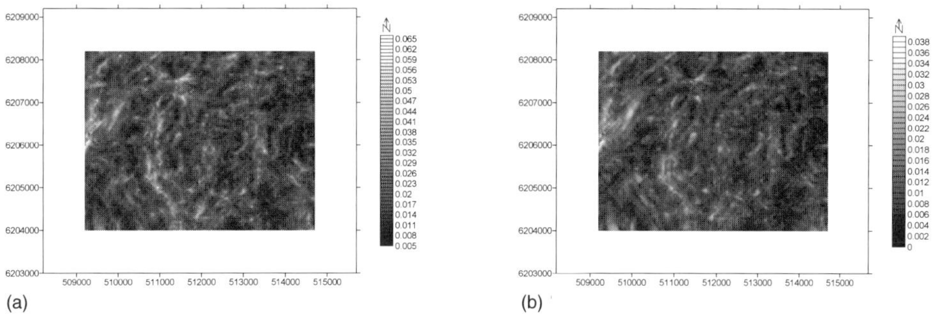

Fig. 15. Maps of the K_1+K_2 spectra integrals from the Fife Field Top Ekofisk surface: (**a**) Raw integral without removing the short-wavelength curvatures (<300 m sample offset) or basin scale curvatures; (**b**) Residual integral after removal of the integrals associated with short wavelengths and basin-scale curvatures (XY scale in metres, integral values in m^{-1}).

with noise, at the grid spacing scale, showed some similar characteristics to the spectra derived from the North Sea chalk surfaces. However, the spectra derived from a plane with noise forms an extreme end member of the range of possible spectral types that could be expected from natural surfaces. In this example the power law exponent was approximately −2. At the other extreme would be a smooth, cylindrically folded surface (see Fig. 1a) with a large radius of curvature compared to the grid spacing (e.g. 2000 m–4000 m for a 50 m grid). Such a surface would show K_1+K_2 trendline power law exponents of zero because the measured curvature is constant at all scales. The trendline power law exponents of the spectra examined from the Fife and Valhall fields were typically in the range of −0.5 to −1.8 and fall within the range of values defined by the extreme examples described above. It is possible that the power law characteristics of spectra for many, if not all, natural surfaces will fall between these two extreme end member values.

Of more practical use is the fact that the examples of spectra from the Fife and Valhall fields could be related to the topography of the surfaces. Although curvature spectra cannot be used in isolation to identify the origins of features on a surface, they can be used to identify the range of sample offsets over which a particular feature influences curvature. For the Fife Field dataset, it was proposed that seismic acquisition- or processing-related noise dominated the curvature spectra at sample offsets of less than 300 m. This cut-off value could be used as a basis for smoothing the dataset prior to future studies involving curvature or other parameters. Conversely, for the Valhall Field data it was apparent that fault-related curvatures dominated the spectra at sample offsets of less than 300 m. In this case smoothing would not improve the dataset and would probably result in loss of information with respect to the faulting. In general, it is recommended that curvature analysis is initially performed on unsmoothed data so that the full range of curvature characteristics can be established. If smoothing is performed prior to curvature analysis then some potentially useful information may be lost without a proper evaluation of the noise-to-signal ratio.

Curvature and strain

As discussed above, the principal reason for mapping the curvature of geological surfaces is to assess the distribution of curvature-related strain affecting a lithological unit or group of units. However, it will have become apparent that care needs to be taken when interpreting curvature variations. In addition to the short-wavelength curvatures produced from the seismic acquisition- and processing-related noise, there will also be some curvature from the pre-deformation topography of the surface. It will not always be possible to deconvolve this curvature from the tectonic curvature purely by examination of the curvature spectra or maps. However, it may be possible to compensate for these curvatures in a strain restoration package prior to calculating the curvature of the surface. Another approach would be to calculate the curvature variables and spectra on different realizations of a strain model. The realizations could be separated either by time or by the strain path.

Curvature and fracture permeability

There are several factors that contribute to the development of fracturing on a fold that can be linked to permeability; only one of these factors is the present-day curvature. The strain path followed by the rock is important as it is possible that the measured curvature of a fold or a portion of a fold is not the maximum curvature that has affected the rock during the deformation history. This will be relevant to fault bend folds formed in overthrust belts where the hanging wall rocks may be 'fed' through fold hinges and then unfolded to some degree as transport progresses (cf. Ramsay & Huber 1987; Price & Cosgrove 1990). Another factor is the timing of the deformation with respect to the diagenetic history of the rocks. Deformation whilst the rocks are undercompacted may result in a more ductile mechanism for strain accommodation and the presence of considerable quantities of formation fluids may result in mineral infills of any brittle fractures that do form. The mineralization of fractures may occur at any time during the history of a structure and could cause a significant reduction in fracture permeability.

In addition to the various curvature parameters studied here, it is also possible to plot the principal curvature directions on a map. If curvature variability can be successfully linked to fracture intensity then this may provide a means to map fracture trends across a surface. If a correlation can also be made between fracture intensity and permeability then the principal curvature orientations could be used for mapping permeability trends across the surface. For this to be successful the relationship of the fracture sets to the fold geometries needs to be known or estimated. The simplest case is for fracturing to strike perpendicular to K_1 as in tensile fractures formed in the outer arcs of tangential longitudinal strain folds. It may be possible to determine other fold geometry to fracture orientation relationships for other tensile fractures or shear fracture sets. The use of the spectral curvature plots and maps in this process would allow an assessment of the scales for which the correlation holds.

As part of a comprehensive study of the deformation style and history of a structure, spectral curvature mapping is a potentially useful tool in identifying regions of higher fracture density. It should also be noted that a comprehensive assessment of the fracture distributions within a reservoir unit or units will require the input of fracture data gathered from core or wellbore imagery. These data will allow the identification of the lithomechanical units and possibly the folding styles. These fracture data can also be used to test and calibrate any models of fracture distribution and density derived from curvature variations.

Conclusions

The spectral curvature mapping techniques described and discussed in this paper are particularly useful in allowing the analysis of curvatures at a variety of wavelengths affecting a point or group of points on a gridded surface.

A variety of variables can be plotted. Spectra of the sum of the absolute curvatures (K_1+K_2) and Gaussian curvature allow the identification of curvatures at a variety of different wavelengths.

The representation of several different wavelengths of curvature can be achieved by using the integral of the K_1+K_2 versus offset plot. This integral can be filtered for short-wavelength noise or features and long-wavelength basin-scale curvatures.

Maps of various curvature types can be used to assess the changing areal distribution of curvature with increasing offset. Particularly useful variables are K_1+K_2 and Gaussian curvature.

Maps of the integral of the K_1+K_2 variable indicate that it can show the combined distribution of curvature from several different wavelengths.

W. Belfield and J. M. Daniel are thanked for thorough reviews that significantly improved the manuscript. W. Belfield is also thanked for providing a pre-print of his manuscript entitled 'Predicting Natural Fracture Distribution in Reservoirs From 3D Seismic Estimates of Structural Curvature', SPE paper 60298. The views expressed in this paper are solely those of the authors and do not necessarily reflect the views of either GeoScience Limited or Amerada Hess Limited.

References

AGARWAL, B., ALLEN, L. R. & FARRELL, H. E. 1997. Ekofisk field reservoir characterisation: mapping permeability through facies and fracture intensity. *In: SPE Formation Evaluation*. SPE Paper **35527**, 227–233.

BROWN, A. R. 1991. *Interpretation of Three-Dimensional Seismic Data* (3rd edn). AAPG Memoir **42**.

CALLADINE, C. R. 1986. Gaussian Curvature and Shell Structures. *In:* Gregory, J. A. (ed.) *The Mathematics of Surfaces*. Clarendon Press, Oxford, 179–196.

COOKE-YARBOROUGH, P. 1994. Analysis of fractures yields improved gas production from Zechstein carbonates, Hewett Field, UKCS. *First Break*, **12**, 243–252.

COUPLES, G. D., LEWIS, H. & TANNER, P. W. G. 1998. Strain partitioning during flexural-slip folding. *In:* COWARD M. P., DALTABAN T. S. & JOHNSON H. (eds) *Structural Geology in Reservoir Characterisation*. Geological Society, London, Special Publications. **127,** 149–165.

ERICSSON, J. B., MCKEAN, H. C. & HOOPER, R. J. 1998. Facies and curvature controlled 3D fracture models in a Cretaceous carbonate reservoir, Arabian Gulf. *In:* JONES G, FISHER Q & KNIPE R.J. (eds) *Faulting, Fault Sealing and Fluid Flow in Hydrocarbon Reservoirs*. Geological Society, London, Special Publications, **147**, 299–312.

GIBBS, A. D. 1984. Structural evolution of extensional basin margins. *Journal of the Geological Society, London,* **141**, 609–620.

GOWERS, M. B., HOLTAR, E. & SWENSSON, E. 1993. The structure of the Norwegian Central Trough (Central Graben area). *In:* PARKER, J. R. (ed.) *Petroleum Geology of Northwest Europe: Proceedings of the 4th Conference*. The Geological Society, London, 1245–1254.

HANKS, C. L., LORENZ, J., TEUFEL, L. & KRUMHARDT, A. P. 1997. Lithological and structural controls on natural fracture distribution and behavior within the Lisburne Group, northeastern Brooks Range and North Slope Subsurface, Alaska. *AAPG Bulletin,* **81**, 1700–1720.

JACKSON, M. P. A., VENDEVILLE, B. C. & SCHULTZ-ELA, D. D. 1994. Structural dynamics of salt systems. *Annual Review of Earth and Planetary Sciences,* **22**, 93–117.

JAMISON, W. R. 1997. Quantitative evaluation of fractures on Monkshood Anticline, a detachment fold in the foothills of western Canada. *AAPG Bulletin,* **81**, 1110–1132.

LISLE, R. J. 1994. Detection of zones of abnormal strains in structures using Gaussian curvature analysis. *AAPG Bulletin,* **78**, 1811–1819.

LISLE, R. J. & ROBINSON, J. M. 1995. The Mohr circle for curvature and its application to fold description. *Journal of Structural Geology,* **17**, 739–750.

MCQUILLAN, H. 1973. Small-scale fracture density in the Asmari formation of S.E. Iran and its relation to bed thickness and structural setting. *AAPG Bulletin,* **57**, 2367–2385.

MCQUILLAN, H. 1985. Fracture-controlled production from the Oligo-Miocene Asmari Formation in Gachsaran and Bibi Hakimeh Fields, southwest Iran. *In:* ROEHL P.O. & CHOQUETTE P.W. (eds) *Carbonate Petroleum Reservoirs.* Case Books in Earth Science. Springer Verlag, New York, 511–523.

MURRAY, G. H. 1968. Quantitative fracture study, Sanish Pool, McKenzie County North Dakota. *AAPG Bulletin,* **52**, 57–65.

NARR, W. 1991. Fracture density in the deep subsurface: Techniques with application to Point Arguello oil field. *AAPG Bulletin,* **75**, 1300–1323.

NUTBOURNE, A. W & MARTIN, R. R. 1988. *Differential Geometry Applied to Curve and Surface Design, Vol. 1, Foundations*. Ellis Horwood, Chichester.

PRICE, N. J. & COSGROVE, J. W. 1990. *Analysis of Geological Structures.* Cambridge University Press, Cambridge.

RAMSAY, J. G. & HUBER, M. 1987. *The Techniques of Modern Structural Geology. Volume 2: Folds and Fractures.* Academic Press, London, 543–559.

STEARNS, D. W. 1964. Macrofracture patterns on Teton anticline, northwest Montana. *American Geophysical Union Transactions,* **45**, 107–108.

STEARNS, D. W. & FRIEDMAN, M. 1972. Reservoirs in fractured rock. *In:* KING R. E. (ed.) *Stratigraphic Oil and Gas Fields – Classification, Exploration Methods and Case Histories. AAPG* Tulsa, Memoirs,**16**, 82–106.

STEEN, O, SVERDRUP, E. & HANSSEN, T. H. 1998. Predicting the distribution of small faults in a hydrocarbon reservoir by combining outcrop, seismic and well data. *In:* JONES, G, FISHER, Q & KNIPE, R. J. (eds) *Faulting, Fault Sealing and Fluid Flow in Hydrocarbon Reservoirs*. Geological Society, London, Special Publications, **147,** 299–312.

STEWART, S. A. & PODOLSKI, R. 1998. Curvature analysis of gridded geological surfaces. *In:* COWARD, M. P., DALTABAN, T. S. & JOHNSON, H. (eds) *Structural Geology in Reservoir Characterisation.* Geological Society, London, Special Publications, **127**. 133–147.

WITHJACK, M. O. & SCHEINER, C. 1982. Fault patterns associated with domes – an experimental and analytical study. *AAPG Bulletin,* **66**, 302–316.

Reservoir characterization for the management of stress-sensitivity

BRIAN G. D. SMART[1], JIM McL. SOMERVILLE[1], MIN JIN[1] & NICOS C. KOUTSABELOULIS[2]

[1]*Department of Petroleum Engineering, Heriot-Watt University, Edinburgh, Scotland EH14 4AS, UK* (*e-mail: brian.smart@pet.hw.ac.uk*)

[2]*VIPS Ltd, Elm Lodge, North Street, Winkfield-Windsor, Berkshire, SL4 4TE, UK*

Abstract: Reservoir stress-sensitivity operates at the near-wellbore and the reservoir scale. While the near-wellbore effects have been recognized and managed for some time, awareness of the significance of the reservoir scale effects is only now developing, aided by extreme examples such as Ekofisk, HP/HT reservoirs and unconsolidated sands. Fractured reservoirs must also be considered to be stress-sensitive, with this possibly being masked in the past by high productivities and abundant reserves. It is proposed that the appropriate conceptual model for all reservoirs is that of intact blocks of rocks bounded by discontinuities (fractures and bedding-parallel), loaded by an anisotropic stress state. This model should be used to drive the collection of stress-sensitive reservoir characterization data, knowing that the new generation of coupled simulators can usefully accommodate this data set. The data set listed enables all aspects of reservoir stress-sensitivity to be examined.

Reservoir rocks and the surrounding strata react mechanically to stress changes induced by drilling and production processes. Examples of rock mechanical reactivity include near-wellbore effects such as borehole collapse and sand production. The increased application of highly deviated and horizontal wells has revealed how well orientation (dip and azimuth) influences the potentially damaging redistribution of *in situ* anisotropic stress states around the wellbore, and well paths can be designed to minimize problems. Drilling in tectonically active ground (J. T. Ford, pers. comm.) or near faults (Maury & Zurdo 1994) has detected meta-stability, and demonstrated the importance of developing a site-specific appreciation of stress-related effects in tectonically active regions. Following completion, the impact of drawdown on the destructive stress concentrations around the wellbore perforations can be modelled and allowed for in sand control work (Morita *et al.* 1987; Vaziri 1995)

Significant reservoir-scale geomechanical phenomena can also be triggered by the production process. Mechanical strains associated with the increase in effective stress as pore pressure is reduced can lead to compaction and subsidence (Wiborg & Jewhurst 1986), causing casing collapse in the overburden and damage to surface installations (Schwall & Denney 1984). In-fill drilling is complicated by the stress-contrasts induced between overburden and reservoir in depleted reservoirs. Faults can be activated altering the compartmentalization of the reservoir (Bayat & Tehrani 1985) activation being detected by the induced seismicity (Segal 1989). Thermal strains associated with water injection interact with the anisotropic *in situ* stress state to impose directionality on flood front migration by creating conduits of flow along fractures which open against the least horizontal stress s_h, i.e. fractures that open along the direction of the regional or local maximum horizontal stress σ_H (Heffer & Koutsabelousis 1990) (see Fig. 1) Furthermore, the manner in which the reservoir reacts at this scale is influenced by its structural setting (Teufel *et al.* 1993).

Near-wellbore effects are usually recognized and managed by drilling and production engineers, e.g. by well design, completion design, shutting off weak zones, or ultimately limiting drawdown, but there is still relatively little attention paid to reservoir-scale geomechanical phenomena. This is probably due to:

- lack of appreciation among a majority of senior reservoir engineers of geomechanical phenomena which can act at this scale, due in

From: AMEEN, M. (ed.) *Fracture and In-Situ Stress Characterization of Hydrocarbon Reservoirs*. Geological Society, London, Special Publications, **209**, 145–153. 0305–8719/03/$15.00
© The Geological Society of London 2003.

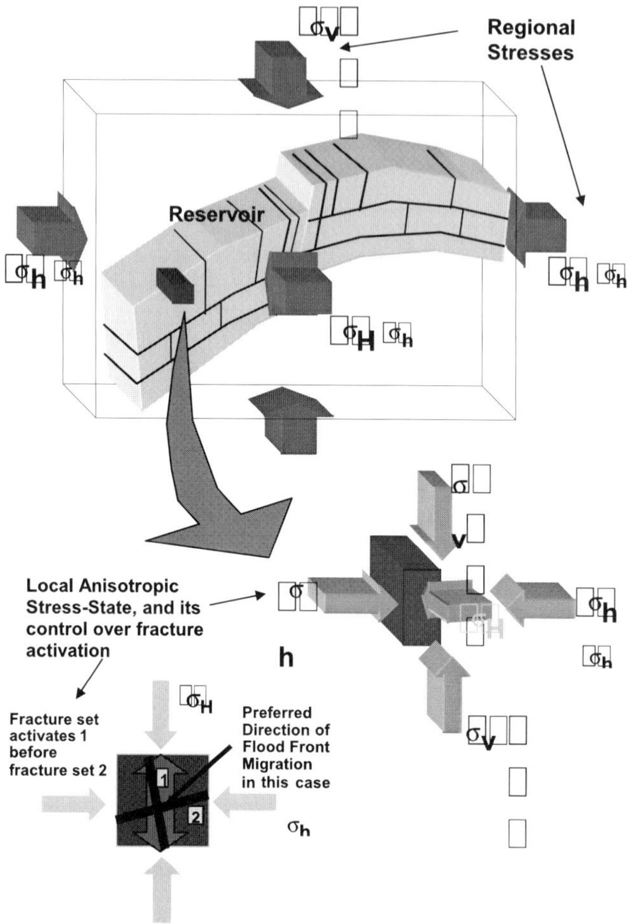

Fig. 1. Anisotropic stress state. Extensional setting $\sigma_v > \sigma_H > \sigma_h$ (as shown). Thrust Setting $\sigma_v < \sigma_H < \sigma_h$.

turn to the relative newness of the stress-sensitivity subject area and hence the dearth of widespread teaching/lifelong learning of the subject;

- related to the above, a credibility problem due to lack of sufficient published case studies which unequivocally demonstrate the occurrence of such phenomena and the threats/opportunities that allow for stress-sensitivity at the reservoir scale regarding recovery;
- until recently, lack of suitable stress-sensitive reservoir simulators or other appropriate management tools, including a methodology for assessing likely stress-sensitivity before or during production;
- as a result of all of the above, an unwillingness on the part of management to commit the extra resources required to characterize reservoirs for stress-sensitivity and run stress-sensitive simulations in what is still seen as a speculative technical advance, with the notable exceptions of some parts of a few multinational and national oil companies who see application of the topic providing a business advantage.

While there is irrefutable evidence from particularly stress-sensitive reservoirs, e.g the Ekofisk chalks, and increasingly from the unconsolidated sands of the Gulf of Mexico, that allowing for geomechanical phenomena is essential if these reservoirs are to be properly managed, the hypothesis presented here is that these reservoirs are high-end members on the stress-sensitivity awareness scale, where the effects are easily recognized. We argue, however, that many reservoirs have a more subtle level of stress-sensitivity, which may not yet have been recognized. i.e. these reservoirs are currently further down the stress-sensitivity awareness scale. Their stress-sensitivity is more subtle, not being derived from an obviously weak reservoir rock. An appreciation of more subtle effects can, however, be used to significantly improve hydrocarbon recovery. For example, in one major field, the relation between water breakthrough, fault sets, and local orientation of the stress state only became apparent after lengthy and careful analysis of the well data. It was estimated that the stress influence made a difference to cumulative production of oil in the order of 1 million barrels per well, as well as causing larger watercuts in wells unsympathetically orientated with the stress state. Correlations of rate histories between pairs of wells indicated that the stress-related effects were influential even in monthly production fluctuations.

Fractured reservoirs must, by their very nature, be stress-sensitive. Apparent stability of the reservoir zone is preserved by a load-sharing between intact discrete blocks of reservoir rock across fractures which are open enough to control flow. The relatively limited mechanical load-bearing contacts between blocks will concentrate stresses, exacerbating compaction and straining of the rock mass induced by reservoir pressure reduction. There will certainly be preferential activation of fracture sets orientated in the most favourable direction with respect to the anisotropic stress state as shown in Figure 1. Possibly the full significance of stress sensitivity in fractured reservoirs has in part been masked in some fractured reservoirs by the abundance of reserves and their high productivity, while in others production problems are attributed to difficulty in encouraging displacement of hydrocarbon from the matrix (intact blocks) into the flowing fractures. There is, however, a growing perception that effective stress changes and associated rock mechanical controls exerted in real time during production are important influences on fluid flow and so reservoir performance, with natural fractures closing with depletion (National Research Council 1996), although care must be taken to consider the production-rate sensitivity, e.g with water production, also observed in fractured reservoirs.

Reservoir stress-sensitivity is not only caused by fractured or low-strength, highly compactable reservoir rocks, but is also dependent on the production scenario applied. Even reservoirs with strong, apparently unfractured rocks, may well exhibit some significant aspects of stress-sensitivity in the latter stages of their life as they are depressurized and feel a significant increase in effective stress. Alternatively, as increasingly larger volumes of rock are influenced by production, differences in the straining and deformation of reservoir compartments can lead to fault activation, e.g. the evidence for production-induced changes in the compartmentalization of Beryl (Buck & Robertson 1996). The stress-sensitivity assessment process must therefore look at the reservoir within the context of its geomechanical characteristics linked inextricably with its production scenario. HP/HT or over-pressured reservoirs must represent an extreme end of stress-sensitivity in this respect, blowdown creating typically a 300% increase in effective stress in the reservoir.

Stress-sensitivity also occurs in the rocks surrounding the reservoir, and in particular in the overburden. The layered nature of sedimentary rock sequences leads to the activation of bedding plane parallel slip along laterally extensive weak layers, e.g. shales or coals sandwiched between thick sandstones, as the subsidence basin develops above a compacting reservoir. The relative shear displacement across these layers can be of sufficient magnitude to cause wells to shear (Hamilton *et al.* 1991). Even if the shear displacements are insufficient to be noticed in wells, the redistribution of stresses in the overburden – linked to the so-called 'arching' effect – can influence fault activation and the selection of in-fill drilling windows, and hence decisions related to reservoir management (Kenter *et al.* 1998).

The redistribution of stresses in the overburden and the reservoir should also be taken into account when interpreting the results of repeat (4D) seismic surveys. As the effective stress on the rock matrix is increased, the seismic velocity increases, and vice versa. The volumetric strain change that the rock experiences as a result of the change in effective stress also alters the density of the rock. Since rock acoustic impedance is given by the product of density and velocity, then acoustic impedance, acoustic impedance contrasts, and hence the seismic survey results are subject to stress-sensitivity. This impedance sensitivity can be of the same order as the more conventionally assumed saturation-only related changes in acoustic impedance, and therefore should be taken into account (Corbett *et al.* 1998).

Management of all scales of reservoir and overburden stress-sensitivity begins with the geomechanical characterization of the reservoir and, if deemed necessary, the surrounding rock, creating a data set which can then be interrogated in its own right to help manage stress-sensitivity, or ultimately can be used as input to a coupled stress analysis/fluid flow simulation as shown in Figure 2. Coupled modelling enables a holistic

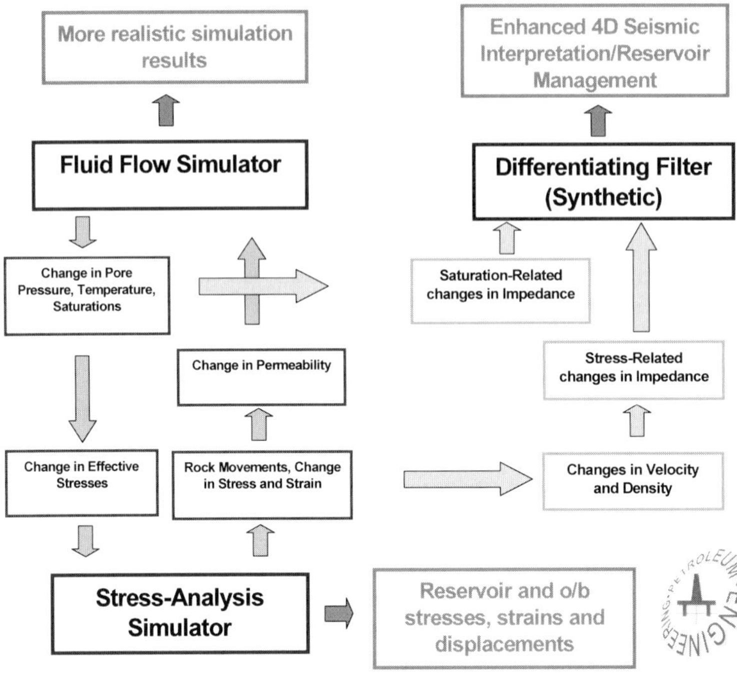

Fig. 2. The Coupled Modelling Process, Enabling the Input of Stress-Sensitivity into Reservoir Management, and Well Design

view to be taken of the prediction and management of the geomechanical phenomena likely to be triggered by hydrocarbon production. This paper introduces the reservoir-specific data set required to achieve full geomechanical characterization, based on the argument that all reservoirs should be assessed in this manner to determine where they lie on the stress-sensitivity scale, and hence what reservoir management threats and opportunities present themselves.

Reservoir characterization for stress-sensitive simulation

The data sought in any reservoir characterization process depend upon the processes thought to have a significant influence on reservoir performance, and therefore to be of importance in understanding and managing the reservoir. Furthermore, there is little point in acquiring data, if within the foreseeable life of the reservoir, it cannot be effectively used. Thus data acquisition is driven by the combination of the realism of the conceptual model being applied and the available simulation capability. Until relatively recently, the conceptual models applied did not allow for stress-sensitivity, and hence reservoir characterization has traditionally been concerned with rock types, poro-perm and fluid distributions. Faulting may also be addressed at several scales, but usually in the context of fixed transmissivity properties, i.e. are the faults permeability barriers or not? Other structural features may be considered in 'fractured' reservoirs, i.e. reservoirs which by virtue of their well in-flow characteristics, have been shown to have open conduits to flow connecting into the well. In this case the spacing, extent and connectivity of the fractures (discontinuities which make a steep angle to bedding) are of importance. The fractured reservoir is typically modelled using conventional reservoir fluid flow simulators with either the so-called 'dual porosity' approach, which captures capillary force controls on fluid flow between fracture and matrix, or the 'dual permeability' approach which captures the viscous force and gravity force controls on flow along the fractures. It is generally felt, however, that in order to capture all of the physics of the flow process in one simulation, discrete modelling of fractures and matrix is required (Edlmann *et al.* 1998). Modelling fractures at this level of detail does have computing resource implications, however, especially for large complex reservoirs.

Furthermore, the rock properties required in the reservoir characterization data set are usually assumed to have constant values which do not change with reservoir stresses, being adjusted only when simulation history matching dictates the need for a change. That said, conventional reservoir characterization provides an essential basis from which a geomechanical characterization can be developed, as will be demonstrated later.

When allowance is made for the rock mechanical reactions caused by production-induced stress changes two enhancements must be made to the conventional characterization data set. Firstly, it should be recognized that production-induced changes in stress in the reservoir – caused by temperature or pore pressure perturbations – initiate systematic changes in all of the rock matrix properties which are linked in any way to pore dimensions and rock compressibility, e.g. permeability, P-wave velocity etc. Thus many of the conventional data are intrinsically stress-sensitive and are not adequately represented by assigning a constant value for the duration of reservoir production life. Secondly, additional attention must be paid to fractures and other discontinuities, extending the dual poro-perm approach. Thus conceptually, all reservoirs should be considered to be composed, as shown in Figure 2 of, 'intact' blocks of rock matrix bounded by two types of discontinuities:

(1) the fractures which make a steep angle to bedding – and these are probably present in all reservoirs from a structural point of view if not from a flow modification point of view;
(2) bedding-plane-parallel discontinuities which are certainly present in all stratified reservoirs from a structural point of view, and may also occasionally influence fluid flow.

The intact blocks and the discontinuities have petrophysical/rock mechanical properties which as well as honouring the heterogeneities of conventional reservoir characterization, are stress-sensitive, and it is these properties which form a major part of the inventory of properties required for stress-sensitive reservoir simulation as shown in Table 1. Furthermore the reservoir structure and the surrounding strata are loaded with a regional stress state which is liable to be anisotropic, with local modification near faults.

The bedding-plane-parallel discontinuities are usually ignored, even in the characterization of fractured reservoirs. Having a relatively low bedding-parallel shear strength, these discontinuities easily strain as they allow shear stress relief, exerting an important control of the lateral and flexural stiffness of the reservoir, and hence of the gross response of the reservoir to stress changes. This influence is in part determined by the separation between low-shear strength bedding planes, e.g. clay-rich layers, or shale bodies or coal seams. In a fluvial depositional environment, the maximum distance between such partings is determined almost on a regional basis by the frequency of flood surfaces which have created laterally extensive clay-rich rocks or coal seams. In carbonates, some stylolites which have produced sufficient non-soluble debris to disengage the opposing teeth of the stylolite may also allow shear stress relief.

In order to complete the stress-sensitive data set, the *in situ* stress state must be described. This presents particular challenges, as it is liable to be anisotropic in three dimensions as shown in Figure 1, with major, intermediate and minor stresses. Usually defined by specifying three orthogonal principal stresses, determination of the magnitude of the intermediate stress presents a particular challenge, although its orientation can be determined, being a member of a mutually orthogonal set of stresses. The complete inventory for data is given in Table 1, while justifications for the data requested are iven in Table 2. Provision of the content of this table requires multidisciplinary collaboration between sequence stratigraphers, geophysicists, structural geologists, rock mechanics laboritories and others. For example, a rationale for the selection of rock mechanical/petrophysical specimens for laboratory testing which honours conventional reservoir characterization is shown in Figure 3.

Spreading the rock mechanical data through the reservoir presents a similar challenge to conventional characterization, although this can be achieved by exploiting correlations between rock types/porosities and rock mechanical properties, via reservoir-specific synthetic rock mechanical property logs (Edlmann *et al.* 1998) Upscaling, or more correctly the specification of *in situ* matrix properties, can be reasonably easily achieved, using a 'critical phenomena' approach. For example, if compaction is the phenomena being considered, then care must be taken to arrive at an effective, integrated *in situ* modulus, while if thermal fracturing is a possible issue, then the lowest measured matrix tensile strength should be used. Upscaling of fracture properties remains a challenge, given the demands on computational resource. The best compromise is probably to use the equivalent material approach for much of the reservoir (Koutsabelousis *et al.* 1991), with certain key discontinuities being

Table 1. *Listing of Intact Rock, Discontinuity and Field Scale Properties Required for Characterisation of a Stress-Sensitive Reservoir*

Property	Management Application						
	Porosity correction	Wellbore stability	Sand production	Hydraulic frac	Thermal frac	Compaction drive	Subsidence and effects
'Intact' Rock							
Young's modulus*	X	X	X	X	X	X	X
Poisson's ratio*	X		X	X	X	X	X
Biot's constant*	X	X	X	X	X	X	X
Tensile strength*		X	X	X	X		
Cohesion*		X	X	X	X	X	X
Angle of friction/ internal friction*		X	X	X	X	X	X
P velocity**		X	X	X	X	X	X
S velocity**		X	X	X	X	X	X
Permeability**		X	X	X	X	X	X
Rel perm**			X	X	X	X	X
Porosity**	X	X	X	X	X	X	X
Discontinuities							
Normal stiffness*	X		X	X	X	X	X
Tensile strength*		X	X	X	X		
Cohesion*		X	X	X	X	X	X
Friction angle*		X	X	X	X	X	X
P velocity**			X			X	X
S velocity**			X			X	X
Permeability**			X		X	X	X
Porosity***						X	X
Surface topography***						X	X
Field Scale							
Virgin *in situ* stress state**	X	X	X	X	X	X	X
Virgin pore pressure*	X	X	X	X	X	X	X
Change in pore pressure*	X	X	X	X	X	X	X
Stratigraphy**		X	X	X	X	X	X
Structure**		X	X	X	X	X	X
Boundary stiffnesses***	X	X	X	X	X	X	X
Temperature*				X	X	X	X
Injected fluid temperature*				X	X	X	X

*Easily available with reasonable accuracy;
**May be available with reasonable accuracy;
***Probably not available with reasonable accuracy.

modelled as discrete entities, e.g. faults and bedding-plane-parallel shear layers.

Having taken the trouble to characterize the reservoir for stress-sensitivity, stress-sensitivity can now be examined and managed using coupled modelling as shown in Figure 2. The production-induced changes in effective stress (via pore pressure and temperature) are simulated, and hence used to determine changes in stress-sensitive reservoir properties such as permeability. The stress-sensitive corrections to the reservoir characterization data embedded in the coupled model are made automatically, enhancing the realism of the predictions made, and justifying the additional effort required. As well as adding realism to the simulation process, this adds another set of output data to that of conventional simulation – stress distributions

Table 2. *Conceptual/analytical/numerical models used, justifying data requirements*

Management Application	Rock Mechanical Model Assumed	Notes
Porosity correction	Linear elastic compression of matrix, changing mainly pore space. Effective stress causes compaction, hence use of Biot's coefficient. The boundary stiffnesses affect the stress condition used in the calculation. Very stiff boundaries usually assumed, dictating the "uniaxial" corection.	Compensates for the difference between measurements made at near ambient stress conditions in the lab and the much higher overburden related compressive stresses experienced by the rock *in situ*. Grains assumed to be very stiff, with most of the rock compression being taken up by pore volume reduction. Other models available allow for grain compressibility
Wellbore stability	Concentration of stresses around the wellbore cause rock failure adjacent to the wellbore when the Mohr-Coulomb failure criterion for the rock or fractures is exceeded. Failure can occur in the matrix or on discontinuities. Sonic velocities used for property correlation.	The mud exerts an outward radial stress on the wellbore, inhibiting failure. Anisotropic *in situ* stress state can be allowed for. Alignment of the wellbore (dip and azimuth) with these stresses is an important control.
Sand production	Concentration of the stresses around the perforation or wellbore case failure as in the wellbore stability case.	Ideally rate dependency should also be modelled.
Hydraulic frac	A fracture is initiated when the wellbore pressure is sufficiently high to overcome the tensile strength of the matrix or a critically orientated discontinuity, and the lowest horizontal principal stress concentrated around the wellbore.	Probably one of the first applications in which the industry recognized the influence of anisotropic stresses.
Thermal frac	A fracture is initiated when the thermally induced tensile stresses caused by the cooling of the rock surrounding the wellbore are sufficient to overcome the tensile strength of the rock around the wellbore, or the strength of a critically orientated discontinuity, whichever is lower.	A relatively recently discovered effect. Again stress anisotropy exerts a major control.
Compaction drive	The reduction in pore pressure increases the effective compressive stress on the reservoir, compressing the matrix and/or fractures, displacing fluid	An important contribution to reservoir energy, even in relatively stiff reservoirs.
Subsidence and effects	Compaction, i.e. strains within the reservoir, cause the redistribution of stresses, strains and displacements in the overburden. Vertical displacements are generally continuous through the overburden, except across activated faults. Horizontal displacements are discontinuous across critical bedding-parallel shear layers. Ideally requires a full coupled model study, hence comprehensive data set.	Well loss due to this now recognised, but impact on 4D seismic just being realised.

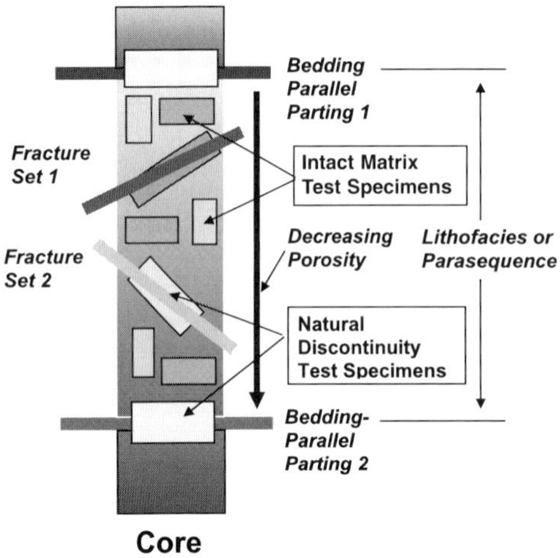

Fig. 3. Selection of test specimens for a fracture reservoir.

and rock strains, displacements and failure state – which can be used in the more comprehensive management process, e.g. in the planning for changes in reservoir compartmentalization, well planning for maximum life in a subsiding overburden, in-fill well path design, and draw-down limitation or flow-unit isolation for sand control. Thus the full power of the data set generated for the characterization of stress-sensitive reservoirs can be realised. For example, Figure 4, which shows reservoir rock displacements in a reservoir layer output from the VISAGE™ stress-analysis simulator, predicts the activation of a major fault during the depletion of a normally pressured UKCS reservoir, enabling the impact of this reservoir-scale geomechanical event to be factored into the reservoir management process. Similarly, the output of stresses and strains can be converted into stress-sensitive acoustic impedance maps and sections, enabling this effect to be factored into the interpretation of 4D seismic surveys, as shown in Figure 2.

Conclusions

The understanding, methodology and tools required for stress-sensitive reservoir simulation are now in place. While additional effort and resources are required compared to conventional simulation, it is likely that recovery in many reservoirs can be improved when due allowance is made for reservoir-scale geomechanical phenomena.

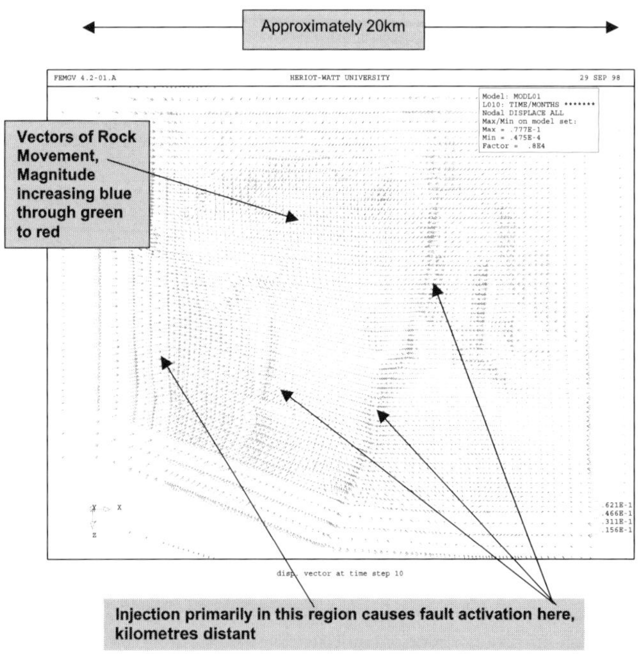

Fig. 4. Output from the VISAGE stress analysis simulator showing rock displacements in one reservoir layer, revealing fault activation.

References

BAYAT, M. G & TEHRANI, D.H. 1985. The Thistle Field – analysis of its past performance and optimisation of its future development. *Offshore Europe, Aberdeen, September 1985*. SPE 13989/1.

BUCK, S. & ROBERTSON, G. 1996. Dynamic behaviour of faults during hydrocarbon production and injection in the highly compartmentalised Beryl Field, North Sea. *In: Faulting, Fault Sealing and Fluid Flow in Hydrocarbon Reservoirs, University of Leeds, Sept 23–25*. Abstracts.

CORBETT, P. W. M. *et al.* 1998. *Final Report on the GUMPA (Geophysics Used in the Management of Petroleum Assets)*. Department of Petroleum Engineering, Heriot-Watt University.

EDLMANN, K., SOMERVILLE, J. M., SMART, B. G. D, HAMILTON, S. A. & CRAWFORD, B.R. 1998. Predicting rock mechanical properties from wireline porosities. *Eurock 98*, Trondheim.

HAMILTON, J. M., MALLER, A. V. & PRINS, M. D. 1992. Subsidence induced shear failures above oil and gas reservoirs. *In:* TILLERSON AND WAWERSIK (eds) *Rock Mechanics, 33rd American Conference*. Balkema, Rotterdam.

HEFFER, K. J. & KOUTSABELOULIS, N. C. 1990. Relationship between azimuths of flood anisotropy and local earth stresses in oil reservoirs. *In: North Sea Oil and Gas Reservoirs II*, Graham and Trotman, London.

KENTER, C. J., BLANTON, T. L., SCHREPPERS, G. M. A., BAAIJENS, M. N. & RAMOS G. G. 1998 Compaction study for the Shearwater Field. *EUrock 98, Trondheim.* SPE 47280.

Koutsabeloulis, N. C., Rylance, M. & Eggington, R. E. 1991. 3–D finite element modelling of the Hope Brook gold mine, *In:* BEER, G., BOOKER, J. R. & CARTER, J. C. (eds) *Proceedings of the 7th International Conference on Computer Methods and Advances in Geomechanics*, 2, A. A. Balkema, Cairns, Australia.

MAURY, V. M. R. & ZURDO, C. 1994 Wellbore instability due to shear displacements along preexisting fractures: are we overlooking a common cause of drilling problems? *IADC/SPE Drilling Conference, Dallas, Feb. 1994*. IADC/SPE 27492

MORITA, N., WHITFIL, D. L., FEDDE O.P. & LOVIK, TH. 1987. Parametric study of sand production prediction: analytical approach. *62nd Annual Conference of the SPE, Dallas Texas.* SPE 16990.

NATIONAL RESEARCH COUNCIL 1996. *Rock fractures and fluid flow: contemporary undersanding and applications*. National Academy Press, Washingon DC.

SCHWALL, G. H. & DENNEY, C. A. 1994. Subsidence induced casig deformation mechanisms in the Ekofisk Field. *Eurock '94.* Balkema, Rotterdam.

SEGALL, P. 1989. Earthquakes triggered by fluid extraction. *Geology* **17**, 942–946.

TEUFEL, L. W, LORENZ, J. C., RHETT, D.W. & FARRELL, H. E. 1993. Control of fractured reservoir permeability by spatial and temporal variations in stress magnitude and orientation. *SPE Annual Conference, Houston, October 19*. SPE 26437.

VAZIRI, H. H. 1995. Analytical and numerical procedures for analysis of flow-induced cavitation in porous media. *Computers and Structures* **54**(2) 223–238.

WIBORG, R. & JEWHURST, J. 1986. Ekofisk subsidence detailed and solutions assessed. *Oil and Gas Journal* (Feb 17), 47–51.

3D fracture network dynamics in reservoirs, faults and salt tectonic systems

K. TUNCAY, A. PARK, D. PAYNE & P. ORTOLEVA

Laboratory for Computational Geodynamics, Chemistry Building, 800 E. Kirkwood Avenue., Indiana University, Bloomington, IN 47405, USA

Abstract: A unique 3D computer simulator is used to predict natural fracture network characteristics in the subsurface. The model is based on the numerical solution of rock deformation processes coupled to the myriad of other basin reaction, transport and mechanical (RTM) processes. The model integrates seismic, well log and surface geological data to arrive at a quantitative picture of the distribution of fractures, stress, petroleum and porosity, grain size and other textural information.

An important component of the model is an incremental stress rheology that accounts for poroelasticity, non-linear viscosity with yield/faulting, pressure solution and fracturing. It couples mechanics to multi-phase flow and diagenesis (through their influence on effective stress and rock rheological properties, respectively). The model is fully 3D in terms of the full range of fracture orientations and the tensorial nature of stress, deformation and permeability. All rock properties (rheologic, multi-phase fluid transport, grain shape, etc.) are coevolved with the other variables. Examples illustrate the relative importance of various overpressuring mechanisms, lithology and flexure on the location and characteristics of a fracture network.

Fracturing in the subsurface is key to the understanding of many fundamental and applied issues in the geosciences. The objective of this study is to show that these rock failure processes are very dynamic and that this dynamic can only be understood via a model that includes a wide range of other crustal processes. Indeed, the evolution of a fracture network is the result of the interplay of internal processes (matrix texture and pore fluid reactions and flow) and overall tectonic factors. We present results of a unique three-dimensional (3D) model of the coevolution of these dynamic phenomena for use in fundamental crustal studies and the petroleum exploration and production analysis. Fractures and faults affect and are affected by many processes. For example, fractures are affected by fluid pressure and, through their influence on permeability, fractures affect fluid flow and hence fluid pressure. As this is just one of many such 'feedback' relations embedded in the network of geological reaction, transport and mechanical (RTM) processes, a predictive fracture model must be fully coupled (see Fig. 1). As lithologic bodies, folds, domes, salt diapirs and other features are fundamentally 3D, a predictive model of fracture and fault phenomena must be 3D.

Here we present a unique 3D RTM model capable of predicting a host of fracture and fault phenomena. Our model is based on the finite element solution of equations of rock deformation, fracture network statistical dynamics, rock failure and gouge, multi-phase flow, organic and inorganic diagenesis, pressure solution and other compaction mechanisms, and heat transfer. The RTM equations are solved consistent with the influences at the boundary of the basin (sedimentation/erosion, basement heat flux, climate, sea level, and extension/compression or uplift/subsidence history) as suggested in Figure 2.

Key to the attainment of predictive power is the coevolution of all variables. Thus, flow changes fluid pressure which can cause fractures and

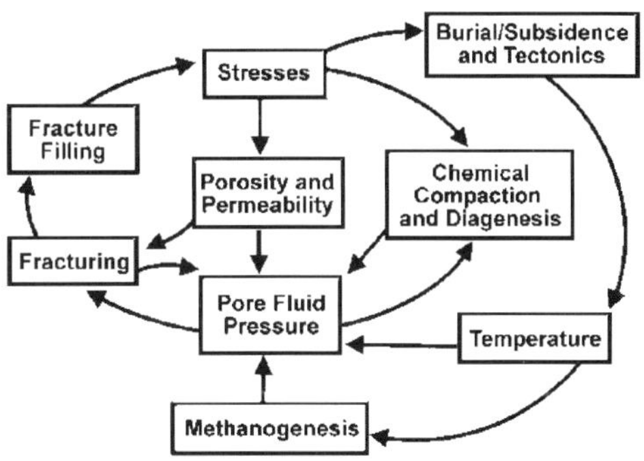

Fig. 1. Complex network of coupled processes underlying the dynamics of a sedimentary basin. These factors and their coupling are accounted for in our unique basin simulator.

From: AMEEN, M. (ed.) *Fracture and In-Situ Stress Characterization of Hydrocarbon Reservoirs*. Geological Society, London, Special Publications, **209**, 155–175. 0305–8719/03/$15.00
© The Geological Society of London 2003.

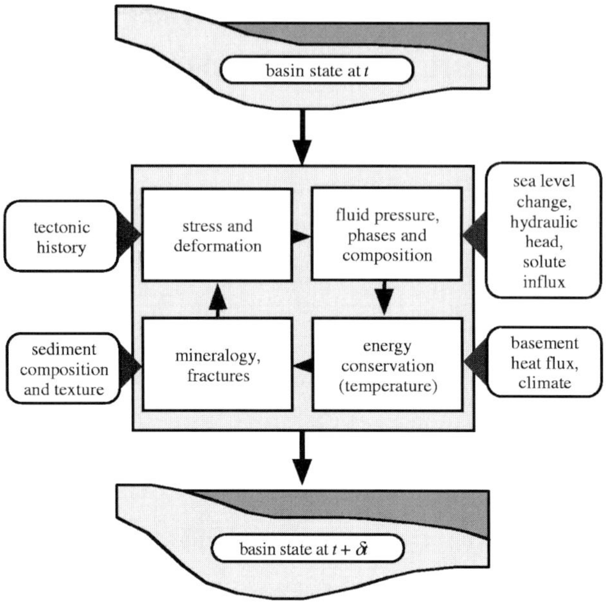

Fig. 2. Schematic Basin RTM flow chart showing the interplay of geological data and the internal RTM processes in evolving a basin over one computational time step.

thereby direct fluid flow along preferred fracture directions. Such coupling and coevolution must be accounted for in their full 3D manifestation. We compute the dynamics of the evolving probability distribution of fracture length, aperture and orientation. Properties such as the permeability tensor are functionals of this probability distribution and constitute a key element of many crustal fluid migration phenomena.

The challenge of modelling fractures and faults in reservoir- or basin-scale systems is one of scale. Typically fractures are metre scale in length and submillimetre scale in aperture. Clearly, it is not feasible to model fractures by introducing a submillimetre scale numerical grid across the reservoir or basin.

The solution is to introduce variables such as average local fracture length, aperture and orientation, characterizing the 'typical' fracture in a macrovolume element of the system. By definition, a macrovolume element is smaller than the scale of the phenomenon of interest (a reservoir) but large enough to contain a statistically significant number of fractures. Such a theory is thus, by nature, a statistical model with the macrovolume element as the sampling volume.

Within such a framework, the variables characterizing the fracture network obey ordinary differential equations in a rock-fixed reference frame (i.e. one moving with the velocity of deformation of the solids). Letting be a collection of texture variables characterizing the fractures as well as the size, shape, orientation, packing, and mineralogy of the grains, we have

$$\frac{D\Theta}{Dt} = \mathrm{A}\left(\Theta, \underline{\underline{\sigma}}, f, T\right) \quad (1)$$

where $\underline{\underline{\sigma}}$ is the stress tensor and f and T are the fluid properties (phase and composition) and temperature in the macrovolume element. In this way fractures (and in general texture dynamics) couple to stress and other factors. Furthermore, rock rheology depends on Θ (as well as f and T), completing a stress/deformation ⇔ texture feedback. This is just one of the feedback loops contained in Figure 1. Introducing the concept of texture dynamics and capturing the many coupling relations driving the dynamic crustal system is at the heart of our fracture and fault prediction approach. In the remainder of this paper, we illustrate the comprehensiveness and power of our approach using a number of examples. Applications to petroleum exploration and production and the fundamentally 3D nature of many crustal RTM phenomena are noted.

The Basin RTM simulator

Purpose

An overview of our Basin RTM simulator used for fracture and fault modelling is as follows. A complex network of geochemical reactions, fluid and energy transport and rock mechanical (RTM) processes underlies the genesis, dynamics and present-day characteristics of petroleum reservoirs or other crustal phenomena in Basin RTM (see Fig. 1). Basin RTM integrates all the relevant geological factors and RTM processes believed to operate in a sedimentary basin. As reservoirs are fundamentally 3D in nature, Basin RTM is fully 3D.

The RTM processes and geological factors accounted for in Basin RTM are outlined in Figure 2. External influences such as sediment input, sea level, thermal and tectonic effects are allowed to influence the progress of internal RTM processes. Within the basin, these RTM processes modify the sediment chemically and mechanically to arrive at petroleum and mineral reserves, basin compartments and other internal features.

Basin RTM provides a platform for integrating all available geological data as suggested in Figure 2 using the framework provided by the laws of physics and chemistry to facilitate exploration or field development. Available information can be divided into geological data and the physico-chemical rate laws and parameters. The former make a simulation tailored to the specific basin. The physico-chemical information gives Basin RTM the power to predict resource location and characteristics and other features of the evolving basin.

Basin RTM can also be used to carry out sensitivity analyses or to identify new phenomena such as self-organization and other non-linear effects that can dramatically affect the disposition of reservoirs (Ortoleva 1990, 1994*a*). Basin RTM simulations show that the sedimentary basin or other crustal system is highly dynamic, exhibiting a strong degree of autonomy, rather than simply responding to the details of the external influences. As Basin RTM uses data on present-day localtime and characteristics of lithologies to extrapolate these properties beyond the locations of these data, it enhances the use and interpretation of seismic, well log, surface geological and other data to understand the present-day and historical state of the crust. Basin RTM can be used to identify windows of time during which formations along a proposed migration pathway were open, and not blocked due to compaction, fracture closure or diagenetic cementation. Alternatively, Basin RTM can predict if and when a seal was breached and some of the hydrocarbons have escaped due to natural fracturing or permeability-enhancing diagenetic reactions (see Figs 3–5).

How basin RTM works

Basin RTM makes its predictions in a completely self-consistent way through a set of multiphase, organic and inorganic, reaction-transport modules. Calculations of all effects are done self-consistently to preserve all cross-couplings

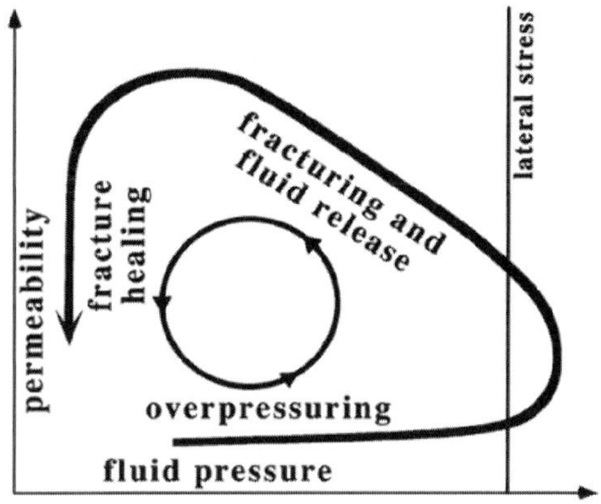

Fig. 3. Fluid pressuring, fracturing and fracture healing feedback cycle, one example of the many feedback mechanisms inherent in the RTM process network. This cycle can repeat many times during a basin's evolution when conditions are appropriate.

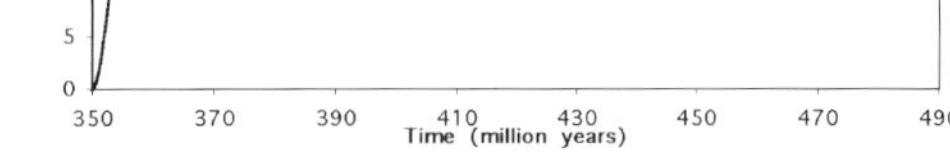

Fig. 4. Overpressure evolution at the bottom of the Ellenburger Formation in Permian Basin, West Texas. Overpressuring starts around 350 million years into the simulation, when fractures in the layer above the source rock disappear. Oscillatory behaviour is a result of cyclic fracturing of the seal driven by petroleum generation. After 470 Ma the cyclic petroleum expulsion ceases and the pressure, oil saturation fracturing and other variables show a more steady behaviour.

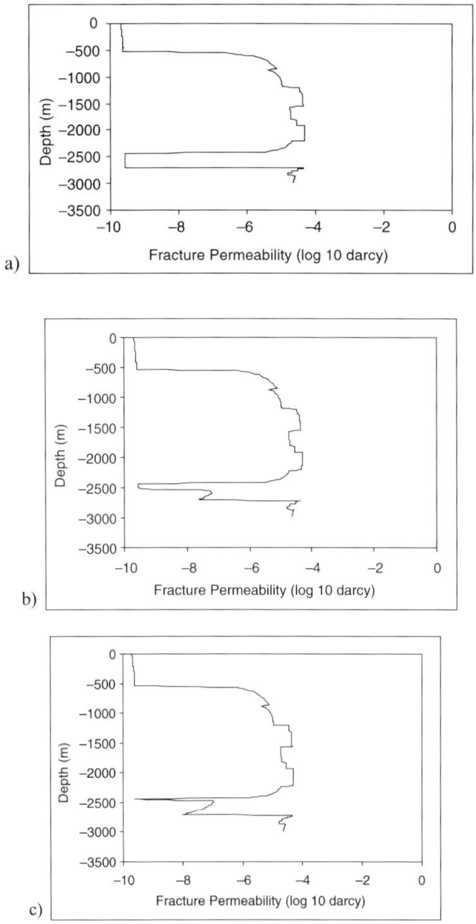

Fig. 5. Fracture permeability profile sequence illustrating the fracture front moving through the seal (between 2450 and 2700 m) Permian Basin, West Texas. Overpressuring of oil and water phases primarily due to oil generation creates a fracture front moving upward through the seal. Once the overpressure is released, the fractures close, which in turn results in descent of the fracture front and overpressuring restarts. This cycle continues until the oil generation rate slows down, or the seal remains fractured due to tectonic effects.

between processes (see Fig. 1). For example, the determination of temperature is affected by transport, which is affected by the changes of porosity that evolve due to temperature-dependent reaction rates. Similarly, the rate of kerogen decomposition depends on temperature which, in turn, depends on thermal transport that is affected, through fluid buoyancy, thermal conductivity, capillarity and relative permeability, by the content of organic material and its thermal decay. All such coupling relations between the full set of RTM processes as in Figure 1 are accounted for in our model.

Predictive power is limited for less rigorous approaches that use statistical correlations. For example, in such methods, porosity history is often based on a formula relating it to mineralogy and depth of burial. However, porosity evolves due to the detailed stress, fluid composition and pressure, and thermal histories. These histories are different for every basin. Thus a simple correlation of porosity with depth and lithologic type does not exist in principle. Basin RTM avoids such problems by solving the fully coupled rock deformation, fluid and mineral reaction, fluid transport and heat transfer problems. Finally, statistical correlations give the average behaviour. As 'on the average' there are no interesting features such as producible pools of petroleum, such approaches can only have a limited interest.

The interplay of geological and physico-chemical information in Basin RTM is suggested in Figure 2. Consider one forward time step in a Basin RTM simulation. The purpose of the incremental evolution step is to advance the state of the basin from a time t to a later time $t+dt$. Two distinct operations take place simultaneously during this time interval dt. The geological information is used to (1) fix the input/output of energy and mass at the basin boundaries and (2) impose the tectonic history (i.e. the overall basin deformation or stress) at the basin boundaries. On the other hand, the physico-chemical processes are used to determine the evolution in dt of the spatial distribution of the local state. The latter describes stress, fluid properties, mineral content, rock texture, permeability, fracture characteristics and temperature.

Basin RTM geological input data are divided into four categories as shown in Figure 2. The tectonic data give the change in the lateral extent and the shape of the basement–sediment interface during δt. As suggested in Figure 2, these data provide the conditions at the basin boundaries needed to calculate the change in the spatial distribution of stress and rock deformation within the basin. This latter physico-chemical calculation is carried out by a stress module that solves equations for incremental stress rock rheology and force balance (see page 167).

The next type of Basin RTM geological data are those affecting the fluid transport, pressure and composition. These fluid data include sea level changes, basin recharge conditions and the composition of fluids injected from the ocean, meteoric and basement sources. This history of boundary input data is then used by the hydrologic and chemical modules to calculate the evolution of the spatial distribution of fluid pressure, fluid composition and fluid phases within the basin. These physico-chemical calculations are based on single or multi-phase flow in a porous medium and on fluid phase molecular species conservation of mass (i.e., the reaction-transport equations). The physico-chemical equations draw on internal data banks for permeability–rock texture relations, relative permeability formulae, chemical reaction rate laws and reaction and phase-equilibrium thermodynamics.

The spatial distribution of heat flux imposed at the bottom of the basin is another geological input/control. These data as well as the temperature imposed at the top of the sediment pile (i.e. climate and ocean-bottom temperature) are used to evolve the spatial distribution of temperature within the basin during the time interval δt. This evolution is computed using the equations of energy conservation and data for mineral and rock thermal properties (conductivities and specific heats).

The sedimentation data provide the detailed textural characteristics such as grain size, shape, mineralogy, mode and organic texture of the sediment being deposited during dt. This history is automatically computed by Basin RTM using interpreted well log, seismic core and surface data. The physico-chemical laws and data are used to calculate the change of the spatial distribution of mineral and organic texture within the basin during dt. These physico-chemical calculations involve the rate laws for free face grain chemical kinetics, pressure solution and grain rotation or breakage, grain nucleation and the laws of kerogen chemical kinetic transformation. Also used are the laws of fracture nucleation, extension, aperture dynamics and the kinetics of cement infilling.

All these geological input data and physico-chemical calculations are integrated in Basin RTM over many time steps to arrive at a prediction of the evolution history and the present-day internal state of a basin or field. In this way, the physico-chemical laws are used to translate the geological input data into a prediction of the changes of internal basin state and its evolution over a basin's history from its inception (or other chosen initial state) to the present.

Numerical approach

We use the updated Lagrangian approach to solve the time-dependent large deformation problem for geological materials satisfying the incremental stress rheology (Tuncay *et al.* 2000a). In our numerical approach, all variables are referred to the updated configuration in each time step. The approach has two major steps. First, the incremental stress rheology equations are solved at the integration points of the finite elements. Second, the displacements are computed by using a global deformation solver. At each time step, iterations of these two steps are performed until the norm of the change in displacements between two consecutive iterations is less than a specified tolerance. The two-step solution technique allows the introduction of new deformation processes with only minor changes in the code. We use the conjugate gradient iterative technique with a simple diagonal preconditioner to solve for the incremental displacements. The finite element code and iterative solver are parallelized. The details of the finite element formulation are provided in Tuncay *et al.* (2000*a*).

In the multi-phase module, the Galerkin-type finite element approximation is used for saturations, concentrations, and pressures. The non-linear terms and boundary conditions are treated in a fully implicit manner. An upwinding method is developed and implemented in the multi-phase module to stabilize the saturation fronts. The mass matrices are lumped to increase the stability as suggested in previous studies (Huyakorn *et al.* 1994; Abriola & Rathfelder 1993). The computer model accommodates a wide variety of boundary conditions. Because of the highly non-linear behaviour of the equations, a Newton–Raphson technique is employed to solve the non-linear algebraic equations arising from the discretization.

The finite element grid accretes as sediment is added. A new grid layer is added when sediment input causes the top grid layer to become too thick. In contrast, when erosion creates a top layer that is locally too thin, the finite element grid is locally reorganized to preserve numerical accuracy. This accreting, reorganizing grid that also adapts to sedimentary features as they are added or subtracted is required to capture sedimentary detail and ensure numerical stability and accuracy.

The interaction of the top of the sediment pile with the overlying fluids (atmosphere or sea bottom) is accounted for by the value of normal stress and the (assumed) absence of tangential shear. A no-shear lateral boundary condition allows for natural compaction at the sides of the basin. Lateral compression/extension and subsidence/upheaval are imposed at the sides and bottom. The sides and bottom are assumed to be impermeable to fluid flow.

All computational modules are packaged in an overall structure that ensures each equation is satisfied at every time step. The time step is allowed to change to ensure accuracy and computational efficiency.

Applications of Basin RTM to sedimentary basins

Other basin models have been used to gain valuable insights into the petroleum system. Typically these models are 2D and treat only two or three processes (Ungerer *et al.* 1990; Larson *et al.* 1993; Maubeuge & Lerche 1993, 1994; Roberts & Nunn 1995; Schneider *et al.* 1996; Luo & Vasseur 1996; Luo *et al.* 1998; Wang & Xie 1998). Consequently, they are not sufficiently comprehensive to accurately simulate natural sedimentary systems and, in particular, fracture networks.

Fracture-mediated petroleum expulsion, migration, and escape from reservoirs are key aspects of the dynamic petroleum system (Figs 3–5). In most of the existing basin evolution models, it is assumed that rocks fracture when the fluid pressure exceeds a specified fraction of lithostatic stress (Maubeuge & Lerche 1993, 1994; Roberts & Nunn 1995; Wang & Xie 1998). This assumption ignores the dependence of fracturing on lithologic properties. In our studies, we find that fracturing strongly depends on rock composition and texture, including mineralogy, grain size, and porosity which indirectly affect the stress tensor (Figs 6 & 7). Another limiting assumption is that there exists a simple dependence of porosity on effective stress (Ungerer *et al.* 1990; Maubeuge & Lerche 1993, 1994; Roberts & Nunn 1995; Luo & Vasseur 1996; Schneider *et al.* 1996; Wang & Xie 1998). This results in a very smooth porosity profile but ignores the dependence of porosity on rock composition and texture. In our approach, porosity is obtained by both solving the mass balance equation for solids and computing rock deformation velocity using a multi-process, incremental stress rheology that contains the elastic and viscous parameters which are functions of texture and composition. By coupling mass balance algorithms with our porosity and stress solvers, porosity and stress are computed self-consistently. As a result, shales usually have lower porosity, and higher least compressive stress than sandstones (Fig. 7). The small grain size combined with low porosity results in very low permeability and thus these layers can form efficient seals. Furthermore, our results show that low shear viscosity/bulk viscosity ratio makes fracturing very unlikely in the absence of flexure or extreme overpressuring mechanisms, such as petroleum generation or fluid thermal expansion.

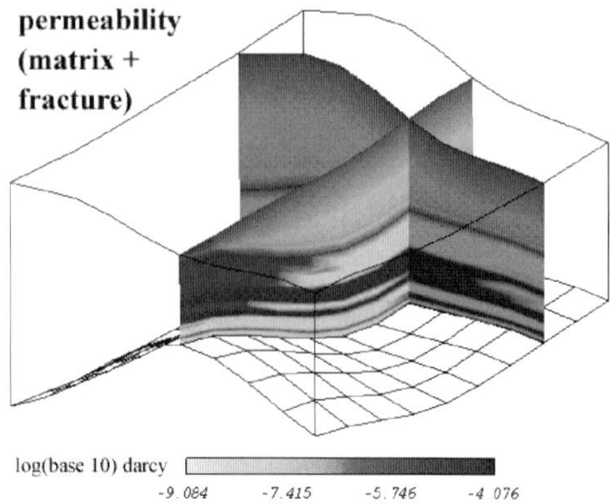

Fig. 6. Predicted cross-sections of permeability at 45 Ma from a simulation of Piceance Basin, Colorado. The system is 50×48 km wide and 1 to 3 km deep. Different permeabilities reflect varying sediment compositions including porosity, texture and mineral grain sizes, plus that resulting from fractures. The computation was carried out on an accreating, lithology-adapted hexahedral grid which at the final time was 20x20x60 elements. The uplift/subsidence, erosion and sedimentation histories were computed using tops, core, and well log data provided by DOE and Barrett Resources. Lateral compression/extension effects were not accounted for in the simulation.

In other basin evolution simulators, fracture permeability is assumed to be isotropic. This simplification ignores the fact that, for an accurate description of fracture orientations, the full stress tensor as well as irreversible (e.g., viscous-like rather than purely elastic) behaviour must be known (Fig. 8). Also disregarded is the interaction between the deformation/stress computation and fracturing. Rocks fracture due to the difference between fluid pressure and least compressive stress. But as fractures open, overall rock volume increases and fluid pressure in the fractures compresses the rock, increasing the compressive stress normal to the fracture plane. Thus, fracturing is a self-limiting process: firstly, as fractures open, they provide a pathway for fluid escape; and secondly, the volumetric strain caused by fractures increases the confining stress that reduces the rate of fracture growth (Tuncay *et al.* 2000*a*,*b*).

Although one- and two-dimensional studies give hints about the dynamics of basin evolution, a three-dimensional basin simulator is necessary to take three-dimensional geometric effects into account (Figs 6 & 9). This becomes extremely important when fracturing is due mainly to flexure and the direction of tectonic compression/extension is changing during the basin's history. Fracture networks provide a pathway for

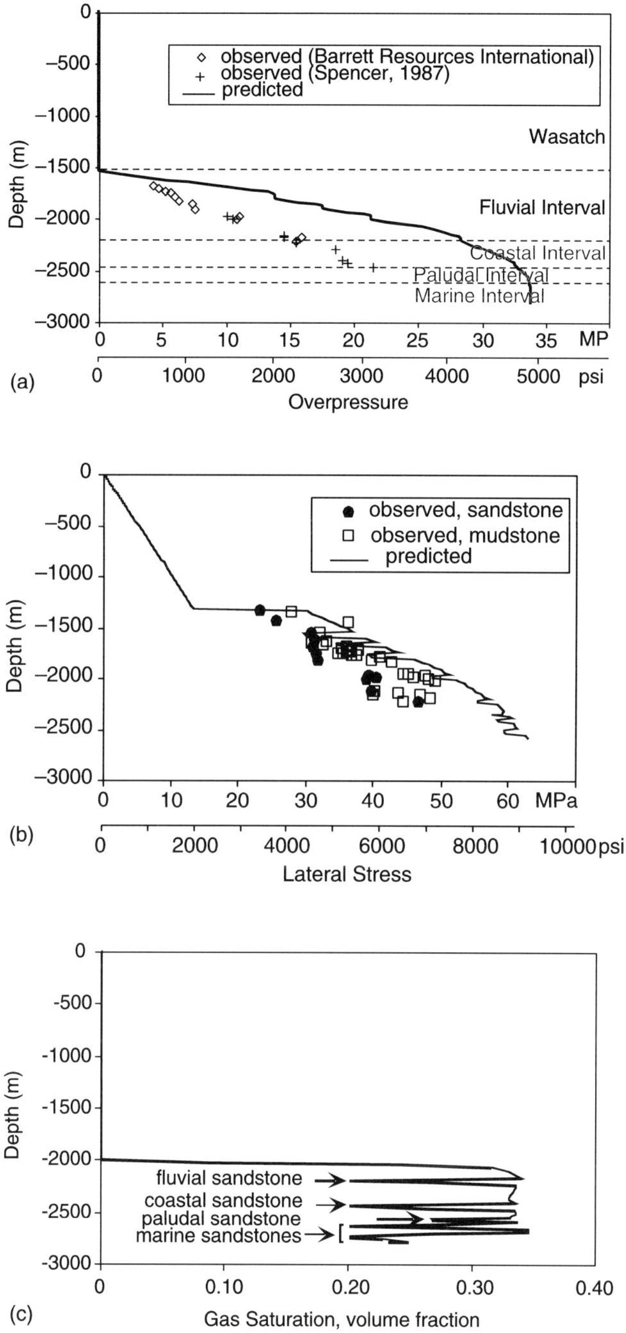

Fig. 7. Simulations by Basin RTM show good agreement with observations. Shown are present-day fluid pressure (**a**) and least compressive stress (**b**) at the MWX site, Piceance Basin, Colorado, compared with observations. In sandstones, the lateral stress and fluid pressures are found to be similar, indicating their vulnerability to fracturing whereas in the mudstones lateral stress exceeds fluid pressure, underscoring the lack of fracturing in them at present day (as observed). Our predictions also show that this situation is dynamic: during an epoch in the past, some of the mudstones were also fractured in some areas in the Piceance Basin as suggested in the literature. (**c**) Predicted natural gas saturation (from Payne *et al.* 2000).

fluid flow and, especially in layered rocks, fluids can move laterally out of the vertical plane. Appropriate simulations can be accomplished only by a three-dimensional basin simulator with a stress/deformation solver that can capture non-planar effects and strong contrasts in rheology from layer to layer.

Although the history of temperature, sedimentation rate, and subsidence rate are among the most important parameters that affect the evolution of a basin, some basin simulators start with a pre-defined grid (Schneider *et al.* 1996). In other words, two basins with the same final thickness and lithology but different time history are assumed to behave similarly. This approach ignores the time-dependence of, for example, overpressuring which often correlates with sedimentation rate and is a key factor in fracturing as well as deformation (Wang & Xie 1998; Ortoleva 1998).

Another limitation of previous models is the absence of shear failure. Only a few models consider the contribution of mechanical shear to rock failure (Luo *et al.* 1998; Larson *et al.* 1993). However, these models are limited to two dimensions. In our model, a Drucker-Prager-type failure function is used to build in a shear failure via a transition in rock viscosity. The coefficients of the failure function have been fitted with experimental data (Sakrani 1996; Ozkan *et al.* 1998). The shear viscosity is assumed to depend on the failure function and texture. As the rock approaches failure, shear viscosity decreases. This feature extends the applicability of our basin simulator to fault formation problems (Fig. 10). We have already made an attempt to include the rate of strain due to gouge (Fig. 11) to show the effect of shearing on the particle size distribution and porosity (Ozkan *et al.* 1998; Ozkan & Ortoleva 2000).

Our model accounts for the changing rock rheological parameters that accompany the changing lithologic and fracture properties. The bulk, shear and effective stress coefficients of the (assumed) isotropic rocks are computed using Berryman's composite medium theory (Berryman 1980, 1986). In our approach, by assuming that the shear and bulk viscosities depend on grain contact areas, rheology is computed as a function of the mechanically and diagenetically modified texture. Thus, dynamic rheologic properties are used to compute basin stress/deformation history using an incremental stress rheology (Zienkiewicz & Cormeau 1974; Rice 1975) extended by the authors(Tuncay *et al.* 2000a,b; Ortoleva 1994*a*, 1998).

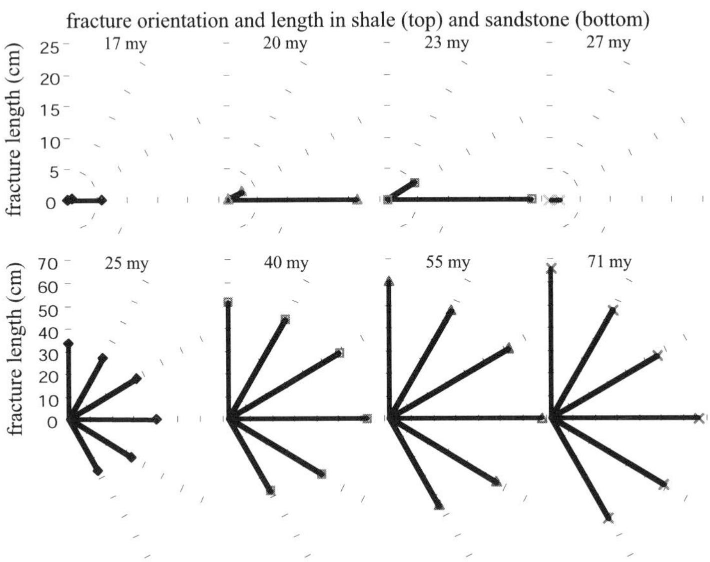

Fig. 8. Predicted fracture orientations and length in a shale (top) and a sandstone (bottom) from the Piceance Basin simulation of Fig 6. Changing sediment properties, stress and fluid pressure during the evolution of the basin result in dynamic fracture patterns, which in turn significantly affect the anisotropy of fracture permeability. If the enhanced permeability from fracturing is significant, it can direct the flow of petroleum; understanding such occurrences of the past, therefore, can be important for identifying or understanding reservoirs in presently unlikely structural and stratigraphic locations.

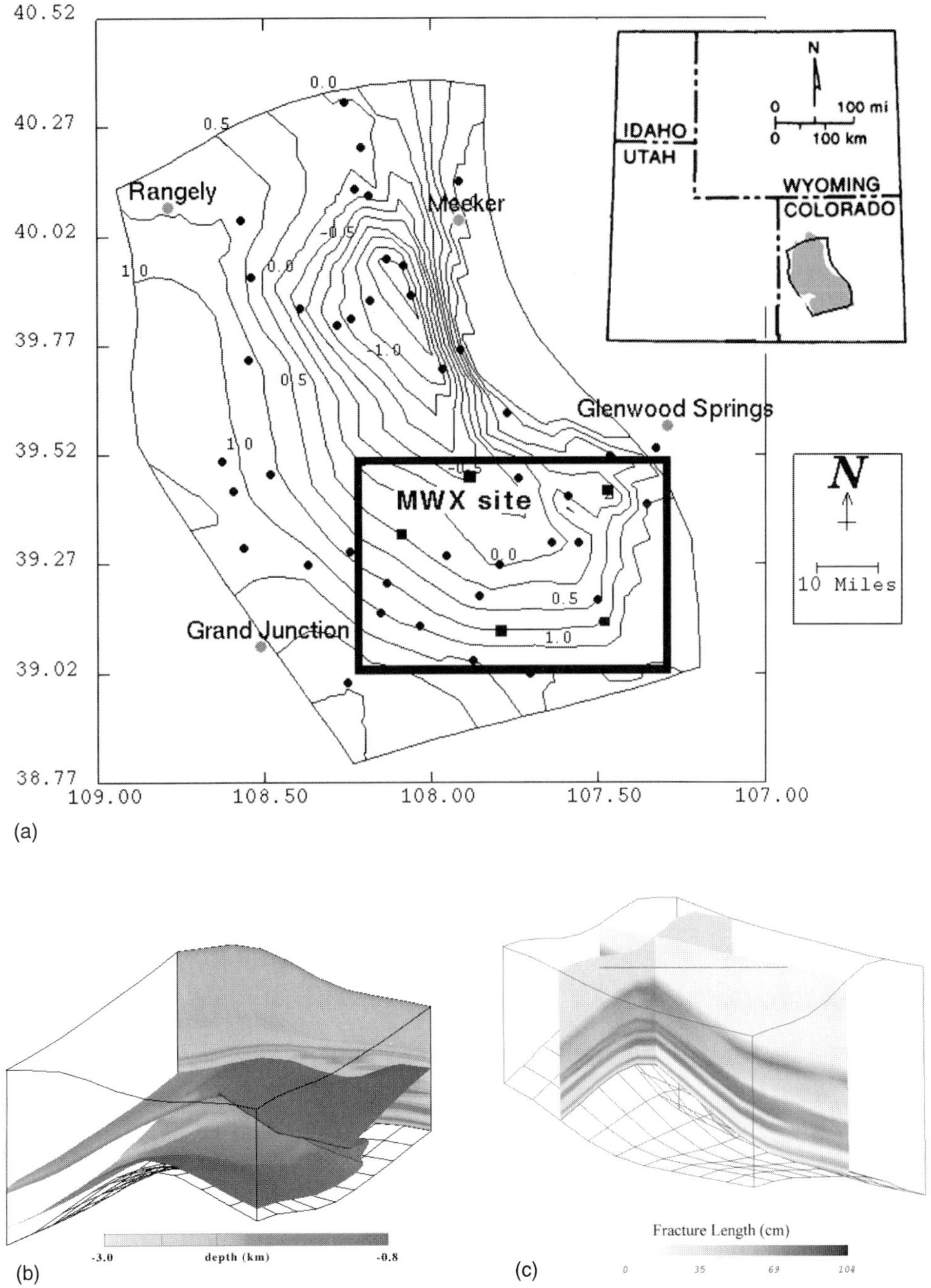

Fig. 9. (**a**) Simulation domains for basin-scale and inter-field studies (thickly outlined box) of Piceance Basin simulations. The Rulison Field is in the upper northwestern area of the latter box. (**b**) Isosurface of overpressure (15 bars) shaded with depth. The folded, multi-layered structure is dictated by the interplay of lithological differences and fracturing and shows the three-dimensional complexity of conductivity of overpressured zones. Thus, stacked overpressured compartments as viewed as a simple pressure–depth curve may hold little insight into the full three-dimensionality of the structure. (**c**) The distribution of fracture length reflects lithologic variation and the topography imposed by the basement tectonics. The layered fracture length structure is closely related to the layering in overpressure surface (from Tuncay *et al.* 2000*a*).

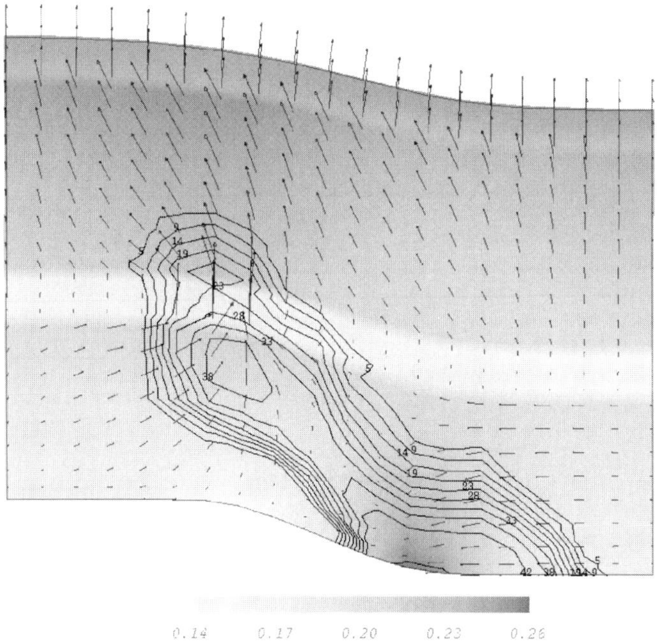

Fig. 10. Cross-section of a 2D, 5×2.5 km normal fault system after 5 My of simulation. The shading indicates porosity and shows differences between the four lithologies; the shales (low porosity) are at the middle and top of the domain. Higher porosity regions (in the lower-right and upper-left corners) and the fracture length (contour lines) arose due to the deformation created by differential subsidence. Both stress field and fracturing are strongly affected by rock composition and texture. The arrows indicate fluid flow toward the region of increasing porosity (lower-right) and through the most extensively fractured shale.

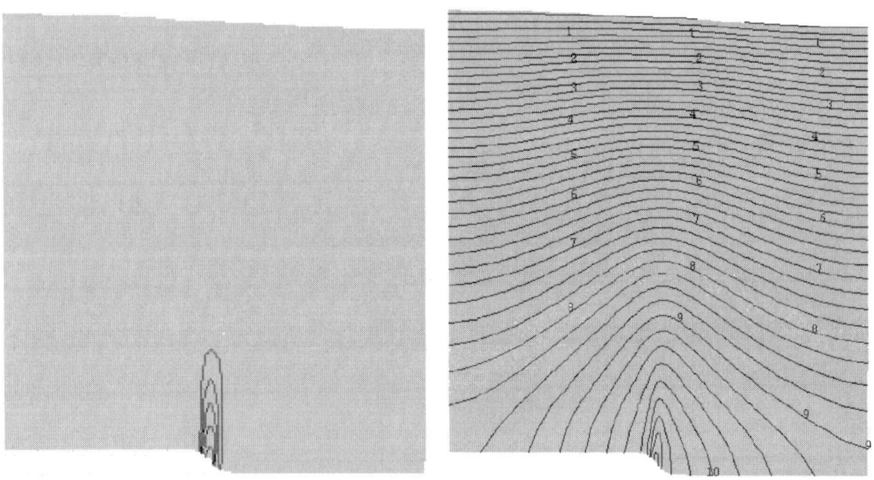

Fig. 11. The effect of gouge on **(a)** porosity and **(b)** overpressure. The size of the domain is 3 km by 3 km in cross-section. The fault width is 0.1 km and the rate of faulting is 0.2 km My^{-1}. To isolate the effect of gouge on the pressure and texture, a very high bulk viscosity is used. Therefore, the only overpressuring mechanism present is the gouge-related changes in the texture. The weakness of the San Andreas Fault is attributed to high fluid pressures in the fault zone (Byerlee 1990; Rice 1992). This example demonstrates fluid flow and deformation interactions in basin environments.

Salt tectonics

Salt movement in the subsurface can dramatically affect the distribution and character of fracture zones and faults for several reasons:

- salt diapirism and related phenomena can cause flexure and otherwise dramatically affect the stress regime in the neighbouring sediments;
- salt movement can affect the topography of the sea floor and thereby the geometry and lithologic character of the salt-adjusted sediments;
- salt is of vanishingly small permeability and therefore can isolate high overpressure (up to lithostatic) and thereby influence effective stress in the underlying sediments; and
- by distorting the heat flux, changing the salinity, and altering the fluid flow patterns, salt can change the diagenetic history and, thereby, rock mechanical properties.

As salt withdrawal is believed to be an important factor in fracturing in some areas, we have incorporated advanced salt tectonics modelling into Basin RTM by calibrating our rheology to be consistent with salt deformation experiments.

With this, our simulator can address the following exploration and production challenges:

- predict the location and geometry of faults or zones of fracturing created by salt motion;
- predict the morphology of sedimentary bodies created by salt deformation;
- locate pools of petroleum or migration pathways created by salt tectonics; and
- assist in the interpretation of seismic data in salt tectonic regimes.

The interplay of salt deformation with the rheology of the surrounding strata is key to understanding the salt deformation and reservoir location and characteristics relationship. The continuous and discontinuous (i.e. fauting and fracturing) responses of the surrounding sediments are induced by salt movement. In turn, the deformation of these sediments promotes or inhibits salt motion.

Figure 12 shows a cross-section of a Basin RTM-simulated salt wave. Note the relationship among deformation, fluid flow and fracturing. The orientation of fracturing follows salt morphology. Fluid flow patterns are strongly influenced by salt morphology and associated fracture patterns. In this case and those to follow, the salt deformation is due to the interplay of salt buoyancy, salt modification of the distribution of the rate of sedimentation and the overall tectonics (i.e. basin extension/compression and upheaval/subsidence).

To predict salt tectonics-related petroleum pools, one must co-evolve petroleum generation/

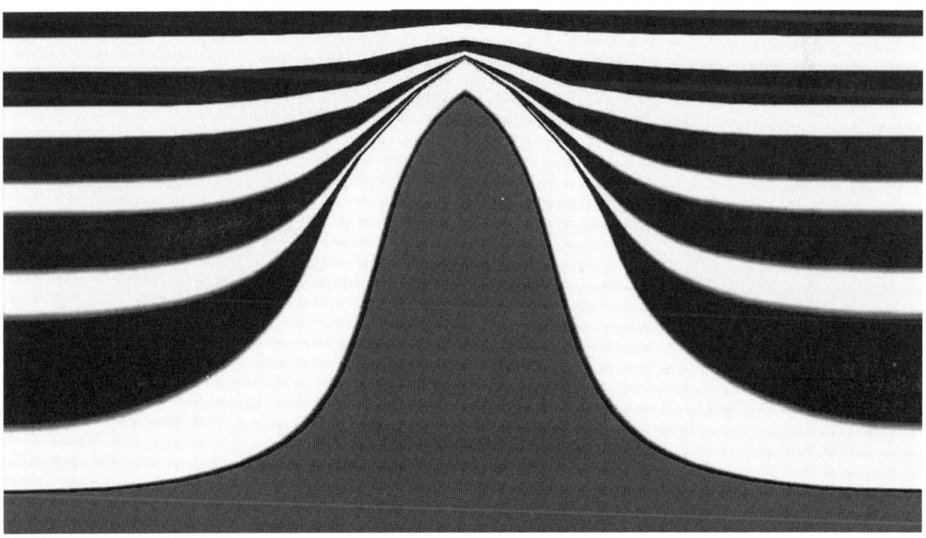

Fig. 12. Basin RTM simulated salt wave showing associated deformation (as indicated by the shape of the evolved salt body). Shown is the state after 8 My of deformation and sedimentation. The salt bed originally only had a 400 m topographic contrast. There are alternating layers of sandstones and shales above the salt at this stage. The salt tectonic – sedimentation interaction induced the sediment geometry and the generation of mini-basins at the flanks of the wave.

expulsion/migration with salt and rock deformation. Figure 13 shows an oil pool developed in association with salt diapirism. The oil was generated in a source rock overlying the salt, both of which were essentially flat in an early stage that preceded diapirism. A similar situation but for a subsalt source rock is seen in Figure 13. In both cases, the influence of salt motion on fracturing and porosity strongly affects migration and ultimate reservoir quality. The escape of subsalt oil at the edges of the salt lens is also seen in Figure 14. Salt tectonics is, for the most part, a fundamentally 3D phenomenon. An example of a simulated salt diapir in 3–D view is seen in

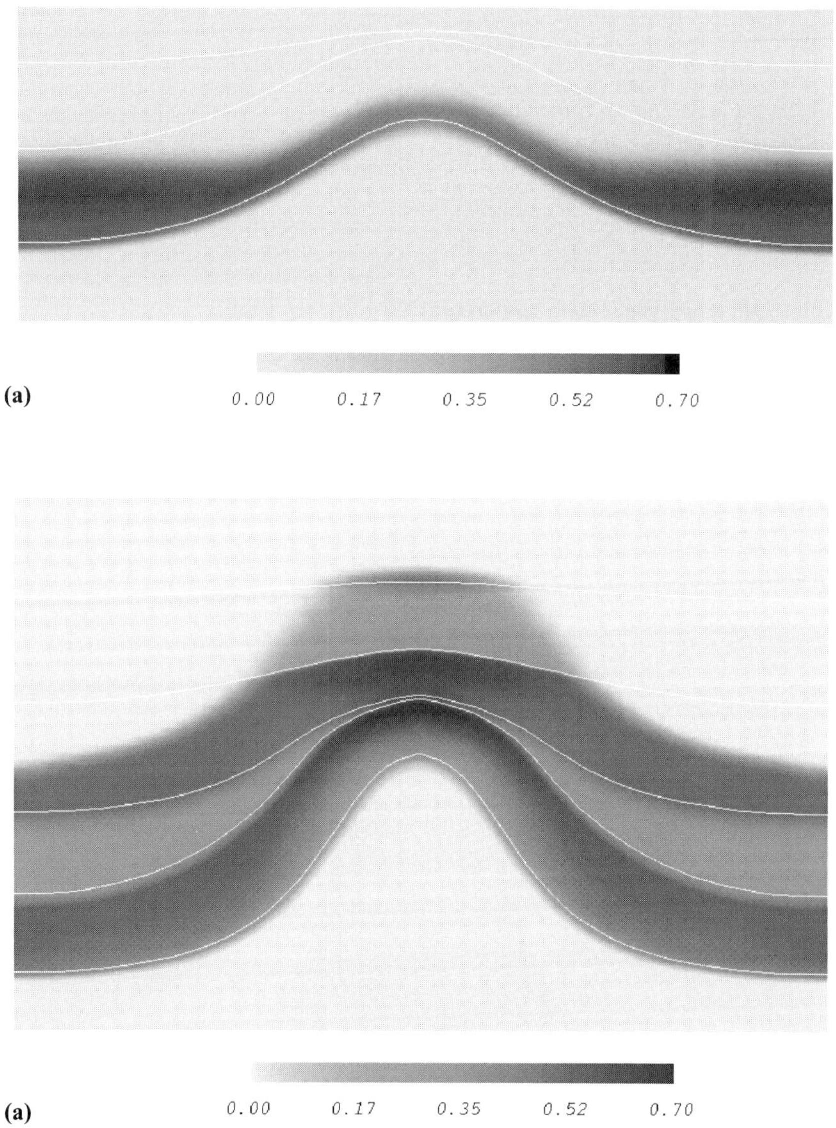

Fig. 13. Simulated time sequence (1.46 (**a**) and (**b**) 3.76 My into the simulation) of oil saturation overlying a rising salt dome. Source rock overlying the dome was transiently overpressured and fractured, facilitating upward oil migration within it and into the overlying layers. Orientations of long-lived fractures (residing in the sandstones) illustrate the relationship between the salt motion and fracture pattern. The location of oil strongly depends on the rate of salt motion and the coevolving mechanical and transport properties of the adjacent sediments.

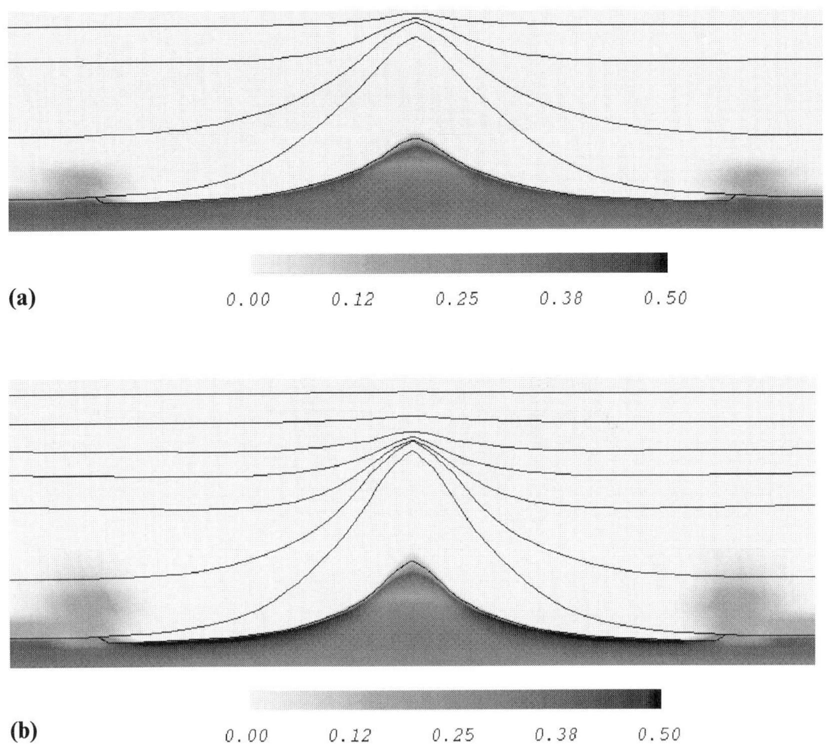

Fig. 14. As in Figure 13 except for an initially finite size (lenticular) salt body and in addition the co-evolution of subsalt petroleum (3.12 (**a**) and 5.56 (**b**) My into the simulation). Shown is the oil saturation with curves indicating lithologic contacts. The overpressure under the salt body and the stress regime on the underlying sediment have preserved porosity in the centre region under the salt while the compaction under the edge of the salt led to the formation of a seal. Thereby a subsalt compartment is formed.

Figure 15. Note the pockets of fracturing associated with the salt diapir. These and other Basin RTM simulations of salt phenomena illustrate the ability of our approach to predict fractures related to salt tectonics. The details of the salt tectonics simulations are provided in Tuncay & Ortoleva (2000).

Incremental Stress Rheology, Statistical Fracture Dynamics and Rock Competency

Incremental stress rheology

The strongly coupled nature of the crustal deformation problem can be captured using an incremental stress rheology. The specific rheology used in our modelling integrates most of the strain mechanisms believed to operate in the crust. It has the form (Tuncay *et al.* 2000*a*,*b*; Ortoleva 1994*a*, 1998; Ortoleva *et al.* 1997)

$$\underline{\underline{\dot{\varepsilon}}} = \underline{\underline{\dot{\varepsilon}}}^{el} + \underline{\underline{\dot{\varepsilon}}}^{vp} + \underline{\underline{\dot{\varepsilon}}}^{ps} + \underline{\underline{\dot{\varepsilon}}}^{fr} + \underline{\underline{\dot{\varepsilon}}}^{go} \qquad (1)$$

Here, $\varepsilon^{\dot{Y}}$ is the net rate of strain while the terms on the right-hand side denote the contributions of five processes: poroelasticity (el), continuous viscoplastic (vp), pressure solution (ps), fracturing (fr) and gouge (go). Specific expressions for each term have been taken from the literature. In this study we consider poroelasticity and continuous viscoplasticity.

The poroelasticity rate of strain $\dot{\varepsilon}^{el}$ may be expressed in terms of stress $\underline{\sigma}$, pressure p of the (assumed single) fluid phase, and rock texture Θ via:

$$\underline{\underline{\dot{\varepsilon}}}^{el} = \underline{\underline{\underline{\underline{C}}}}^{-1}(\Theta)\frac{D}{Dt}\left(\underline{\underline{\sigma}} + \alpha(\Theta)p\underline{\underline{I}}\right) \qquad (2)$$

where $\underline{\underline{\underline{\underline{C}}}}$ is the fourth rank tensor of poroelastic coefficients, α is the effective stress coefficient, and I is the second rank identity matrix. Here, D/Dt represents a material time derivative measuring the rate of change of a tensor in time

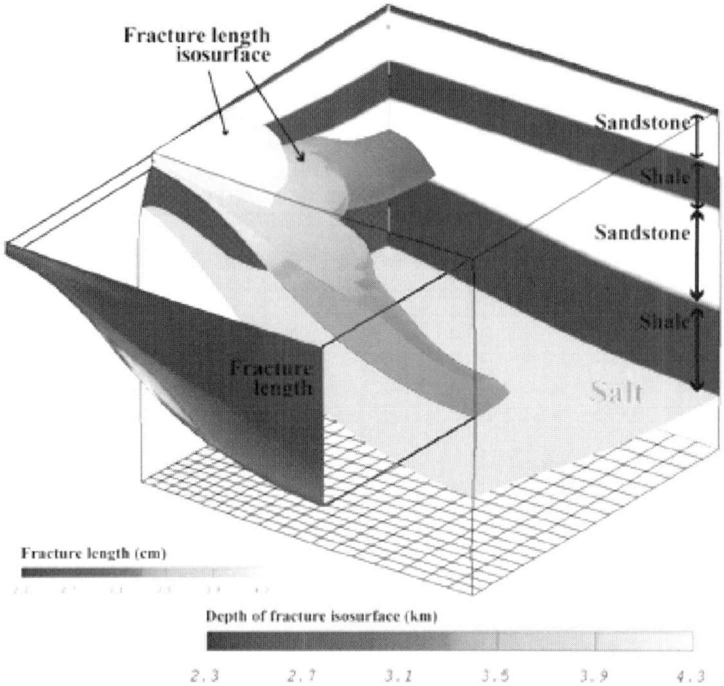

Fig. 15. Simulated quarter section of a salt diapir showing the relationship to fracturing in the overlying sandstones after 3 My of deformation. Extracted partial front cross-section shows the fracture length.

with respect to a local reference frame fixed to a translating, rotating material volume element. The texture Θ represents a set of variables characterizing the mineralogy, shape, size, orientation and packing of the grains of each mineral. The bulk and shear moduli of the drained porous medium, and the effective stress coefficient of the medium can be computed in terms of the Θ and mineral elastic properties using Berryman's (1980, 1986) approach.

The inelastic mechanical contribution to $\dot{\varepsilon}$ is cast in the present approach as a non-linear viscosity law in the form:

$$\dot{\underline{\underline{\varepsilon}}}^{(vp)} = \underline{\underline{\underline{\eta}}}^{-1}\left(\underline{\underline{\sigma}} + \tilde{\alpha} p \underline{\underline{I}}\right) \tag{3}$$

The fourth rank viscosity tensor $\underline{\underline{\underline{\eta}}}$ depends on stress, fluid pressure and texture. The second term in the effective stress involves a coefficient that is usually taken in the literature to be unity. The shear viscosity is assumed to be a strong function of rock competency. Sleep & Blanpied (1994) assumed that the viscosities depend on the intrinsic viscosity of grains and the volume fractions of pores and fractures. However, they did not consider the sudden drop in the shear viscosity due to failure, which is one of the key parameters required to model the stress drop following rupture. We expect that in the case of failure, the change in the shear viscosity is greater than the change in the bulk viscosity. Since the state of stress is strongly dependent on the viscosity ratio, this feedback might play an important role in the reorganization of fault zones after failure.

The texture must be further augmented when rock competency is considered. This is described in the following section. Again, in the spirit of our Markov approach, only when it is sufficiently comprehensive can a self-consistent deformation model be set forth. One expects that $\dot{\varepsilon}^{(j)}$ (j=poroelasticity, viscosity, pressure solution, fracturing, gouge, etc.) should depend on all the aforementioned variables . With this:

$$\dot{\underline{\underline{\varepsilon}}} = \sum_{j=1}^{N_d} \dot{\underline{\underline{\varepsilon}}}^{(j)}(\Theta, \underline{\underline{\sigma}}, p, c, T) \tag{4}$$

for a system with N_d deformation mechanisms. The dependence of the strain rates on state clarifies the central role of incremental stress theory in integrating all the crustal RTM processes into a unified model. It is the coupling allowed by this integration that underlies fault dynamics.

The total rate of strain $\dot{\varepsilon}$ is defined as:

$$\dot{\varepsilon}_{ii'} = \frac{1}{2}\left(\frac{\partial u_i}{\partial x_{i'}} + \frac{\partial u_{i'}}{\partial x_i}\right) \qquad (5)$$

for rock deformation velocity $\underline{u}$ and Cartesian coordinates *xi, xii, xiii*. The six independent components of the symmetric second rank tensor (Equation 2) must be supplemented with three additional equations so that the three deformation velocity components ($\underline{u}=u_1,u_2,u_3$) and the six independent stress components can be determined. The required condition arises from force balance:

$$\sum_{i'=1}^{3} \frac{\partial \sigma_{ii'}}{\partial x_{i'}} + f_i = 0 \qquad (6)$$

for body force f_i which is given by:

$$f_i = g\rho_m \delta_{i3} \qquad (7)$$

Here g is the gravitational acceleration, ρ_m is the crustal mass density, and the subscript $_3$ denotes upward direction.

In addition to the coupling of deformation to other phenomena through the incremental stress rheology, there are numerous indirect couplings. For example, rock properties such as permeability, multi-phase flow parameters, reactive grain surface area and thermal conductivity depend strongly on texture. As the latter is affected by stress and deformation, a complex network of coupling relations is thereby expressed. For further discussion of the consequences of this network, see Ortoleva (1994*a*,*b*, 1998), Tuncay *et al.* (2000*a*,*b*) and Dewers & Ortoleva (1994).

Rock Competency

Predicting faulting and other rock failure phenomena requires a model that accounts for rock competency. We have introduced a measure of rock competency that accounts for the degree to which grains are attached to each other (Tuncay *et al.* 2001). Schematically, the theory is as follows.

Let Γ, rock competency, measure the fraction of grain surface that is attached to other grains. Thus Γ is in the range $0 \leq \Gamma \leq 1$. Large Γ implies competency while in a low Γ rock there are few intact grain–grain contacts. Thus rheologic quantities such as rock strength or viscosity are strongly dependent on Γ.

Schematically, our model is as follows. The equation of motion of Γ is taken in the form:

$$\frac{D\Gamma}{Dt} = R(\Gamma, F) \qquad (8)$$

where F is a failure function that depends on macroscopic stress, fluid pressure, rock texture, mineralogy and temperature. In three dimensions, the failure function is assumed to take the form (Drucker & Prager 1952):

$$F = aJ_1 + \sqrt{J_2} - b \qquad (9)$$

where J_1 is the first invariant of the effective stress tensor and J_2 is the second invariant of the deviatoric effective stress tensor. The coefficients a and b can be expressed in terms of angle of internal friction φ and cohesion C determined from conventional triaxial compression experiments (Desai & Siriwardane 1984):

$$a = \frac{2\sin\varphi}{\sqrt{3}(3-\sin\varphi)} \qquad (10)$$

$$b = \frac{6C\cos\varphi}{\sqrt{3}(3-\sin\varphi)} \qquad (11)$$

Here, we assume that cohesion depends on rock competency. For an intact rock ($\Gamma=1$), b is large. As the rock competency is lost, the cohesion-like term b vanishes. Therefore b is a strong function of Γ taken here to be $b=b^*\Gamma^n$ where b^* refers to the value when $\Gamma=1$ and n is a phenomenological exponent.

If the dynamics of Γ is relatively fast, its evolution is closely related to the shape of the curve $R(\Gamma,F)=0$ (see chapters 2 and 3 of Ortoleva (1992) for further discussion). The Γ dynamics is, in a sense, a cooperative phenomenon, i.e. a decrease in competence fosters more rapid decline. This is captured by the qualitative picture of Figure 16 where a schematic evolution path in the F,Γ plane is shown.

Through this model, failure is rapid while healing can be a much slower process. This follows if R is relatively small when Γ is small and F is less than a healing value F_h. This type of effect gives geological materials the memory they

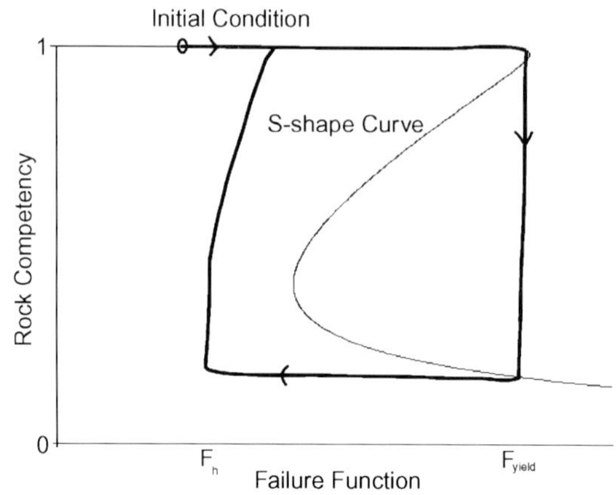

Fig. 16. Schematic competence (Γ) and failure function (*F*) plane illustrating the cooperative aspects of rock failure. When Γ is near unity, the rock is competent but when *F* exceeds F_y, it is compromised. However, for $F<F_h$, competence is regained through chemical healing processes.

have of zones of earlier faulting. Here, $R(\Gamma,F)$ is taken in the form

$$R(\Gamma,F)=k(\Gamma,F)\left(\frac{-F+d_3(\frac{1}{2}-\Gamma)+}{\frac{d_2}{\Gamma}+\frac{d_1}{1-\Gamma}+d_4}\right)\Gamma(1-\Gamma) \quad 12$$

where d_1,d_2,d_3 and d_4 are material constants. The function k is chosen such that when Γ is small and $F<F_h$, k is small, ensuring that healing is slow. Finally, F_h can be determined in terms of d_1,d_2,d_3 and d_4. Conceptually, these parameters depend on mineralogy, grain size, shape and packing.

We now show that the feedback associated with the coupling of shear stress (via incremental stress rheology) and rock competency naturally supports autonomous oscillation. Consider a simple shear system with the total rate of strain given by the sum of poroelastic and non-linear viscous contributions. The rate of strain due to poroelasticity can be written in terms of the total rate of strain and rate of strain due to viscosity as $\dot{\varepsilon}^{cl}=\dot{\varepsilon}-\dot{\varepsilon}^{vp}$. Assuming isotropic simple shear, the following ordinary differential equation for the shear stress is written:

$$\frac{D\tau}{Dt}=\frac{G_f}{\mu(\Gamma)}(2\dot{\varepsilon}\mu(\Gamma)-\tau) \quad (13)$$

where τ, $\dot{\varepsilon}$, G_f and μ are the shear stress, total rate of shear strain, elastic shear modulus and shear viscosity, respectively. Shear viscosity is taken to be an increasing function of rock competency $\mu=\mu^*\Gamma^m$. The exponent m is taken as 8 to capture the large change in the order of magnitude of shear viscosity between intact and failed rock. For simplicity, assume that the failure function is in the form $F=|\tau|-c$. In this case c is taken to be $c=-a(\sigma_n+p)+b^*\Gamma$ where σ_n and p are the normal stress (positive in tension) and fluid pressure, respectively. The coefficient b^* is taken to be 50 MPa within the range of typical rock properties. Note that for small values of Γ, $F \approx |\tau|+a(\sigma_n+p)$, capturing the usual friction law. The term $-a(\sigma_n+p)$ is taken as 30 MPa. The rate of shear strain is either taken to be a specified function of time or determined by the energy dissipation condition

$$\tau\dot{\varepsilon}=E \quad (14)$$

where $\dot{E}$ is the rate of energy input.

Equations (8) and (13) form a strongly coupled set of non-linear ordinary differential equations for the shear stress and rock competency that were integrated numerically by the fourth-order Runga-Kutta technique with adaptive time stepping. Setting the RHS (Equation 13) to zero yields for constant $\dot{E}$ and the Drucker Prager failure function (Equation 9):

$$\Gamma=\left(\frac{(F+c)^2}{2\mu^*\dot{E}}\right)^{1/m} \quad (15)$$

Fig. 17 illustrates the null curves for different values of $\dot{E}$. For small $\dot{E}$ the stress null curve (Equation 15) intersects the S-shaped null curve $R=0$ on the upper stable branch near $\Gamma=1$ (competent rock), i.e. the competent rock viscosity μ^* is sufficiently low that the rate of energy input is equal to the rate of viscous energy dissipation (ductile flow). For very large $\dot{E}$, rock fails but the null curves intersect at the lower stable branch, i.e. the rock remains failed because of the very low shear viscosity needed to dissipate the required energy (aseismic faults). For intermediate values of $\dot{E}$, the null curves intersect at the unstable branch of $R=0$. In this case, for a constant $\dot{E}$, rock fails and heals cyclicly (seismic faults). It is also possible that the null curves intersect at three distinct points. In this case, depending on the initial conditions of shear stress and rock competency, rock will either never fail or never heal. If a point F,Γ is above the null curve given by Equation 15, the shear stress increases, otherwise it decreases. Similarly, if a point (F,Γ) is to the right of the null curve $R=0$, the rock competency decreases. These four cases suggest a classification of fault dynamics: intact stable sliding, failed stable sliding, multiple state and oscillatory sliding.

The above model can support autonomous oscillatory failure reminiscent of the earthquake cycle when the null curves intersect on the unstable (middle) branch. Figure 18 shows the F,Γ plane for an oscillation-supporting set of parameter values ($\dot{E}=0.06$ GPa Ma^{-1}). The arrows show the direction of movement. Figure 18 illustrates the predicted time dependence of rock competency (Fig. 18a) , failure function (Fig. 18b) and shear stress (Fig. 18c). As seen in Figure 18, after a short transient period, the system finds its limit cycle. The stress drop after failure is quite large because of the use of Equation (14). It is likely that in two- and three-dimensional simulations, the stress drop will be less due to the interaction of failed and unfailed regions. Furthermore, the frequency of failure events is expected to increase since the energy release at each failure (stress drop) will be less.

Fracture Network Statistical dynamics

We have developed a model of the probability for fracture length, aperture and orientation (Tuncay *et al.* 2000*b*). The model predicts the evolution of this probability in response to the changing stress, fluid pressure and rock properties as the basin changes. The fracture probability is used to compute the permeability tensor. The latter affects the direction of petroleum migration, information central to planning infill drilling spacing and likely directions for field extension. It is a key to the design of horizontal wells and the optimum rate of production in stress-sensitive reservoirs. Finally, the predicted dis-

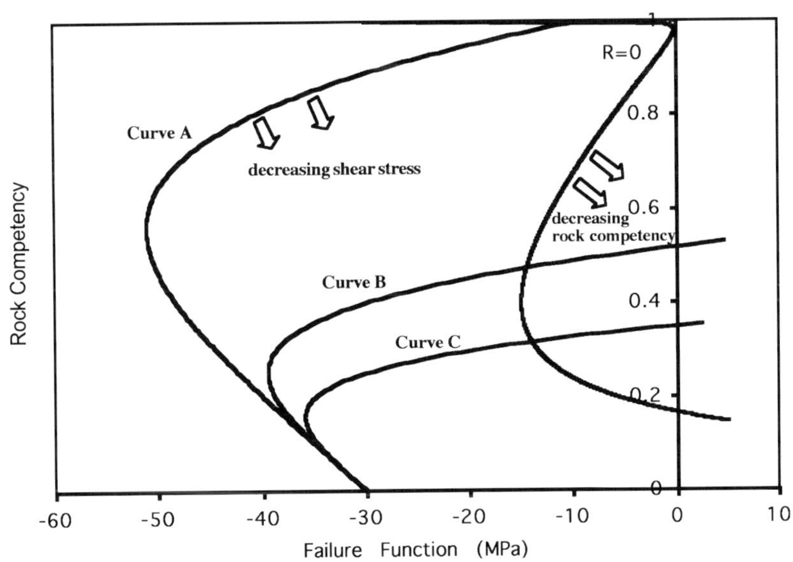

FIG. 17. The null curves and phase diagram for a time-dependent $\dot{E}$ (0–100 years, $\dot{E}=1$ GPa Ma^{-1}; 100–200 years, $\dot{E}=100$ GPa Ma^{-1}; 200–300 years, $\dot{E}=50$ GPa Ma^{-1}; 300–400 years, $\dot{E}=100$ GPa Ma^{-1}; 400–600 years, $\dot{E}=1$ GPa Ma^{-1}).

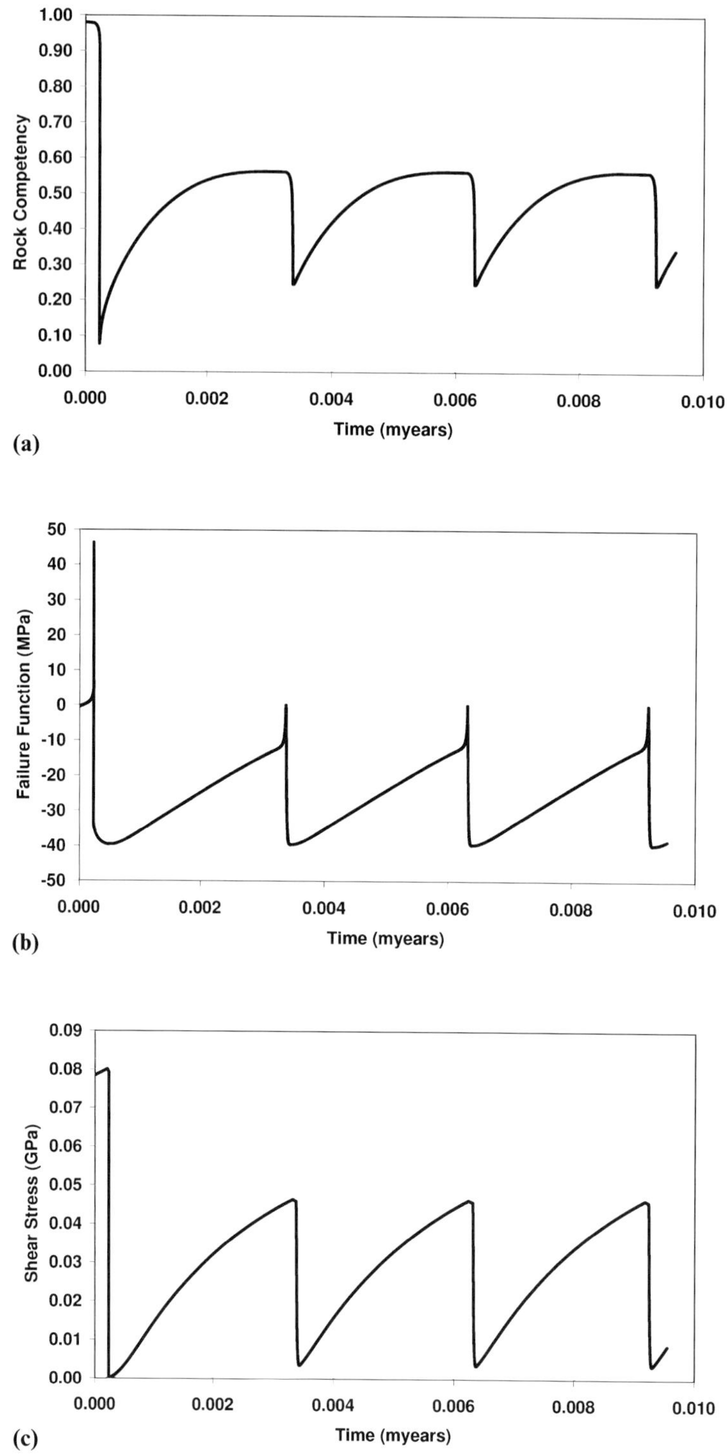

Fig. 18. History of **(a)** rock competency, **(b)** failure function and **(c)** shear stress. After a short transient period, the system finds its limit cycle.

tribution of fracture network statistics across a field is a necessary input to reservoir simulators used to optimize production.

The dynamics of the fracture network in our model is based on a statistical representation. For example, consider a set of fractures of length L with normal $\underline{n}$ for a 3D spectrum of normal orientations. Then the rate of change for L in the rock-fixed frame takes the form:

$$\frac{dL}{dt} = R(L, a, p, \Theta, \underline{\underline{\sigma}}) \tag{16}$$

where the fracture extension rate R depends on the normal stress σ, the wetting phase fluid pressure p and the texture Θ of the surrounding rock, and c is the aperture of the n-fracture. A similar equation for the fracture aperture is developed (see Tuncay *et al.* (2000*b*) for further details).

Let η' be the number density of sites at which fractures may nucleate. By definition of the undeformed rock $\eta'=\eta$ (being the original, depositional value of η') but η' can differ from due to changes in rock texture from diagenesis or mechanical processes. In the simplest case where fracture nucleation sites are not created or destroyed, η' obeys the conservation equation $\partial\eta'/\partial t+\nabla\cdot(\eta' u)=0$. In a macrovolume element of volume V there are $V\eta'$ fractures and hence a fracture void space $V\eta'\pi L^2{}_a$ where a and L are the aperture and length (radius) of the assumed penny-shaped fractures, respectively. To compute the dilatation, we focus on a fixed volume V_m of solids and follow its change in a time δt. The volume of the unfractured rock V_{unfr} is related to V_m and the porosity ϕ_m of the unfractured rock via $V_{unfr}=V_m+\phi_m V_{unfr}$. Hence, $V_{unfr}=V_m/(1-\phi_m)$. The total volume V of the sample of rock containing V_m is then:

$$V = (1-\phi_m)^{-1} V_m + V\Delta \tag{17}$$

where $\Delta=\eta'\pi L^2 a$. With this, the volume of rock $V(t)$ at time t for fixed volume of solids V_m (considered incompressible and not to expand thermally or react) is given by:

$$V(t) = V_m(1-\phi_m)^{-1}(1-\Delta) \tag{18}$$

Noting that

$$tr\dot{\underline{\underline{\varepsilon}}}^{fr} = \lim_{\delta t\to 0}\frac{V(t+\delta t)-V(t)}{V(t)\,\delta t} \tag{19}$$

one obtains

$$tr\dot{\underline{\underline{\varepsilon}}}^{fr} = [1-\Delta]^{-1}\frac{D\Delta}{Dt} \tag{20}$$

The tensor character of the fracture-mediated deformation is related to the directions of each fracture through its normal to the fracture plane. Consider the expression:

$$\dot{\varepsilon}^{fr}_{kl} = [1-\Delta]^{-1}\frac{D}{Dt}(\Delta n_k n_l) \tag{21}$$

Here D/Dt represents a material time derivative; however, now, it also must account for the rotation of the fracture normals as they change direction with flexure, shearing or other deformation. Note that the trace of this expression agrees with the earlier result for the dilatation. Finally, this expression agrees with simple cases wherein all fractures are parallel.

In our model, a finite (but representative) number of fracture orientations is accounted for. We use the fracture kinetics formulation of Ortoleva 1994*a*, and Sonnenthal & Ortoleva (1994). However, here we replace the least compressive stress in the formulation by the stress component normal to each fracture plane. This allows calculation of fracture length and aperture for each fracture orientation. For example, if we assume that only vertical fractures can occur as for a one-dimensional problem, since the stress component normal to any vertical plane is the same because of the symmetry, an isotropic fracture network develops. In three-dimensional problems, our proposed algorithm has the power to predict a complex fracture network with preferential orientations dictated by the structure of the stress tensor.

Since the fracture network is well defined, the anisotropic fracture permeability can be calculated approximately. The anisotropic fracture permeability of a fracture network consisting of a single fracture orientation is given by:

$$K_{ij}{}^{fr} = \lambda(\delta_{ij} - n_i n_j) \tag{22}$$

where $\underline{n}$ is the unit normal to the fracture plane and $\underline{\underline{K}}^{fr}$ is the fracture permeability. The parameter λ can be approximated by:

$$\lambda = \beta\phi_{fr}\frac{a^2}{12} \tag{23}$$

Here β is a factor accounting for the connectivity of fractures. For large fracture lengths and dense networks β approaches unity whereas for small fracture lengths and low fracture densities it vanishes (Oda 1986). Oda (1985, 1986) proposed that this coefficient should be a function of a dimensionless second-order tensor of fracture geometry. He called this tensor the fabric tensor (Oda 1982). In this study β is taken as unity. In our computation, the total fracture permeability is taken to be the sum of fracture permeabilities for all orientations. It is assumed that fluid flow is slow and the disturbance at fracture intersections is negligible. Summation of fracture permeability of different sets is inadequate when the fracture density is lower than the percolation threshold (Berkowitz 1995; Odling 1992; Bour & Davy 1998). Another limitation is due to the surface roughness of fractures. In this study fracture aperture is assumed to be constant in a particular fracture. The spatial distribution of fracture aperture alters the fracture permeability. Waite *et al.* (1998) measured water flow through a sinusoidal fracture to compare sinusoidal flow with parallel-plate flow. Their experimental and numerical results showed that a sinusoidal fracture has a significantly lower permeability and for the sinusoidal geometry the effective aperture is very close to the minimum value of the normal aperture. Thomson & Brown (1991) showed that the directional non-uniformities in the fracture surface are more important than the degree of surface roughness.

Conclusions

The examples presented above illustrate the complex set of factors that yield the disposition of the present-day fracture system. The latter is the product of a history of these factors and thereby often has little correlation with one or a few other present-day factors. The extensive set of rock and fluid parameters at each point within the crust makes inversion of seismic data difficult except for the simplest cases. Therefore the integration of standard seismic techniques with comprehensive modelling presented here should provide a new oppurtunity for predicting the location and characterization of fractured zones in the subsurface. We believe that our Basin RTM simulator is uniquely suited for this purpose.

This project was supported by the Office of Science of the US Department of Energy (grant no. DE-FG02–91ER14175) and the Gas Research Institute (contract no. 5097–260–3779).

References

ABRIOLA, L. M. & RATHFELDER, K. 1993. Mass-balance errors in modeling 2–phase immiscible flows – causes and remedies. *Advances in Water Resources*, **16**, 223–239.

BERKOWITZ, B. 1995. Analysis of fracture network connectivity using percolation theory. *Mathematical Geology*, **27**, 467–483.

BERRYMAN, J. G. 1980. Long-wavelength propagation in composite elastic media I. Spherical inclusions, *Journal of the Acoustics Society of America*, **68**, 1809–1819.

BERRYMAN, J. G. 1986, Effective medium approximation for elastic constants of porous solids with microscopic heterogeneity. *Journal of Applied Physics*, **59**, 1136–1140.

BOUR, O. & DAVY, P. 1998. On the connectivity of three-dimensional fault networks. *Water Resources Research*, **34**, 2611–2622.

BYERLEE, J. D. 1990. Friction, overpressure and fault normal compression. *Geophysical Research Letters*, **17**, 2109–2112.

DESAI, C. S. & SIRIWARDANE, H. J. 1984. *Constitutive laws for engineering material*. Prentice-Hall, New Jersey.

DEWERS, T, & ORTOLEVA, P. 1994. Nonlinear dynamical aspects of deep basin hydrology: Fluid compartment formation and episodic fluid release. *American Journal of Science*, **294**, 713–755.

DRÜKER-PRAGER, D. C. & PRAGER, W. 1952. Soil mechanics and plastic analysis or limit design. *Quarterly of Applied Mathematics*. **10**, 157–165.

HUYAKORN, P. S., PANDAY, S. & WU, Y. S. 1994. A three-dimensional multiphase flow for assessing NAPL contamination in porous and fractured media, 1. Formulation. *Journal of Contaminant Hydrology*, **16**, 109–130.

LARSON, K. W., WAPLES, D. W. FU, H. & KODAMA, K. 1993. Predicting tectonic fractures and fluid flow through fractures in basin modeling. *In:* DORE, A. G. (ed.) *Basin Modeling: Advances and Applications*.Norwegian Petroleum Society. Special Publications, **3**, Elsevier, Amsterdam, 373–383.

LUO, X. & G. VASSEUR. 1996. Geopressuring mechanism of organic matter cracking: numerical modeling. *AAPG Bulletin*, **80**, 856–873.

LUO, X., VASSEUR, G., POUYA, A., LAMOUREUX-VAR, AND, V. & POLIAKOV, A. 1998. Elastoplastic deformation of porous medium applied to the modeling of compaction at basin scale. *Marine and Petroleum Geology*, **15**, 145–162.

MAUBEUGE, F. & LERCHE, I. 1993. A north Indonesian basin: geo, thermal and hydrocarbon generation histories. *Marine and Petroleum Geology*, **10**, 231–245.

MAUBEUGE, F. & LERCHE, I. 1994. Geopressure evolution and hydrocarbon generation in a north Indonesian basin: two-dimensional quantitative modeling. *Marine and Petroleum Geology*, **104**, 104–115.

ODA, M. 1982. Fabric tensor for discontinous geological materials. *Soils and Foundations*, **22**, 96–108.

ODA, M. 1985. Permeability tensor for discontinous rock masses. *Geotechnique*, **35**, 483–495.

ODA, M. 1986. An equivalent continuum model for coupled stress and fluid flow analysis in jointed rock masses. *Water Resources Research*, **22**, 1845–1856.

ODLING, N.E. 1992. Network properties of a two-dimensional natural fracture pattern. *Pure and Applied Geophysics*, 138, 95–114.

ORTOLEVA, P. (ed.) 1990. Self-organization in geological systems. *Earth-Science Reviews*, **29**(1–4),

ORTOLEVA, P. 1992. *Nonlinear Chemical Waves*. John Wiley and Sons: New York.

ORTOLEVA, P. 1994*a*. *Geochemical Self-Organization*. Oxford University Press, Oxford.

ORTOLEVA, P. 1994*b*. Basin compartmentation: definitions and mechanisms. *In:* ORTOLEVA, P. (ed.) *Basin Compartments and Seals*. AAPG Memoir **61**, 39–51.

ORTOLEVA, P. 1998. *Basin compartment fundamentals*. GRI Topical Report (Contract No. GRI-97/0097). Gas Research Institute:Chicago.

ORTOLEVA, P., MAXWELL, J. M., PAYNE, D., SIBO, W. & COMER, J. 1997. Naturally fractured reservoirs and compartments: a pedictive basin modeling approach. *In:* HOAK, T. E., KLAWITTER, A. L. & BLOMQUIST, P. K. (eds) *Fractured Reservoirs: Characterization and Modeling*, Rock Mountain Association of Geologist Guidebooks, 67–102, Denver, RMAG.

OZKAN G, & ORTOLEVA, P. 2000. Evolution of the gouge particle size distribution: A Markov model. *Pure and Applied Geophysics*, **157**(3), 449–468.

OZKAN, G., TUNCAY, K. & ORTOLEVA, P. 1998. Process-based fault seal/conduit prediction. In: *AAPG Annual Convention Abstracts, Salt Lake City, UT, May 17–28, 1998* (CD-ROM format).

PAYNE, D. F., TUNCAY, K., PARK, A., COMER, J. & ORTOLEVA, P. 2000. A reaction-transport-mechanical approach to modelling the interrelationships between gas generation, overpressuring, and fracturing –Implications for the Upper Cretaceous natural gas reservoirs of the Piceance Basin, Colorado. *AAPG Bulletin*, **84**, 545–565.

RICE, J.R. 1975. Continuum mechanics and thermodynaimcs of plasticity in relation to microscale deformation mechanisms. *In:* ARGON, A. S. (ed.), *Constitutive Equations in Plasticity*. MIT Press: Cambridge, MA.

RICE, J. R. 1992. Fault stress states, pore pressure distributions, and the weakness of the San Andreas Fault. *In:* EVANS, B. & WONG, T.-F. (eds) *Fault Mechanics and Transport Properties in Rocks*. Academic Press, London, 475–503.

ROBERTS, S. J. & NUNN, J. A. 1995. Episodic fluid expulsion from geopressured sediments. *Marine and Petroleum Geology*, **12**, 195–204.

SAKRANI, K. 1996. *A unified texture and mineralogy dependent model for rock deformation*. PhD dissertation, Indiana University, Bloomington.

SCHNEIDER, F., POTDEVIN, J. L., WOLF, S. & FAILLE, I. 1996. Mechanical and chemical compaction model for sedimentary basin simulators. *Tectonophysics*, **263**, 307–317.

SLEEP, N. H. & BLANPIED, M. L. 1994. Ductile creep and compaction: a mechanism for transiently increasing fluid pressure in mosly sealed fault zones. *Pure and Applied Geophysics*, **143**, 9–40.

SONNENTHAL, E. & ORTOLEVA, P. 1994. Numerical simulations of overpressured compartments in sedimenatary basins. *In:* ORTOLEVA, P. (ed.) *Basin compartments and seals*. AAPG Memoir **61**, 403–416.

THOMSON, M. E. & BROWN, S. R. 1991. The effect of anisotropic surface roughness on flow and transport in fractures. *Journal of Geophysical Research*, *96*, 21923–21932.

TUNCAY, K. & ORTOLEVA, P. 2000. Salt tectonics as a self-organizing process: a three dimensional reaction, transport and mechanics model. *Journal of Geophysical Research*, **106**, 803–818.

TUNCAY, K., PARK, A. & ORTOLEVA, P. 2000*a*. A forward model of three dimensional fracture orientation and characteristics. *Journal of Geophysical Research*, **105**, 16719–16735.

TUNCAY, K., PARK, A. & ORTOLEVA, P 2000*b*. Sedimentary basin deformation: An incremental stress rheology approach. *Tectonophysics*, **323**, 77–104.

TUNCAY, K., KHALIL, A. & ORTOLEVA, P. 2001. Failure, memory and cyclic fault movement. *Bulletin of Seismological Society of America*, **91**, 538–552.

UNGERER, P., BURRUS, J., DOLIGEZ, B., CHENET, P. Y. & BESSIS, F. 1990. Basin evaluation by integrated two-dimensional modeling of heat transfer, fluid flow, hydrocarbon generation, and migration. *AAPG Bulletin*, **74**, 309–335.

WAITE, M.E., GE, S., SPETZLER, H. & BAHR, D. B. 1998. The effect of surface geometry on fracture permeability: A case study using a sinusoidal fracture. *Geophysical Research Letters*, **25**, 813–816.

WANG, C. & XIE, X. 1998. Hydrofracturing and episodic fluid flow in shale-rich basins –A numerical study. *AAPG Bulletin*, **82**, 1857–1869.

ZIENKIEWICZ, O. C. & CORMEAU, I. C. 1974. Visco-plasticity–plasticity and creep in elastic solids—A unified numerical solution approach. International *Journal for Numerical Methods for Engineering*, **8**, 821–845.

Geological evidence of patterns of fluid flow through fracture networks: examination using random realizations and connectivity analysis

RICHARD J. H. JOLLY[1] & JOHN W. COSGROVE[2]

[1]*Golders Associates (UK) Ltd. Clyde House, Reform Road., Maidenhead, Berkshire, SL6 8BY, UK. Present address: BP Amoco Exploration, Chertsey Road, Sunbury-on-Thames, Middlesex, TX16 7LN.*
[2]*Department of Earth Sciences and Engineering, Royal School of Mines, Imperial College of Science, Technology & Medicine, London SW7 2BP, UK*

Abstract: Fluid flow through fractured media is dependent upon a variety of parameters including fracture length, orientation and density, and also on the relative magnitude of the lithostatic stress and the fluid pressure. Because of this it is often difficult to determine the fluid pathway through a particular fracture system even when the geometry of the network is known. Although many fluids (e.g. gas, water etc.) leave little or no evidence of their passage through the rock others such as magmas and fluidized sediments preserve the pathways they follow by forming dykes and sills.

It is found that the pathways preserved by the two types of fluids are different i.e. the spacing of the clastic dykes follows a power-law distribution and that of the igneous dykes a log-normal distribution. It is suggested that this in part might reflect the different properties of the dyke material (specifically its permeability) which determines whether or not the fracture containing the dyke can continue to act as a channel of easy fluid migration once the dyke has been emplaced.

Fractures are abundant in the upper crust and are known to exert a significant influence on the migration of fluids through the rock mass. This is particularly true when the rock matrix has a low permeability and as a result the bulk of the flow must be within fracture systems. As a result the task of understanding the migration of hydrocarbons into areas of accumulation and their subsequent extraction from these sites is often directly linked to the problem of fluid flow through a fractured rock mass in response to regional or local stress gradients. The problem can be subdivided into three parts.

(1) The determination of the potential fluid pathways through the fracture network. This involves percolation theory and the determination of the degree of connectivity of the network which depends upon the length, orientation and density of the fractures of the various fracture sets making up the network.
(2) The subsequent determination of which, if any, of these connected pathways through the fracture network will be open and will therefore allow the percolation of the fluids along them. This is determined primarily by the relative magnitude of the fluid pressure and the lithostatic stress (see Delaney *et al* 1986; Barton *et al.* 1995; Jolly & Sanderson 1995; Zhang & Sanderson 1999).
(3) The magnitude and orientation of the mean stress gradients across the fracture network.

In attempting to understand the migration of hydrocarbons through fracture networks within the crust geologists are frequently frustrated by having to study the problem indirectly or at best remotely by examining the results of well tests. In addition, details of the fracture networks through which the migration has and/or is occurring are generally poorly known. In an attempt to circumvent these problems geologists have turned to percolation theory and to field observations of other fluids that have migrated through fracture networks but which, unlike many hydrocarbons, have left evidence of the pathways they followed. Examples of such fluids are magmas and fluidized sediments. With these fluids the pathways are preserved by igneous or sedimentary dykes and sills.

In this paper computer realizations are used to examine the influence of fracture length on connectivity of the fracture system. In addition to examining the factors that control the onset of connectivity between one boundary and another,

From: AMEEN, M. (ed.) *Fracture and In-Situ Stress Characterization of Hydrocarbon Reservoirs*. Geological Society, London, Special Publications, **209**, 177–186. 0305–8719/03/$15.00
© The Geological Society of London 2003.

the way in which the pattern of connectivity, particularly the spatial distribution of the connected fractures, varies as the connectivity increases, is also considered.

The spatial distribution displayed by igneous and clastic dykes, together with models for their formation are then presented. The paper concludes with a comparison of the theoretically predicted and the observed spatial distribution for clastic and igneous dykes and a discussion of the implications that these results have on hydrocarbon migration through fracture networks.

Percolation theory and the styles of connectivity

Percolation theory is a mathematical theory that examines the likelihood of connectivity, through a generated fracture network, from one boundary to the opposing boundary (Stauffer & Aharony 1994). There has been much work examining the connectivity of these models, and the onset of the connectivity of these types of fracture networks (Balberg *et al.* 1991; Berkowitz 1995, Bour & Davy 1999) In this paper, simple programs are used to generate networks in which the number of fracture sets, and the fracture length, orientation and density can be prescribed. The program places the fractures of each set randomly between the boundaries of the model (e.g. Fig. 1). The program then seeks out the shortest connected route or routes between prescribed boundaries. Only those fractures that are connected via a fracture network to the lower boundary of the model are considered and displayed (Fig. 1a$_2$, b$_2$, c$_2$). In that a connectivity to a specific boundary is considered this approach differs from the more traditional approach of percolation theory where the connectivity of the networks is considered independently of a boundary connection (Balberg *et al.* 1991; Hestir & Long 1990).

Figure 1 illustrates the change in connectivity across a fracture network made up of two orthogonal fracture sets, as the fracture length is gradually increased. For a set of input parameters the simulation is repeated a number (35) of times to ensure that the characteristic connectivity for those input parameters can be established. By plotting percentage of simulations connected against fracture length (Fig. 2), it can be seen that the onset of connectivity does not occur gradually. It occurs instantly as soon as some critical fracture length is reached, this is consistent with the findings of percolation theory (Stauffer & Aharony 1994). In the example shown in Figure 2, which relates to a network with a particular value of individual fracture length with likelihood that there is connectivity between the upper and lower boundaries of the model, connectivity first occurs when the fractures have a length of around 0.065 of the distance

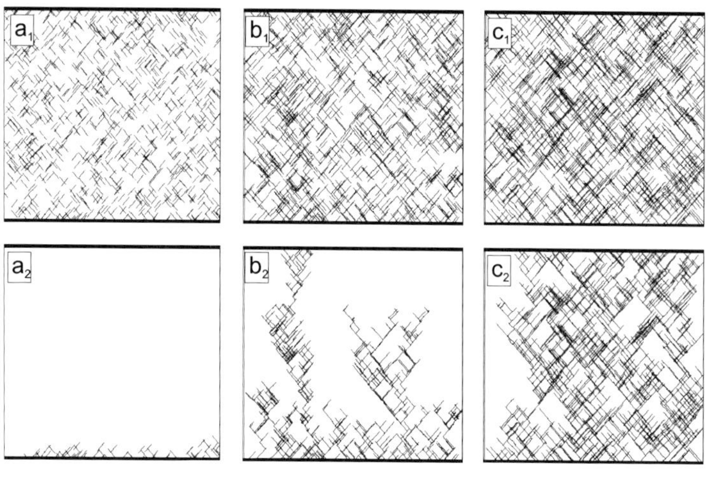

Fig. 1. (a$_1$, b$_1$ & c$_1$) Fracture networks made up of two orthogonal fracture sets. The orientation and density of the fracture sets in the three models are identical and they differ only in the length of the fractures which increases from models a to c. (**a$_2$**, **b$_2$** and **c$_2$**) The shortest fracture connections between the lower and upper boundaries, and where there is no connection between the upper and lower boundary only those fractures connected to the lower boundary. The models show the styles of connectivity varies from none (a) through channelized (b) to pervasive (c).

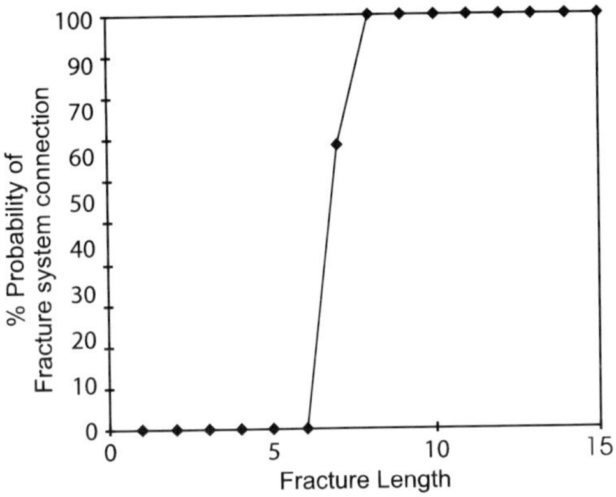

Fig. 2. Critical effect of changing fracture length on connectivity.

between the upper and lower boundary of the model, i.e. the distance across which the value of connectivity is required.

Connectivity and styles of connectivity

In the two orthogonal sets of fractures shown in Figure 1a the fractures in both sets are 0.05 of the distance from the lower to the upper boundary. It is found that when the fractures are relatively short there is no connectivity between the upper and lower boundaries of the model (Fig. 1a$_2$). For fluids to migrate through such a system they would have to utilize the intrinsic permeability of the rock matrix. In Figure 1b the fractures are 0.08 of the distance from the lower to the upper boundary and in this model some of the fractures do link to form a connection between the upper and lower boundary. However, not all of the fractures are connected and it can be seen that the connected fractures (emboldened) form channels within the network separated from each other by zones within the network where no connectivity exists, Fig. 1b$_2$. As the length of the fractures is further increased the connectivity increases and the style of connectivity changes from channelized to pervasive where most of the fractures within the model are connected. This is illustrated in Figure 1c$_2$ in which the length of the fractures has been increased to 0.11 of the distance from the lower to the upper boundary.

By examining the relationship between connectivity and fracture density a threshold value for the onset of connectivity, similar to the one shown in Figure 2, is also obtained. As the density of fractures is increased so the connectivity changes from channelized to pervasive.

Spatial distribution of connected fractures

There is a fundamental difference between the transition between no connectivity and channelized connectivity and channelized connectivity and pervasive connectivity. As mentioned in the previous section the first transition is abrupt and occurs as shown in Figure 1. The second transition is more gradual and in order to study this transition in more detail a series of models whose widths are considerably greater than their heights were examined. The extra width of these models allows enough connected fracture pathways to be established to enable an examination of their spatial distribution to be carried out (Fig. 3).

The fracture network examined consists of two orthogonal fracture sets, one parallel to the direction in which the connectivity was being investigated (vertical fractures Fig. 3) and the other at right angles to it. Several 'experiments' were carried out in which the length of the vertical fractures was progressively increased and that of the horizontal fractures kept the same. Three of these simulations are shown in Figure 3. They show that initially the network has channels of connectivity, but with increasing length of the vertical fractures there is a gradual change from this channelized connectivity to one that is more pervasive.

In order to quantify the change in spatial organization of the channels of connectivity as the connectivity became progressively more pervasive, their spacing was measured along the upper boundary of each experiment. For the experiments carried out in which the length of the vertical fractures was just above the threshold value (i.e. were relatively short), the spacings of the connected fractures conform best to a power-law distribution (Fig. 4, fracture length 16). An increase in the length of the vertical fractures results in the fracture network having a higher degree of connectivity. In addition the spacing of the connected fractures drifts away from a power law distribution towards a negative-exponential distribution (Fig. 4, fracture length 18). Further increases in the vertical fracture length increases the degree of connectivity of the fracture system, and the spacing distribution moves closer and closer to a negative-exponential (Fig. 4, fracture length 20).

This drift from a power-law distribution of the channels of connectivity towards one that is negative-exponential can be calculated as an R-squared fit of the data. As the length of the vertical fractures is increased the change in R-squared fit is shown quantitatively in Figure 5.

The change in the spatial distribution of the connected fractures indicates an important change

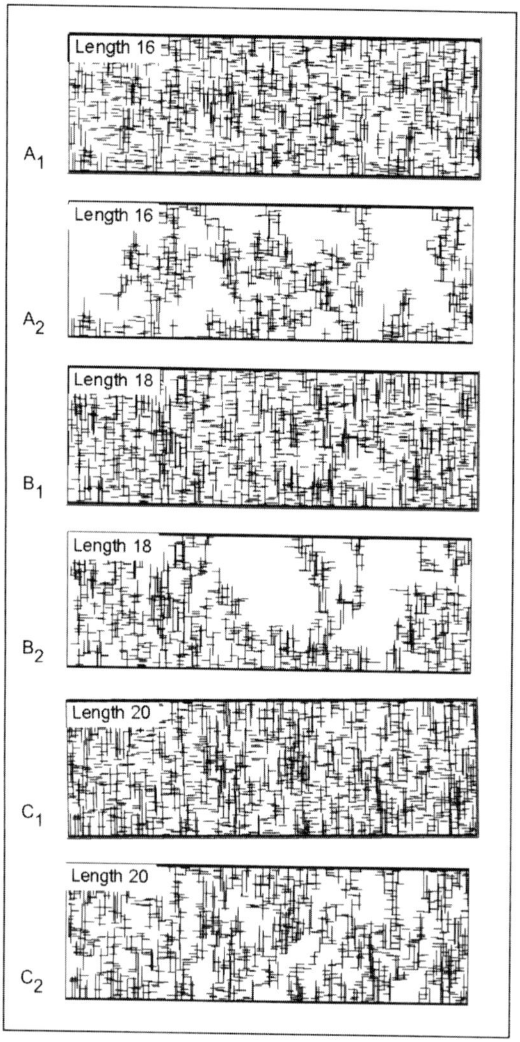

Fig. 3. ($\mathbf{A_1}$, $\mathbf{B_1}$ and $\mathbf{C_1}$) Three percolation models made up of two orthogonal fracture sets. ($\mathbf{A_2}$, $\mathbf{B_2}$ and $\mathbf{C_2}$) The effect on connectivity of increasing the length of the vertical fractures. The spacing of the connected channelways with increasing fracture length changes, and these changes are shown quantitatively in Figure 4.

in the style of the connectivity of the fracture system. The conformity of the spacing to a power-law distribution when the vertical fractures have a length of 0.16 of the distance from the lower to the upper boundary, reflects the clustered nature of the connectivity of the fracture system close to the percolation threshold. At this stage in the evolution of connectivity, only those fractures located close to the flow channel are connected to it. With increasing connectivity of the fracture system more and more of the fractures are connected to the potential fluid pathways so the spacing of the connected fractures tends towards the spacing of the fracture network that is being exploited. In the examples discussed in this paper the spacing of the fractures in both fracture sets is random and as a result the distribution of the connected fractures as the connectivity becomes pervasive tends to a negative exponential thus reflecting the total distribution of the host fracture network. It is likely that the resulting degree of clustering seen just after the boundaries become connected through the fracture network within this model would vary as a result of the size of the

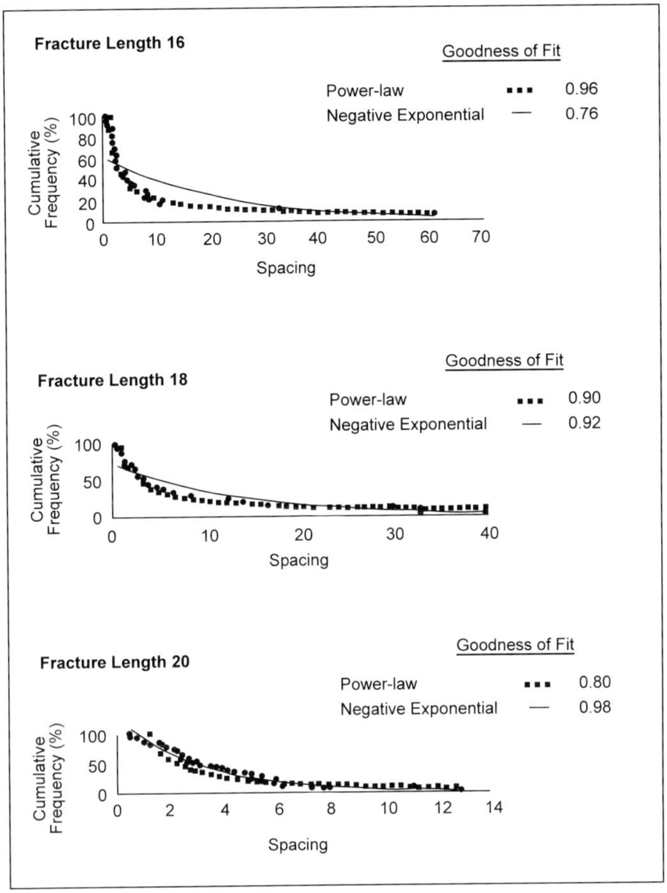

Fig. 4. A plot of the spatial distributions of channels of connectivity against cumulative frequency for the fracture networks shown in the models in Figure 3. Note that as the fracture length increases, so the spacing drifts from a power law distribution to that of a negative exponential.

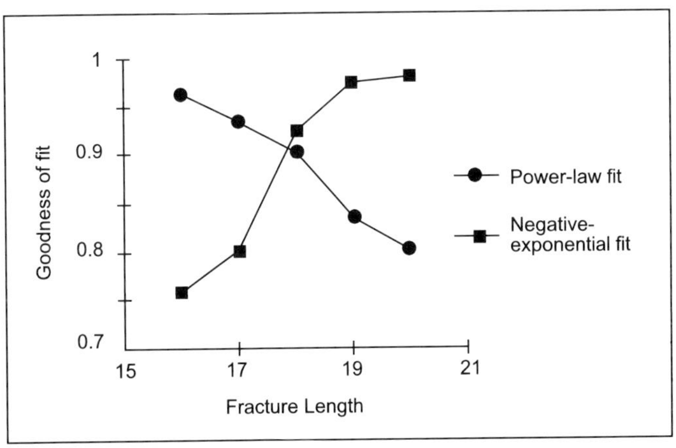

Fig. 5. The change in goodness of fit to power-law and negative-exponential distribution with fracture length.

horizontal connecting fractures; however, the connectivity would resemble the channelized flow yielded from these simulations.

Although the geometrical boundary connectedness can be used to determine the connectivity within a fracture network, this geometric analysis cannot be used to determine whether the fluids will be able to exploit the connectivity. This will be determined principally by the relative magnitudes of the lithostatic stress and fluid pressure (Delaney *et al.* 1986, Barton *et al.* 1995, Jolly & Sanderson 1995).

Thus, even in a pervasively connected fracture network, at low fluid pressures (i.e. when the fluid pressure is much less than the least principal stress, s_3), the fluid will be unable to shear or dilate any of the fractures. The fractures will therefore be unlikely to transmit fluids easily and fluid migration will have to exploit the intrinsic permeability of the rock matrix, assuming fractures can be closed under compressive stresses. When the fluid pressure reaches a certain magnitude relative to the regional stress, fractures of certain orientations will shear (Barton *et al.* 1995); this will cause local areas of dilation to occur (Zhang and Sanderson 1999). For fluid flow within the rock mass that is explosive, as it is for some geological phenomena such as igneous and clastic dykes and the fluid pressure is greater than the minimum principal stress but less than the maximum principal stress, a limited range of fracture orientations will be able to dilate, causing channelized fluid flow. Under extreme conditions when the fluid pressure is greater than the maximum principal stress, all fractures will be able to dilate and there will be pervasive fluid flow through the fracture network.

To summarize, because the connectivity of a fracture network made up of two or more fracture sets is controlled by several parameters (including fracture length, orientation and density) of the individual fracture sets and because the possible fluid pathways through the network are further controlled by the relative magnitudes of the fluid pressure and the lithostatic stress, it is not always possible to determine the pathways by which the fluids would pass through a system from the geometry of the fracture network alone. In nature, the patterns of fluid flow though a fracture network for fluids such as gas and water, are generally not preserved. However, certain fluids such as magmas and fluidized sediments leave evidence of the pathways they followed in the form of dykes and sills. In the following section the patterns of flow preserved in this way are examined and their relevance to the possible migration patterns of hydrocarbons discussed. It can be shown that in addition to the potential flow paths through a fracture network being controlled by the fluid pressure, the type of fluid can also play a role in determining the degree to which the fractures in the network are exploited during subsequent episodes of migration.

The intrusions of igneous and clastic dykes

The formation of dykes and sills (either igneous or sedimentary) is the result of the movement of fluid, either magma or fluidized sediment, through a fracture network. The present distribution of these intrusions is a direct record of the fracture paths followed by these fluids. In this section the spatial distribution of these two types of intrusion is considered. Both these types of intrusion exploit a pre-existing fracture network, and the propagation of new fractures is not required for fluid flow through the rock mass.

Igneous dykes

The west coast of Scotland provides excellent exposures of the Tertiary dyke swarms that emanate from the igneous centres of Mull, Skye and Ardnamuchan and the distribution of the Mull dyke swarm has been studied by Jolly & Sanderson (1995). The coastal areas in this region provide unbroken exposure that allows complete sampling of *all* the dykes within the traversed section. These sections are approximately perpendicular to the main trend of the swarm and this maximizes the number of dykes recorded and allows direct measurement of the spacing. The data were collected by laying out scanlines and linking these scanlines together to form traverses. Where the scanline intersects a dyke the distance along the scanline is recorded. The spacing of the dykes is recorded as the distance between the dyke centres. The dykes at Easdale utilize a previously existing joint set and the cleavage in the Dalradian sediments by which to intrude, resulting in irregular stepping geometries (for a more detailed description of the dykes see Jolly & Sanderson 1995).

The recorded spacing distribution of the dykes was compared with the normal (Gaussian), log-normal, negative-exponential and power-law distributions by using cumulative frequency plots (Fig. 6). These enable a visual determination of the data fit to one of the distributions as a straight line fit indicates conformity to that distribution. The spacing of the dykes was found to fit well to a log-normal distribution (Jolly & Sanderson 1995) (Fig. 6), and showed a poor fit

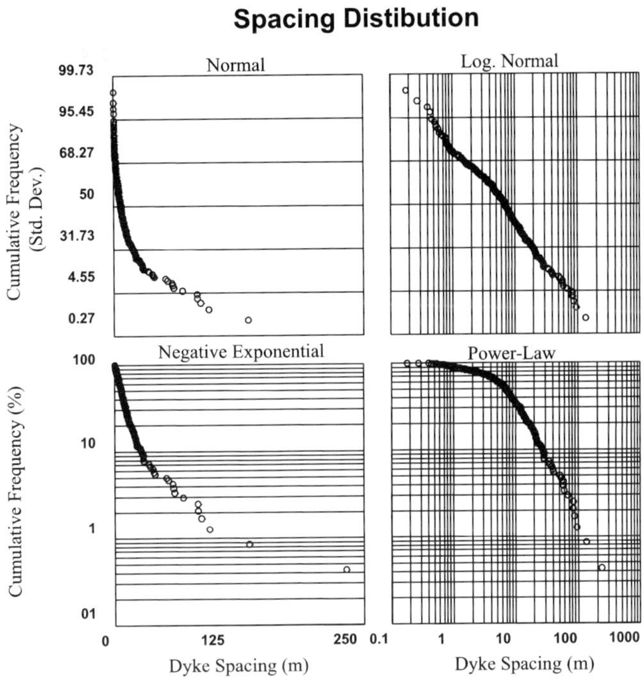

Fig. 6. Cumulative frequency against log dyke spacing for the Tertiary dykes of the Mull swarm, western Scotland. A straight line shows conformity of the data to a log-normal distribution (after Jolly & Sanderson 1995).

to the normal, negative-exponential and power-law distributions. This result is similar to the analyses of the spacing of joints and faults (Priest & Hudson 1976; Gillespie *et al.* 1993). The complete exposure of the coast line and the resulting confidence that all of the dykes were sampled suggest that the distributions represent real distributions of spacings of dykes and certainly at the small spacing are not subject to resolution truncation issues. However, at the large spacing scale the distributions are subject to sample line length.

Clastic dykes

A similar study of spacing of clastic dykes has been carried out by Jolly *et al.* (1998).

Clastic dykes occur where there is rapid injection of fluid carrying grains of sand into a fracture. This typically occurs when the rock surrounding the fluidized sand is of low permeability. One of the most well known clastic dykes swarms is found in the Ono district, North Sacramento Valley, California (Diller 1890, Peterson 1966). Here sand is injected into shales containing vertical joints (for further detail see Jolly *et al.* 1998). The spatial data on the distribution of these clastic dykes can be easily obtained because the swarm is very well exposed along a series of dry river beds. Data were collected by laying out scanlines in the same way as described above for the igneous dykes of the Mull swarm.

The spacing data of these sedimentary dykes were compared with four distributions, normal (Gaussian), log-normal, negative-exponential and power-law. It was found that the data conformed best to a power-law distribution (Fig. 7). The gradient of this distribution is extremely low, less than -0.4 which indicates that the dykes in the swarm are spatially clustered.

A relationship was also noted between the minimum dyke spacing and dyke thickness (Fig. 8) which indicates that thin dykes tend to be close to adjacent clastic dykes whereas thick dykes (>0.04 m), are more widely spaced (i.e. >30 m away for the nearest dyke). The relatively close spacing of the thin dykes suggests that there are zones where there is local interconnectivity of fractures that allow the branching of the flow into multiple fractures.

A model that best explains the spatial distribution and the thickness/spacing relationship of these clastic dykes is a process of repeated branching of the flow at the time of emplace-

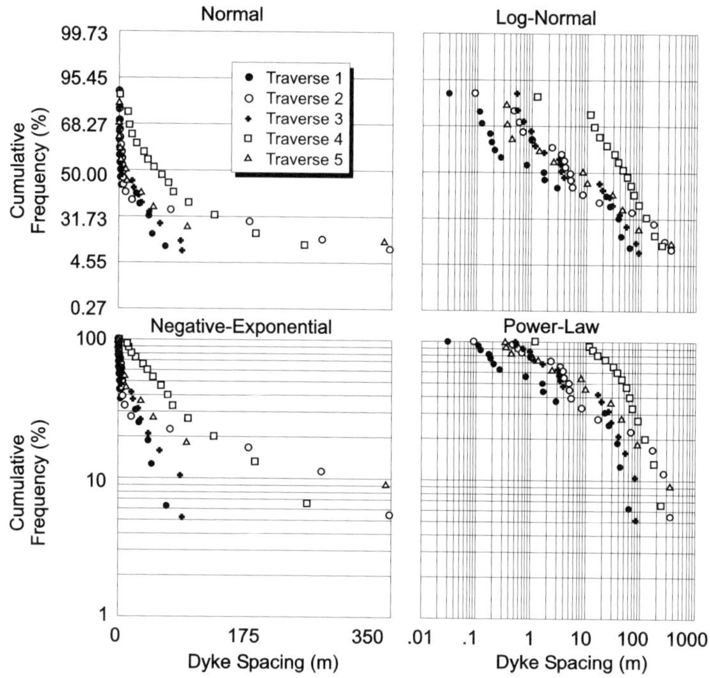

Fig. 7. Cumulative frequency against log dyke spacing for the clastic dykes of the Ono district, California. A straight line shows conformity to a power-law distribution (after Jolly *et al.* 1998).

ment (Fig. 9) (Jolly *et al.* 1998). This produces localized zones of flow.

Discussion

It is suggested that the different spatial distributions of the igneous and sedimentary dykes shown in Figures 6 & 7 reflect the difference in properties between the two types of dykes. In both of the examples of intrusion mentioned above the fluids are exploiting an already existing fracture network, and are not subject to the propagation of new fractures. In the case of igneous dykes, as the magma cools it crystallizes within the fractures. This freezing prevent any subsequent flow within the fracture and forces subsequent pulses of magma to utilize other routes through the fracture network. As a result the flow through the network becomes more pervasive.

In contrast the clastic dykes do not completely block the fractures along which they form. They are more permeable than the low permeability host rock and therefore allow the subsequent flow of fluid along the fractures they occupy. The initial connectivity of the fracture network does not need to be increased for any further fluid flow through the rock mass.

Thus it is argued that the difference in the spatial distribution of the igneous and sedimentary dykes relates to difference in nature of the infilling material of the two dykes. It follows that in the example of clastic dykes, the fracture network they invade need only have channelled connectivity in order for the dykes to flow across the system forming a connecting pathway from source to sink that will remain permeable and permit sustained dewatering of the overpressured sands from which the dykes emanate. The only way for the magma to exploit an existing pathway would be to propagate a new fracture along the previously exploited pathway otherwise as noted above, the igneous dykes in contrast block any further flow along the fractures and impel later magma to seek other pathways. Thus the fracture network becomes progressively invaded by new dykes and sills and any initial channelling of the magma flow tends to become obscured as the magma progressively invades more and more fractures. The pattern of clastic dykes is therefore likely to preserve the original channelized pattern of flow whereas the pattern of igneous dykes is likely to preserve a more pervasive flow. This study is concerned with the exploitation of existing fractures and has not

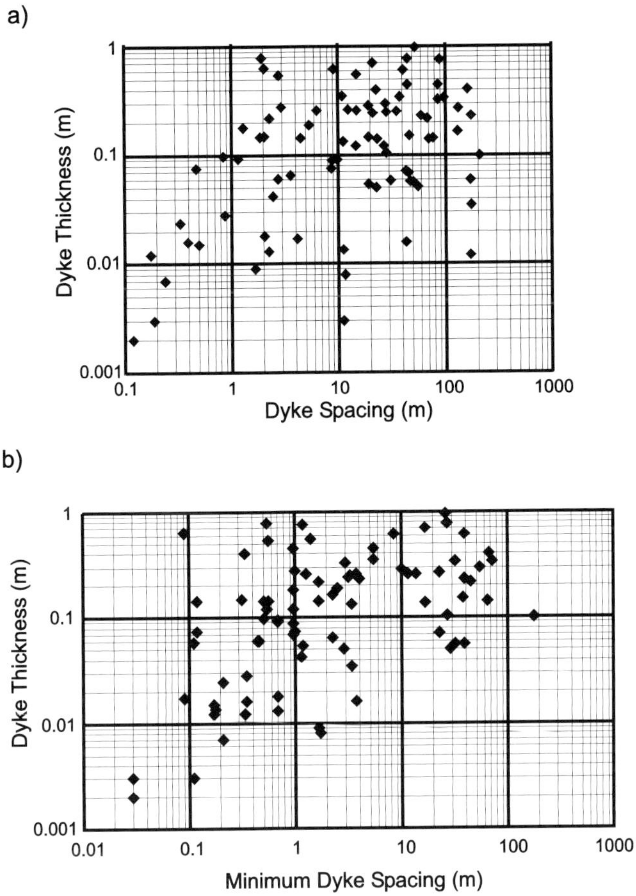

Fig. 8. Thickness–spacing data for the clastic dykes of the Ono district, California. **(a)** Dyke thickness against dyke spacing shows no clear relationship. However, for dyke thickness against minimum dyke spacing **(b)** there is a relationship, suggesting that thin dykes are unable to occur in isolation (after Jolly *et al.* 1998).

considered the impact of propagation fractures through an already existing network. The propagation of a fracture will of course dramatically change the connectivity network from the connectivity prior to the propagation of the fracture.

The migration of hydrocarbons through fracture networks, due to their non-sealing nature, is more like the clastic dyke model, forming local channels (Fig. 3a) flowing through the fracture system (Fig. 9).

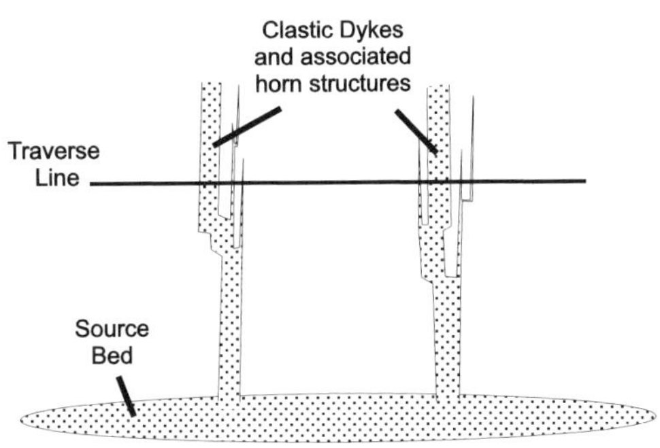

Fig. 9. Model for the clastic dyke distribution in the Ono district California, showing the influence of the process of branching and dyke sensitivity on spatial and thickness distributions seen in the district (from Jolly *et al.* 1998).

Conclusions

This paper shows that once the system is connected there is a change in the style of connectivity from channelled to pervasive.

- The spacing of the connected fractures changes from a power-law (clustered) distribution to the distribution of the fracture network that it is exploiting.

- The spatial distribution of igneous dykes from the Mull dyke swarm is log-normal.
- The spatial distribution of clastic dykes from the Sacramento Valley, California, is power-law suggesting the flow within the fracture network is extremely clustered.
- The spatial distribution of the clastic dykes is attributed to local branching associated with connected flow zones.
- The difference between the spatial distributions observed for the clastic and igneous dykes is attributed to the ability of the fluids to re-utilize existing flow pathways.
- It is likely that migrating hydrocarbons, because they do not seal their flow pathways, have channelized flow through fracture networks. Thus the geometry flow pathways of migrating hydrocarbons are similar to the clastic dyke model.

The migration of hydrocarbons through fracture networks should not be considered as flow within isolated linked fractures, but as localized zones of flow within the fracture network (Figs. 3a & 9).

This work was funded by a NERC Ropa award while R.J.H.J. was at Imperial College. D. Dewhurst, L. Lonergan and D. Sanderson are thanked for many discussions concerning this work. Also thanked are two anonymous reviews for their helpful comments on the manuscript.

References

BALBERG, I., BERKOWITZ, B. & DRACHSLER, G. E. 1991. Application of a percolation model to flow in fractured hard rock. *Journal of Geophysical Research*, **96**, 10,015–10,021.

BARTON, C. A., ZOBACK, M. D. & MOOS, D. 1995. Fluid flow along potentially active faults in crystalline rock. *Geology*, **23**, 683–686.

BERKOWITZ, B. 1995. Analysis of fracture network connectivity using percolation theory. *Mathematical Geology*, **27**, 467–483.

BOUR, O. & DAVY, P. 1999. Clustering and size distribution of fault patterns: theory and measurements. *Geophysical Research Letters*, **26**, 2001–2004.

DELANEY, P. T., POLLARD, D. D., ZIONEY, J. I. & MCKEE, E. H. 1986. Field relations between dikes and joints: emplacement processes and paleostress analysis. *Journal of Geophysical Research*, **91**, 4920–4938.

DILLER, J. S. 1890. Sandstone dikes. *Geological Society of America Bulletin*, **1**, 411–442.

GILLESPIE, P. A., HOWARD, C. B., WALSH, J. J. & WATTERSON, J. 1993. Measurement and characterisation of spacial distribution of Fractures. *Tectonophysics*, **226**, 113–141.

HESTIR, K. & LONG, C. J. S. 1990. Analytical expressions for the permeability of random two-dimensional Poisson networks based on regular lattice percolation and equivalent media theories. *Journal of Geophysical Research*. **92**(B), 21,565–21,581.

JOLLY, R. J. H. & SANDERSON, D. J., 1995. The variation in form and distribution of dykes in the Mull swarm. *Journal of Structural Geology*, **17**, 1543–1557.

JOLLY, R. J. H., COSGROVE, J. W. & DEWHURST, D. N., 1998. Thickness and spatial distributions of clastic dykes, northwest Sacramento Valley, California. *Journal of Structural Geology*, **20**, 1663–1672.

PETERSON, G. L. 1966. Structural interpretation of sandstone dikes, Northwest Sacramento Valley, California. *Geological Society of America Bulletin*, **77**, 833–842.

PRIEST, S. D. & HUDSON, J. A. 1976. Discontinuity spacings in rock. *International Journal of Rock Mechanics and Mining Sciences & Geomechanics Abstracts*, **13**, 135–148.

STAUFFER, D. & AHARONY, A. 1994. *Percolation Theory*. Taylor & Francis, London.

ZHANG, X. & SANDERSON, D. J. 1999. Critical stress localization of flow associated with deformation of well fractured rock masses, with implications for mineral deposits. *In:* McKaffrey, K. J. W., Lonergan, L. & Wilkinson, J. J. (eds) *Fracture Fluid Flow and Mineralisation*. Geological Society, London, Special Publications, **155**, 69–81.

Difficulties and uncertainty in mathematical/ numerical modelling of fluid flow in fractured media

RICHARD E. EWING & ANNA M. SPAGNUOLO

Texas A&M University, College Station, TX 77843-3404, USA
(e-mail: richard-ewing@tamu.edu)

Abstract: The ability to numerically model single-phase and multiphase flow of fluids in porous or fractured media is extremely important in developing an understanding of the complex phenomena governing the flow. The flow is complicated by the presence of heterogeneities in the reservoir at many different length scales by special flow features such as fractures or faults and by phenomena such as diffusion and dispersion. These effects must be effectively modelled by terms in coupled systems of non-linear partial differential equations which form the basis of the simulator. The simulator must be able to model both single and multiphase flows and the transition regimes between the two in unsaturated flow applications. A discussion of some of the aspects of modelling unsaturated and multiphase flows in the presence of heterogeneities and severe channelling is presented along with directions for future work. Simulators are severely hampered by the lack of knowledge of reservoir properties, heterogeneities, fracture dimension and orientation, and relevant length scales and other important mechanisms. Simulations can be performed either deterministically, to predict the outcome of a single realization of reservoir and flow properties, or via stochastic techniques to incorporate uncertainties of flow directly. Due to the extreme difficulties in using stochastic differential equation models for non-linear multiphase flows, we will concentrate on the potential of deterministic models. Recent developments have been made in homogenization, scaled averaging, and the use of the simulator as an experimental tool to develop methods to model the interrelations between localized and larger-scale media effects. Monte Carlo techniques using simulators with effective parameters can generate statistics for multiphase flow.

The understanding and prediction of the behaviour of the flow of multiphase or multicomponent fluids through porous or fractured media are often strongly influenced by heterogeneities, either large-scale lithological discontinuities such as fractures or faults or quite localized phenomena. Considerable information can be gained about the physics of multiphase flow of fluids through fractured media via laboratory experiments and pore-scale models; however, the length scales of these data are quite different from those required from field-scale understanding. The coupled fluid–fluid interactions are highly non-linear and quite complex. The presence of heterogeneities in the medium at all length scales greatly complicates this flow. We must understand the effects of heterogeneities coupled with non-linear parameters and functions on different length scales. We can then use the simulators as 'experimental tools' in the laboratory of high performance computing to simulate the process on increasingly larger length scales to develop intuition on how to model the effects of heterogeneities at various levels.

The partial differential equation models used in the simulators are convection dominated. Mixed finite-element methods are described to treat the strong variation in coefficients arising from heterogeneities. An operator-splitting technique is then used to address the disparate temporal scales of conviction, diffusion, and reaction terms. Convection is treated by time stepping along the characteristics of the associated pure convection problem, and diffusion is modelled via a Galerkin method for single-phase flow and a Petrov Galerkin technique for multiphase regimes. Eulerian Lagrangian techniques, MMOC (modified method of characteristics) described by Douglas & Russell (1982) or Ewing *et al.* (1983), MMOCAA (modified method of characteristics with adjusted advection) introduced for a waterflooding problem (Douglas *et al.* 1977, in press) and for a multicomponent miscible displacement problem (Spagnuolo in press) or ELLAM (Eulerian Lagrangian localized adjoint methods) introduced by Celia *et al.* (1990), effectively treat the advection-dominated processes. Extensions of

From: AMEEN, M. (ed.) *Fracture and In-Situ Stress Characterization of Hydrocarbon Reservoirs*. Geological Society, London, Special Publications, **209**, 187–200. 0305–8719/03/$15.00
© The Geological Society of London 2003.

ELLA to the multiphase regime appear in Ewing (1991). Accurate approximations of the fluid velocities needed in the Eulerian Lagrangian time-stepping procedure are obtained by mixed finite-element methods. When reaction terms are present, their rapid effects in relation to convection or diffusion must be treated carefully since they strongly affect the conditioning of the resulting systems.

In order to scale the highly localized behaviour of fine-scale fingering generated by heterogeneous fractured media up to computational and field scales, we must develop techniques to obtain effective parameters for coarse-grid models which match fine-grid simulations. Ewing *et al.* (1989*b*) presented a coarse-grid dispersion model of heterogeneity and viscous fingering to match fine-grid simulation of miscible displacement processes. They adjusted longitudinal and transverse dispersivities in a dispersion tensor to match recovery curves for simulations of viscous fingering on fine grids. Although they were able to match production from various simulations, they pointed out that permeability averages, variances, and standard deviations along are not able to determine dispersivities, since the age specific permeability distribution in each realization can have significant impact upon the flow. Espedal *et al.* (1991) considered similar dispersion models to describe fingering processes in immiscible, two-phase flow. Neuman and coworkers (Neuman *et al.* 1987; Neuman & Zhang 1990) also presented various models for both Fickian and non-Fickian dispersion. In this paper we combine these ideas for multiphase and multicomponent flow, using dispersion models coupled with accurate treatment of first-order transport effects for both models. This coupling is very important for saturated unsaturated models, which possess aspects of each process. The dispersion models are presented for both single and multiphase cases. Then accurate high-resolution numerical simulators are introduced and used as our experimental tools. Numerical results have illustrated the success of dispersion models for all of these problems.

We discuss some simulations from dual porosity models that were derived through homogenization theory (Douglas *et al.* 1987; Arbogast 1989*a*; Douglas & Arbogast 1990). In Douglas & Arbogast (1990), recovery curves for some simulations of petroleum reservoir waterflooding were presented to give a gross indication of the flow of fluids within the reservoir. These curves sufficiently illustrated important features of naturally fractured reservoir simulation. When comparing a simulation of a fractured reservoir to a simulation of an unfractured reservoir, it was seen that there is a significant difference in response between the two reservoirs. It should be noted that the unfractured reservoir simulation possessed the matrix properties of the fractured one, except for some minor modifications to take into account fracture spacing. In the unfractured reservoir, better sweep efficiency, resulted, but it took a much higher pressure gradient across the reservoir to obtain this recovery. In the unfractured case, the curve had a breakthrough that was fairly obvious to see, but beyond it, recovery dropped dramatically. In comparison, the recovery curve for the fractured case was seen to vary more gently. Studies on the block sizes in fractured reservoirs were also performed. It was seen that sweep efficiency increases with block size.

Even at the finest grid level available using effectively scaled models, many phenomena cannot be resolved. Understanding and modelling of these localized phenomena require the use of adaptive or local grid refinement. Usual implementation of local grid refinement techniques destroys the efficiency of large-scale simulation codes. Techniques which involve a relatively coarse macro-mesh are the basis for domain decomposition methods (Bramble *et al.* 1988; Ewing 1989*a*) and associated parallel solution algorithms. Accuracy, efficiency of implementation, and adaptivity of these techniques are discussed (see also Ewing 1989*b*; Ewing *et al.* 1991, 1994). Experiences obtained whilst using these ideas in three-dimensional multiphase industrial petroleum simulators are briefly described in Ewing *et al.* (1989*a*) and Boyett *et al.* (1992).

In this paper, we discuss some of the difficulties involved in describing a reservoir with features of many different length scales. We also discuss techniques for multilevel homogenization to model fractured media with fractures at several length scales. Next, we present model equations for field-scale simulations followed by a description of operator-splitting techniques. Concepts of local grid refinement and upscaling are presented. Some numerical experiments are also discussed. Finally, we present some conclusions and directions for future study.

Reservoir characterization and description

The processes of both single and multiphase flow involve convection, or physical transport, of the fluids through a heterogeneous porous medium. The equations used to simulate this flow at a macroscopic level are variations of Darcy's law. Darcy's law has been derived for both single and multiphase regimes by Slattery (1969, 1970) via a volume averaging of the Navier Stokes equations, which govern flow through the porous medium at a microscopic or pore-volume level. The length scale for Navier Stokes flow (10^{-4} to 10^{-3} m) is very different from the scale required by normal reservoir simulation ($10–10^3$ m). Reservoirs themselves have scales of heterogeneity ranging from pore level to field scale. In the standard averaging process for Darcy's law, many important physical phenomena which may eventually govern the macroscopic flow could be lost. The continued averaging of reservoir and fluid properties necessary to use grid blocks of the size of $10–10^2$ m in field-scale simulators further complicates the modelling process. We discuss certain techniques to try to address these scaling problems.

Diffusion and dispersion are often critical to the flow processes and must be understood and modelled. Molecular diffusion is typically fairly small. However, dispersion, or the mechanical mixing caused by velocity variations and flow through heterogeneous rock, can be extremely important and should be incorporated in some way in our models.

Since the mixing and velocity variations are influenced at all relevant length scales by the heterogeneous properties of the reservoir, much work must be done in volume averaging terms like porosity and permeability. Gelhar & Axness (1983), Furtado *et al.* (1991) and Neuman and co-workers (Neuman & Yakowitz 1979; Neuman 1980; Neuman *et al.* 1980) have shown that statistical methods show promise in this area. Statistical techniques are currently being considered to obtain effective permeability tensors for large-scale models of flow through anisotropic or fractured media.

Due to the scarcity of direct data measurements, many researchers (Kitanidis & Vomvoris 1983; Carrera & Neuman 1986; Yeh 1986; Ginn & Cushman 1990) have included pore pressure measurements as part of the inverse problem for parameter estimation. In order to enhance these techniques, newer parameter estimation techniques (Rubin *et al.* 1992) also incorporate seismic velocity data in addition to the hydrologic flow data. Stochastic models have been used effectively to determine various properties for single-phase flow (Dagan 1985; Rubin & Dagan 1987*a*, *b*; Dagan & Rubin 1988).

The effects of dispersion in various flow processes have been discussed extensively in the literature (e.g. Perkins & Johnston 1963; Warren & Skiba 1964; Russell & Wheeler (1983) and Young (1984) have given excellent surveys of the influence of dispersion and the attempts to incorporate it in present reservoir simulators. Various terms which affect the length of the dispersive mixing zone include viscosity and velocity variations, and reservoir heterogeneity. Much work is needed to quantify these effects and to obtain useful effective dispersion coefficients for field-scale simulators. The dispersion tensor has strong velocity dependence. The longitudinal dispersion is often an order of magnitude larger than the transverse dispersion. This variation enhances unstable flow regimes induced by viscosity differences and reservoir heterogeneity. Initial work on correlation of dispersion coefficients presented with statistical simulations by Ewing *et al.* (1989*b*) will be discussed below.

In both single and multiphase regimes, the process of flows of fluids with very different viscosities through a heterogeneous porous medium can be unstable. If the flow rate is sufficiently high, the interface between the contaminant and the water may become unstable and may form fingers which grow in length in a non-linear fashion. This phenomenon, termed 'viscous fingering', is well known; different techniques for understanding and modelling it were surveyed by Ewing & George (1984).

Ewing *et al.* (1989*b*) indicated the ability of coarse-grid dispersion models to match results of both laboratory experiments and fine-grid simulations on highly heterogeneous meshes in a single-phase flow context. The use of a dispersion tensor in petroleum recovery applications avoided the optimistic recovery predictions often attributed to standard convection diffusion models. Heterogeneity by itself, when not highly correlated, was shown to be less of an in influence on recovery than viscous fingering from adverse mobility ratios; however, effects of heterogeneity are important and must be incorporated. The mixing parameter approach of Todd & Longstaff (1972) predicted too early breakthrough and optimistic ultimate recoveries compared with experiments and the dispersion models. More research is needed in this area, and this is currently underway. Studies are being carried out to extend the global effective dispersion concept to multiphase flow in an analogous manner.

Multi-level homogenization of fractured media

The investigation of fluid flow in networks of fractures is complex and has been the object of many recent researches. Two approaches have been employed: the discrete fracture approach and the equivalent porous medium approach. The first considers flows through a relatively small number of fractures. The second treats the system of fractures as a continuous porous medium on a macroscopic scale; such models, which are more appropriate when there is a very large number of fractures, are normally derived via some type of averaging process, such as homogenization.

For the case of two scales in the resulting model, a concept of dual porosity was first introduced on heuristic grounds by Barenblatt *et al.* (1960) and Warren & Root (1963). There, the authors treated the matrix-fracture flow interaction by a transmissibility function proportional to the difference between the matrix and the fracture pressures (see Barenblatt *et al.* 1990).

The basic concept underlying compaction/compression-driven flows in fractured reservoirs and related overpressures (Welte & Yukler 1981; Bethke 1985; Nakayama & Lerche 1987) is effective stress:

$$\delta=\mathbf{S}-\mathbf{P}$$

where δ=effective stress, $\mathbf{S}$=total stress, assumed vertical, and $\mathbf{P}$=pore pressure.

Similarly to Smith (1971), we can link the vertical component of effective stress with porosity through lithology-dependent empirical relationships. In some situations, the pore pressures tend to exceed the weight of the overburden, necessitating hydraulic fracturing. In some models (Unger *et al.* 1990) permeability is increased to simulate the opening of cracks when pressure exceeds minimum stress. Hydraulic fracturing has been discussed fairly extensively (e.g. Meissner 1978; Palcrauskas & Domenico 1980; Du Rouchet 1981). We will try to incorporate these effects in our models in later publications. We have coupled compaction in elastic porous media with flow and transport models for single phase flows (Chen *et al.* 1999).

In the past decade, a new way to treat the coupling of the flow through the fracture system with that in the matrix rock system was introduced (Douglas *et al.* 1987, 1991, 1993; Arbogast 1988, 1989*a*, *b*; Arbogast *et al.* 1988, 1990; Douglas & Arbogast 1990); the physical problems discussed therein arise in petroleum reservoir engineering. There, the matrix–fracture interaction is treated on each representative matrix block through boundary conditions that conserve momentum across the block faces and in the fracture system by introducing source terms that impose conservation of mass on the complete system. Several types of flow have been considered, including single-phase compressible flow, immiscible (multiphase) flow, and miscible (multicomponent) flow. Based on these ideas, Douglas *et al.* (1998) consider models in multiply fractured porous media; that is, the fractures occur at any finite number of length scales. These models shall be discussed below.

Here, we restrict ourselves to studying a single-phase, constant-compressibility fluid flowing in a geometrically more complicated structure given (initially) by a naturally fractured reservoir that has a hierarchy of fracture systems, with the first being defined by an interconnected system of planar fractures dividing the reservoir into a collection of disjoint blocks. A second system of fractures divides each of the previous blocks into a collection of equally sized smaller blocks. And so forth, until a last level is reached in which the disjoint blocks behave as a collection of disjoint unfractured matrix (rock) blocks. The geometric structure will be idealized by the assumption that each fracture system is periodic. See Figure 1 for a cross-sectional view of the idealized reservoir in the case of two levels of fractures. In the case of two levels of fractures, posing the flow equations on the three different scales of the domain involves using three different porosity and permeability functions. Homogenizing the flow equations on the small level of fractures and matrix blocks first gives an overall fracture flow in each of the blocks related to the larger-scale fractures similar to that predicted by a double-porosity model. That is, the (porous rock) matrix blocks will provide a source term to the surrounding system of small-scale fractures which, after homogenization, are treated as a continuous pore medium. Thus, a continuous medium relationship exists between the smaller level of fractures and the matrix blocks. Following this average, we can couple the equations for flow in the larger level of fractures and the first level of blocks, each of which is represented now by a double-porosity system. At this point, the set-up will again be analogous to that of a double-porosity model. Averaging the flow in the first level of fractures then gives a macroscopic description of the flow in a three-sheeted covering of the domain; this extends the concepts of the model of Douglas & Arbogast (1990). Thus, the first-level fractures are now smoothed out to cover the entire domain and the first-level blocks interact with the first-level fractures as sources, while the behaviour of the flow on a first-level block is that of a double-porosity system. Overall, the system can be characterized as a triple-porosity model. These ideas can be extended to model the flow in certain porous media that have a finite number of fracture systems. In the final mathematical model for such media, the systems of fractures are treated as a hierarchy of continuous porous media with the permeability

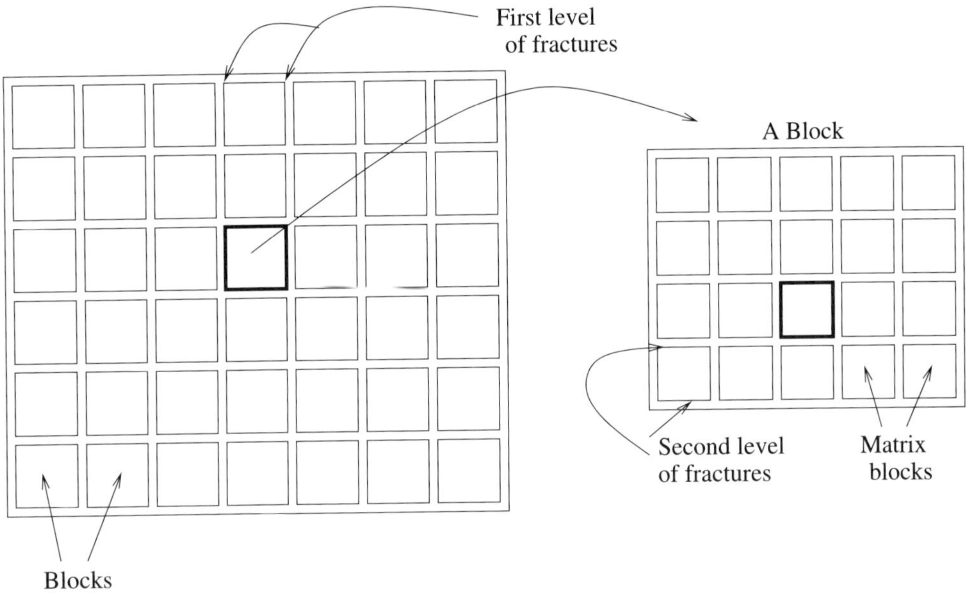

Fig. 1. The periodic structure in a reservoir exhibiting two levels of fractures.

tensors and porosities for the several scales of fractures being derived through homogenization. In addition, there is at the smallest scale of the model a collection of matrix blocks that behave as standard porous media. See Figure 2 for a topological view of the resulting model for N levels of fractures.

Mathematical models for N levels of fractures

The final equations that model the flow are given below. For i=0, . . . , $N-1$, let Y_{iB} be the block at the $(i+1)^{st}$ level of fractures, and let Y_{iF} be half of its surrounding fractures. Set $Y_i = Y_{iB} \cup Y_{iF}$. Then, for i=0, . . ., $N-2$, define the $\{m, n\}^{th}$ component of the 3 x 3 $\tilde{\mathbf{K}}$ matrix by

$$\left(\tilde{\mathbf{K}}_i\right)_{mn} \equiv \int_{Y_{iF}} \frac{\partial \omega_n^i}{\partial \alpha_{i+1}^m} d\alpha_{i+1}$$

where x_{i+1}^m is the mth component of the vector x_{i+1} and the functions w_k^i, or k=1, 2, 3, are Y_{iF}-period auxiliary functions (modulo constants) that satisfy

$$-\Delta_{\alpha_{i+1}} \omega_k^i = 0 \qquad \text{in } Y_{iF}$$

$$\nabla_{\alpha_{i+1}} \omega_k^i \, n_{Y_{iB}} = -e_k n_{Y_{iB}} \qquad \text{on } \partial Y_{iB}$$

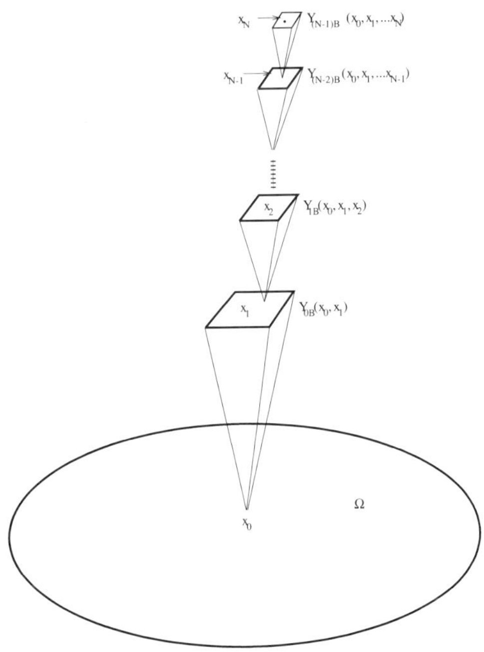

Fig. 2. The pop-up topology of Ω for the (N+1)-scale model.

where $n_{Y_{iB}}$ is the unit outward normal to Y_{iB}.

Let $\mathbf{K}_i=\mathbf{K}_i\,(x_0,\ldots,x_i)$ and $\phi_i=\phi_i\,(x_0,\ldots,x_i)$ be the tensor permeability and the porosity on Y_{iF}, respectively for $i=0,\ldots,N=1$, and let $\mathbf{K}_m=\mathbf{K}_m\,(x_0,\ldots,x_i)$ and $\phi_m=\phi_m\,(x_0,\ldots,x_N)$ be the tensor permeability and the porosity defined on $Y_{(N-1)B}$ (the smallest matrix block), respectively. Through the $\tilde{K}_i$'s, we can then define the following homogenized matrices for $i=0,\ldots,N-1$:

$$\mathbf{K}_i^h \equiv \frac{\mathbf{K}_i}{|Y_i|}\left(|Y_{iF}|\mathbf{I}+\tilde{\mathbf{K}}_i\right)$$

and

$$\mathbf{K}_m^h \equiv \mathbf{K}_m$$

Also, for $i=0,\ldots,N-1$ let

$$\phi_i^h \equiv \frac{|Y_{iF}|}{|Y_i|}\phi_i$$

and

$$\phi_m^h \equiv \phi_m$$

Let Ω be the domain and let $J=(0,T)$ be the time interval of interest. Assume that the source term S_0 is one that originally entered at only the first level of fractures. Let u and c be the viscosity and the compressibility of the fluid, respectively. Then, the macroscopic density ρ_0^h at each point $x_0\in\Omega$ satisfies the following initial boundary value problem:

$$\phi_0^h\frac{\partial\rho_0^h}{\partial t}-\nabla_{\alpha 0}\left(\frac{K_0^h}{\mu c}\nabla_{x0}\rho_0^h\right) = \frac{|Y_{0F}|}{|Y_0|}S_0-\frac{1}{|Y_0|}\int_{\partial Y_{0B}}\frac{\mathbf{K}_i^h}{\mu c}\nabla_{\alpha 1}\rho_1^h\, n_{Y_{0B}}\,da(\alpha_1) \quad \text{in } \Omega\times J$$

$$\frac{\mathbf{K}_0^h}{\mu c}\nabla_{\alpha 0}\rho_0^h\, n_\Omega=0 \qquad \text{on } \partial\Omega\times J$$

$$\rho_0^h(\alpha_0,0)=\rho_{init}(\alpha_0) \quad \text{in } \Omega\times\{0\}$$

In general, for $i=1, N-1$, we associate a block Y_{0B} to each $x_0\in\Omega$, a block Y_{1B} to each $x_1\in Y_{0B}$, $\ldots$, a block, $\ldots$, $Y_{(i-1)B}$ to each $x_1\in Y_{(i-2)B}$ such that for each $x_i\in Y_{(i-1)B}$, ρ_0^h satisfies

$$\phi_i^h\frac{\partial\rho_i^h}{\partial t}-\nabla_{x_i}\left(\frac{\mathbf{K}_i^h}{\mu c}\nabla_{x_i}\rho_i^h\right) = -\frac{1}{|Y_i|}\int_{\partial Y_{iB}}\frac{K_{(i+1)}^h}{\mu c}\nabla_{\alpha(i+1)}\rho_{(i+1)}^h\, n_{Y_{iB}}\,da\left(\alpha_{(i+1)}\right) \quad \text{in } \Omega\times\cdots\times Y_{(i-1)B\times J}$$

$$\rho_i^h(\alpha_0,\cdots,\alpha_i,t)=\rho_{i-1}^h(\alpha_0,\cdots\alpha_{i=1},t) \text{ on } \Omega\times\cdots\partial Y_{(i-1)B}\times J$$

$$\rho_i^h(\alpha_0,\cdots,\alpha_i,0)=\rho_{init}(\alpha_0) \qquad \text{in } \Omega\times\cdots\times Y_{(i-1)B}\times\{0\}$$

where $\mathbf{K}_N^h$ and $\mathbf{K}_m^h$.

Finally, we associate a block Y_{oB} to each $x_0\in\Omega$, a block Y_{1B} to each $x_1\in Y_{0B},\ldots$, a block $Y_{(N-1)B}$ to each $x_{N-1}\in Y_{(N-2)B}$, such that, for $x_N\in Y_{(N-1)B}$, ρ_0^h is the solution of

$$\phi_m^h\frac{\partial_m^h}{\partial t}-\nabla_{\alpha N}\left(\frac{\mathbf{K}_m^h}{\mu c}\nabla_{\alpha_N}\rho_m^h\right)=0 \quad \text{in } \Omega\times\cdots\times Y_{(N-1)B}\times J$$

$$\rho_m^h(\alpha_0,\cdots,\alpha_N,t)=\rho_{N-1}^h(\alpha_0,\cdots,\alpha_{N-1},t) \quad \text{on } \Omega\times\cdots\times\partial Y_{(N-1)B}\times J$$

$$\rho_m^h(\alpha_0,\cdots,\alpha_N,0)=\rho_{init}(\alpha_0) \quad \text{in } \Omega\times\cdots\times Y_{(N-1)B}\times\{0\}$$

Dispersion models

In order to ensure that the information passed from scale to scale is dependent upon the physical properties of the flow and not upon the numerics of the specific simulator, we have extensively studied the codes used in Ewing *et al.* (1983) and Dahle *et al.* (1990) and have shown them to be essentially free of numerical dispersion and grid orientation effects. There are no Courant number restrictions, and the only grid restrictions are sufficiently fine to resolve waves of length XXX. Otherwise, interpolation errors could cause difficulties. Dahle *et al.* (1990) local grid refinement around the interface allowed significantly greater grid spacings elsewhere. The codes utilize mixed finite-element methods for accurate fluid velocities in the presence of heterogeneities and Eulerian Lagrangian techniques for accurate fluid transport without numerical dispersion.

These techniques apply well to the transport of contaminants through the unsaturated or unsaturated soil zones. We first consider for simplicity a two-dimensional horizontal reservoir where gravity effects are negligible. The single-phase flow of an incompressible fluid with a dissolved solute in a horizontal porous reservoir $\Omega \subset IR^2$, over a time period $J=[T_0, T_1]$, is given by

$$-\nabla\left(\frac{\mathbf{K}}{\mu}\nabla p\right)\equiv\nabla\,\mathrm{u}=q \quad x\in\Omega, t\in J \tag{1}$$

$$\phi\frac{\partial c}{\partial t}-\nabla\left(\mathbf{D}\nabla c-\mathbf{u}c\right)=q\tilde{c} \quad x\in\Omega,\ t\in J \tag{2}$$

where p and u are the pressure and Darcy velocity respectively of the fluid mixture, ϕ and $\mathbf{K}$ are the porosity and the permeability respectively of the medium, u is the concentration-dependent viscosity of the mixture, c is the concentration of the contaminant solute, q is the external rate of flow, and $\bar{c}$ is the measurable inlet or outlet concentration. The form of the diffusion–dispersion tensor $\mathbf{D}$ is given by

$$\mathbf{D}=\phi(x)\left\{d_m I=|\mathbf{u}|\left(d_\ell E(\mathbf{u})+d_t E^{\perp}(\mathbf{u})\right)\right\} \tag{3}$$

where

$$E_{ij}(\mathbf{u})=\frac{\mathbf{u}_i\mathbf{u}_j}{|\mathbf{u}|^2} \tag{4}$$

$E^{\perp}=I-E$, d_m is the molecular diffusion coefficient, and d_l and d_t are the longitudinal and transverse dispersion coefficients, respectively. In general, $d_l\approx 10d_t$, but this may vary greatly with different soils, fractured media, etc. The viscosity u in Equation 1 is assumed to be determined by some empirical relationship or mixing rule based on contaminant concentration. In addition to Equations 1 and 2, initial and boundary conditions are specified. The flow at wells is modelled in Equations 1 and 2 via point or line sources and sinks.

When either an air phase or a non-aqueous phase liquid contaminant (NAPL) is present, the equations describing two-phase, immiscible flow in a horizontal porous medium are given by

$$\frac{\partial(\phi\rho_w S_w)}{\partial t}-\nabla\left(\mathbf{K}\frac{\rho_w k_{wi}}{\mu_w}\nabla p_w\right)= q_w\rho_{w\,x\in\Omega,\,t\in J} \tag{5}$$

$$\frac{\partial(\phi\rho_a S_a)}{\partial t}-\nabla\left(\mathbf{K}\frac{\rho_a k_{ai}}{\mu_a}\nabla p_a\right)= q_a\rho_{a\,x\in\Omega,\,t\in J} \tag{6}$$

where the subscripts w and a refer to water and air contaminant, respectively, S_i is the saturation, p_i is the pressure, XXX$_i$ is the density, k_{ri} is the relative permeability, u_i is the viscosity, and q_i is the external flow rate, each with respect to the ith phase. The saturations sum to unity.

One of the saturations can be eliminated; let $S=S_w=1-S_a$. The pressure between the two phases is described by the capillary pressure

$$p_c(S)=p_a-p_w \tag{7}$$

Although formally Equations 1 and 2 seem quite different from Equations 5 and 6, the latter system may be rearranged in a form which very closely resembles the former system. In order to use the same basic simulation techniques in our sample computations to treat both miscible and immiscible displacement, we will follow the ideas of Chavent (1976), and utilize a miscible/immiscible flow analogy.

Let Ω in $\mathbf{R}^3$ represent a porous medium. The global pressure p and total velocity $\mathbf{v}$ formulation of a two-phase (w) and air (a) flow model in XXX is given by the following equations (Chen *et al.* 1994).

$$S_a c_a\frac{dp}{dt}+\nabla\,v=-\frac{\partial\phi(p)}{\partial t}+ q(x,S_w) \quad x\in\Omega,\ t>0 \tag{8}$$

$$\mathbf{v}=-\mathbf{K}\lambda\left(\nabla_p-\mathbf{G}_\lambda\right) \quad x\in\Omega,\ t>0 \tag{9}$$

$$\phi\frac{\partial S_w}{\partial t}+\nabla\begin{pmatrix} f_w\mathbf{v}-\mathbf{K}\lambda_a q_w\delta\rho\mathbf{g}- \\ \mathbf{D}(S_w)\nabla S_w\end{pmatrix}= -S_w\frac{\partial\phi(p)}{\partial t}+q_w \quad x\in\Omega,\ t>0 \tag{10}$$

The global pressure and total velocity are defined by

$$p=\frac{1}{2}(p_w+p_a)+\frac{1}{2}\int_{S_c}^{S}\frac{\lambda_a-\lambda_w}{\lambda}\frac{dp_c}{d\xi}d\xi \text{ and } \mathbf{v}=\mathbf{v}_w+\mathbf{v}_a \tag{11}$$

Further, $\frac{d}{dt} \equiv = \frac{\partial}{\partial t} = \frac{v_a}{Sa}\ \nabla$, $\lambda = \lambda_w + \lambda_a$ is the total mobility, $\lambda_i = \frac{k_{ri}}{mi}$, $i = w$, a is the mobility for water and air, and $\mathbf{K}$ is the absolute permeability tensor.

The gravitational forces G_λ and capillary diffusion term $D(S)$ are expressed as

$$\mathbf{G}_\lambda = \frac{\lambda_w \rho_w + \lambda_a \rho_a}{\lambda} \mathbf{g}$$

$$\textit{and} \quad \mathbf{D}(S) = \mathbf{K}\lambda_a f_w \frac{dp_c}{dS} \tag{12}$$

and the compressibility c_a and fractional flow of water f_w are defined by

$$c_a = \frac{1}{\rho_a} \frac{d\rho_a}{dp_a} \quad \text{and} \quad f_w = \frac{\lambda_w}{\lambda} \tag{13}$$

We note that, in this formulation, the only diffusion/dispersion term is capillary mixing described by Equation 12.

The phase velocities for water and air, which are needed in transport calculations, are given by

$$\begin{aligned} \mathbf{v}_w &= f_w \mathbf{v} + \mathbf{K}\lambda_a f_w \nabla p_c - \mathbf{K}\lambda_a f_w \delta\rho \mathbf{g} \\ \mathbf{v}_a &= f_a \mathbf{v} + \mathbf{K}\lambda_w f_a \nabla p_c - \mathbf{K}\lambda_w f_a \delta\rho \mathbf{g} \end{aligned} \tag{14}$$

where $f_a = \lambda_\alpha / \lambda$, $\alpha = w$, α, and $\alpha = \omega$, and $\rho_a - \rho_w$. Within the groundwater literature, the pressure is normally scaled by the gravity potential function. Equation 8 would then be given in terms of the pressure head. We should also note that if the Richards approximation, infinite mobility of air, or $p_a = 0$ are valid, Equation 10 can be replaced by $p_c(S_w) = -p_w$. We may note that the phase velocity for air is given by Equation 14 even if the Richards approximation is used.

If Γ is the boundary of Ω, general boundary conditions for Equation 8–10 can be given by a combination of the following expressions:

$$p = p_\Gamma(x,t) \qquad x \in \Gamma_1,\ t > 0 \tag{15}$$

$$\mathbf{v}\,\nu + b(x,t,S_w)p = G(x,t,S_w) \qquad x \in \Gamma_2,\ t > 0 \tag{16}$$

$$\int_{\Gamma_3} \mathbf{v}\,\nu = g(t) \quad \text{and} \quad p = p_\Gamma(x,t) + d(t) \qquad x \in \Gamma_3,\ t > 0 \tag{17}$$

$$S_w = S_\Gamma(x,t) \qquad x \in \Gamma_4 \tag{18}$$

$$\left(f_w \mathbf{v} + \mathbf{K}\lambda_a f_w (\nabla p_c - \delta\rho \mathbf{g})\right)\nu + b_w(x,t,S_w)p = G_w(x,t,S_w) \quad x \in \Gamma_5, t > 0 \tag{19}$$

where Γ_i, $i = 1, \ldots, 5$ are given partitions of Γ.

Normally, the boundary conditions will be non-linear functions of the physical boundary conditions for the original two-pressure formulation (Chen *et al.* 1994). This means that we have to iterate on the boundary conditions as a part of the solution process. Our experience is that this does not cause problems.

Since the transport term and the diffusion/dispersion term in Equation 10 are governed by the fluid velocity, accurate simulation requires an accurate approximation of the velocity **v**. Because the lithology in the reservoir can change abruptly, causing rapid changes in the flow capacities of the rock, the tensor **K** in Equations 9 and 10 can have discontinuous entries. In this case, in order for the flow to remain relatively smooth, the pressure changes extremely rapidly. Thus, standard procedures of solving Equations 8 and 9 as a parabolic partial differential equation for pressure, differentiating or differencing the result to approximate the pressure gradient, and then multiplying by the discontinuous $K\lambda$ can produce very poor approximations to the velocity **v**. A mixed finite-element method for approximating **v** and p simultaneously, via a coupled system of first-order differential equations, has been used. This formulation allows the removal of singular terms as in Ewing *et al.* (1985) and accurately treats the problems of rapidly changing flow properties in the reservoir.

Both single-phase and multiphase codes used in our simulation utilize a physical dispersion tensor with different longitudinal and transverse terms. Although this is clearly natural for single-phase-contaminant modelling, the local physics of multiphase flows does not normally involve a dispersion phenomenon. However, via perturbation analysis, Espedal and Langlo have developed a natural dispersion tensor arising from heterogeneous flow at larger length scales following the single-phase work of Dagan (1989). Descriptions of these concepts appear in Espedal *et al.* (1990, 1991). Furtado *et al.* (1991) have arrived stochastically at a dispersion phenomenon with effects somewhere between transport and diffusion in origin. We feel that this corresponds to the need to match the gross permeability effects with first-order transport concepts and the finer-scale fingering with dispersion models.

Operator-splitting techniques

In finite-difference simulators, the convection is stabilized via upstream-weighting techniques. In a finite-element setting, we use a possible combination of a modified method of characteristics and Petrov Galerkin techniques to treat the transport separately in an operator-splitting mode.

In miscible or multicomponent flow models, the convective part is a linear function of the velocity. An operator-splitting technique has been developed to solve the purely hyperbolic part by time stepping along the associated characteristics (Douglas & Russell, 1982; Ewing *et al.* 1984, 1985; Russell 1985).

In immiscible or multiphase flow, the convective part is non-linear. A similar operator-splitting technique to solve this equation needs reduced time steps because the pure hyperbolic part may develop shocks. Recently, an operator-splitting technique has been developed for immiscible flows (Espedal & Ewing, 1987; Dahle *et al.* 1990) which retains the long time steps in the characteristic solution without introducing serious discretization errors.

The splitting of the convective part of Equation 10 into two parts $f^m(S)S+\mathbf{b}(S)S$ is constructed (Espedal & Ewing 1987) such that $f^m(S)$ is linear in the shock region, $0 \leq S \leq S_1 \leq 1$, and $\mathbf{b}(S) \equiv$ for $S^1 \leq S \leq 1$.

The operator splitting is defined by the following set of equations:

$$\phi \frac{\partial \overline{S}_w}{\partial t} + \frac{d}{dS}\mathbf{f}^m\left(\overline{S}_w\right)\nabla \overline{S}_w \equiv \phi \frac{d}{dr}\overline{S}_w = 0 \tag{20}$$

$$\phi \frac{\partial S_w}{\partial \tau} + \nabla\left(\mathbf{b}^m\left(S_w\right)S_w\right) - \nabla\left(D\left(S_w\right)\nabla S_w\right) = -S_w \frac{\partial \phi(p)}{\partial t} + q_w \tag{21}$$

$t_m \leq t \leq t_{m+1}$, together with proper initial and boundary conditions. As noted earlier, the saturation S_w is coupled to the pressure/velocity equations, which will be solved by mixed finite-element methods (Douglas *et al.* 1983; Ewing & Heinemann 1984; Ewing *et al.* 1984, 1985).

For a fully developed shock, the characteristic solution of Equation 20 will always produce a unique solution and, as in the miscible or single-phase case, we may use long time steps Δ_t without loss of accuracy. Equation 21 is solved by Petrov Galerkin variational methods, where the time derivative and the non-linear constants are approximated by the solution from Equation 20 (Espedal & Ewing 1987). An iterative solution procedure based on domain decomposition methods (Bramble *et al.* 1988) is used in the solution of the variational form of Equation 20.

Local grid refinement

It seems natural to relate the size of the coarse domains to the solution of the pressure velocity equation (Espedal & Ewing 1987), since the velocity varies slowly and defines a natural long length scale compared with the variation of the saturation S at a front. A simple local error estimate, which determines if a coarse-grid block must be refined, is given in Espedal & Ewing (1987). Normally, local refinement must be performed if a fluid interface is located within the coarse-grid block in order to resolve the solution there. A slightly different strategy is to make the region of local refinement big enough such that we can use the same refinements for several of the large time steps allowed by the method. The local grid refinement strategy combined with operator splitting is defined in the literature (Espedal & Ewing 1987; Dahle *et al.* 1990). The solution at groups of the coarse-grid vertices and the local refinement calculations may be sent to separate processors to achieve a high level of parallelism in the solution process.

The difficulty with these techniques is the communication of the solution between the fine and coarse grids. The use of local grid refinement in large-scale simulators often destroys the vectorization and efficient solution capabilities of the codes. Patch approximation techniques coupled with domain decomposition iterative solution methods (Bramble *et al.* 1988) have proven to be very effective in developing accurate and efficient local grid refinement in the context of existing simulators. Mass balance considerations are very important for accuracy. Approximation techniques for cell-centred finite-difference methods appearing in Ewing *et al.* (1991, 1994) are discussed and compared with other methods.

These techniques can be extended to local time-stepping schemes (Ewing *et al.* 1989*a*) and to algorithms for mixed finite-element methods (Ewing *et al.* 1990). Mixed methods are being incorporated into existing finite-difference simulators to address the need for accurate approximation of fluid velocities in the context of heterogeneous media.

Upscaling

The solution procedure described above represents an excellent tool for handling multiscale phenomena. Often data such as permeability, porosity, and capillary forces will have a multiscale dependence. Our multilevel solution procedure fits into this very well. However, the local refinement capabilities also mean that we must be able to give the appropriate model equations for different computational scales. Given a local computational grid, subgrid information has to be incorporated properly into the data representation.

Large-scale groundwater or oil reservoirs may have a very complex structure, and the geological description is normally a subject of great uncertainty. A two-stage geostatistical model has been proposed (Haldorsen & Damseth 1990).

(1) Large-scale heterogeneities associated with facies are modelled from the information achieved from seismic data, well data, and analogous outcrops.
(2) Rock properties of the facies are modelled by a continuous multivariate Gaussian field or other statistical models. Seismic and well data can be used by a conditioning technique, and core data and other available data should be used to determine the statistical properties of the random field (mean, variance, correlation, etc.).

The coarsest computational domains should coincide with the facies of the model. The level of refinement of these domains and the grid within a given domain have to be decided from the knowledge of geometry, permeability variation, pressure gradients, etc.

In both groundwater and petroleum modelling, a substantial amount of research has been done on the upscaling of the permeability field to give a grid-block permeability (Dagan, 1989; Rubin, 1990), which could be used within our computational framework. The homogenization type of upscaling (Amaziane & Bourgeat 1988; Amaziane *et al.* 1991), which leads to a symmetric block-tensor for the permeability, seems to be especially well suited. The additive Schwarz type of domain decomposition methods leads to zero-Dirichlet boundary conditions for the local computations, consistent with the periodic boundary condition needed for the homogenization technique. One should note that within our computational framework, we need only the assumption of a periodic medium locally on a given domain. We want to extend this single-scale homogenization technique to a multiscale model. Based on a wavelet representation of permeability, we have started research within this area, and so far the results look promising.

The upscaling of the saturation equation gives a new macrodispersion term in Equation 10, originating from the subgrid permeability variation. For a single-phase model this has been successfully studied (Lasseter *et al.* 1986; Dagan 1989; Ewing *et al.* 1989*b*; Rubin 1990; Binning 1994).

Within two-phase models little work has been done. Using a multifractal hypothesis, scaling laws for macrodispersion terms have recently been presented in the literature (Glimm & Lindquist 1992; Glimm *et al.* 1993). Also, macrodispersion models have recently been derived for a model where the permeability has a lognormal distribution (Langlo & Espedal 1994). The derivation is based on the solution technique given above. It gives a saturation-dependent block-tensor dispersion coefficient. From the numerical experiments that are performed, we can conclude that the weakly correlated saturation fluctuations, on average, can be adequately described by this dispersion term.

Upscaling, leading to block-tensor dispersion terms, falls naturally into our computational setup, and we will continue our work based on this kind of modelling.

Numerical experiments

A wide variety of numerical experiments have been performed using the techniques and concepts described in this chapter. A variety of numerical results for single-phase problems were presented by Ewing *et al.* (1989*b*). There the entire heterogeneity distribution and fingering phenomenon were described, with some lack of success, by the dispersion tensor alone. We feel that the highly correlated heterogeneities must be described by first-order effective permeability via homogenization techniques and the subgrid effects by dispersion methods. Continued research is underway for these problems.

In the multiphase flow case, experiments have been described in Espedal *et al.* (1990, 1991). These computations indicate that fingering instabilities initiated by fine-grid heterogeneities can effectively be modelled via a dispersion tensor. However, highly correlated heterogeneities must be treated by effective coarse-grid permeabilities. The non-linearities limit the growth of the mixing zone in this study so that an isotropic tensor can be used for the fine grid effects. Future work will consider a continuum of scales of heterogeneity.

In all the numerical experiments, we systematically varied mobility ratio, longitudinal and transverse dispersivity, and heterogeneity on fine grids. We used lognormal permeability distributions, considering the effect of variance. We also simulated several different randomly generated permeabilities with the same statistical properties to see whether the gross fingering behaviour and recovery are similar. Then we sought relationships between the fine-grid parameters and those in the coarse-grid models to use effective parameters which match 'averaged' properties of many fine-grid simulations. The computations for both the multicomponent and multiphase models on fine grids have been matched to some extent via dispersion models for limited examples. Additional work for several scales of heterogeneity is essential.

Conclusions and directions for further study

Given the importance of dispersion and fingering in the modelling of many flow processes, research must progress in several directions. First, the averaging processes used to change length scales must be improved, perhaps via more statistical techniques, to obtain better effective reservoir coefficients for the macroscopic models. Simultaneously, the effective length scales of dispersion and its effect upon dispersive phenomena must be better understood. Also, better macroscopic techniques for including the effects (perhaps statistical) of channelling and dispersion must be developed.

Even if the information known about the reservoir properties in a highly heterogeneous medium is complete, the problem of how to represent this medium on coarse-grid blocks of different length scales still remains. The power of supercomputers must be brought to bear for simulation studies using homogenization and statistical averaging to represent fine-scale phenomena on coarser grids. The estimated viscosity, permeability, or dispersion coefficients must be modified to incorporate the important effects of fingering. Describing fingering on a grid size that resolves the fingers is impossible, even on the largest supercomputers, but since its presence can dominate flow, its effects must be included in large-scale simulators on coarse grids. We are developing effective equations and effective parameters to model this important effect. Much work is needed in this research area. Additional directions for research have also been given. Progress will depend upon the combination of stochastic techniques and deterministic simulation with accurate numerical techniques that do not possess artificial grid-size-related diffusion or dispersion phenomena that may mask the true phenomena.

References

AMAZIANE, B. & BOURGEAT, A. 1988. Effective behaviour of two-phase flow in heterogeneous reservoir. *In*: WHEELER, M. F. (ed.) *Numerical Simulation in Oil Recovery*. IMA Volumes in Mathematics and its Application **11**. Springer-Verlag, Berlin, 1–22.

AMAZIANE, B., BOURGEAT, A. & KOEBBE, J. 1991. Numerical simulation and homogenization of two-phase flow in heterogeneous porous media. *In*: HORNUNG, U., DOGA, Z. M. & KNABER, P. (eds) *Transport in Porous Media*. Kluwer Academic Publishers, Dordrecht, 519–548.

ARBOGAST, T. 1988. The double porosity model for single phase flow in naturally-fractured reservoirs. *In*: WHEELER, M. F. (ed.) *Numerical Simulation in Oil Recovery*. IMA Volumes in Mathematics and its Applications, **11**. Springer-Verlag, Berlin, 23–45.

ARBOGAST, T. 1989*a*. On the simulation of incompressible miscible displacement in a naturally-fractured reservoir. *RAIRO Modélisation Mathématique et Analyse Numérique*, **23**, 5–51.

ARBOGAST, T. 1989*b*. Analysis of the simulation of single phase flow through a naturally-fractured reservoir. *SIAM Journal of Numerical Analysis*, **26**, 12–29.

ARBOGAST, T., DOUGLAS, J. JR. & HORNUNG, U. 1990. Derivation of the double porosity model of single phase flow via homogenization theory. *SIAM Journal of Mathematical Analysis*, **21**, 823–836.

ARBOGAST, T., DOUGLAS, J. JR. & SANTOS, J. E. 1988. Two-phase immiscible flow in naturally-fractured reservoirs. *In*: WHEELER, M. F. (ed.) *Numerical Simulation in Oil Recovery*. IMA Volumes in Mathematics and its Applications, **11**. Springer-Verlag, Berlin, 47–66.

BARENBLATT, V. M., ZHELTOV, I. P. & KOCHINA, I. N. 1960. Basic concepts in the theory of seepage of homogeneous liquids in fissured rocks [strata]. *Prikladnaja Matematika i Mehanika*, **24**, 852–864.

BARENBLATT, V. M., ENTOV, V. M. & RYZHIK, V. M. 1990. Theory of fluid flows through natural rocks. *In*: BEAR, J. (ed.) *Theory and Applications of Transport in Porous Media*. Kluwer Academic Publishers, Dordrecht.

BETHKE, C. M. 1985. A numerical model of compaction-driven groundwater flow and heat transfer and its application to the paleohydrology of intracratonic sedimentary basins. *Journal of Geophysical Research*, **90**, 6817–6828.

BINNING, P. J. 1994. *Modeling Unsaturated Zone Flow and Contaminant Transport in the Air and Water Phase*. PhD Dissertation, Department of Civil Engineering and Operations Research, Princeton University.

BOYETT, B. A., EL-MANDOUH, M. S. & EWING, R. E. 1992. Local grid refinement for reservoir simulation. *In*: FITZGIBBON, W. E. & WHEELER, M. F. (eds) *Computational Methods in Geosciences, Frontiers in Applied Mathematics*. SIAM, Philadelphia, 15–28.

BRAMBLE, J. H., EWING, R. E., PASCIAK, J. E. & SCHATZ, A. H. 1988. A preconditioning technique for the efficient solution of problems with local grid refinement. *Computer Methods in Applied Mechanics and Engineering*, **67**, 149–159.

CARRERA, J. & NEUMAN, S. P. 1986. Estimation of aquifer parameters under transient and steady state conditions, 1. Maximum likelihood method incorporating prior information. *Water Resources Research*, **22** (2), 199–210.

CELIA, M. A., RUSSELL, T. F., HERRERA, I. & EWING, R. E. 1990. A Eulerian-Lagrangian localized adjoint method for the advection-diffusion equation. *Advances in Water Resources*, **13**, 187–206.

CHAVENT, G. 1976. A new formulation of diphasic incompressible flows in porous media. *Applications of Methods of Functional Analysis to Problems in Mechanics*, **503**, Springer-Verlag, Berlin.

CHEN, H., EWING, R. E., LYONS, S. L. ZIN, G., SUN, T. & YALE, D. P. 1999. A numerical algorithm for single-phase fluid flow in elastic porous media. *Lecture Notes in Physics*. Springer-Verlag, Berlin, 1–44.

CHEN, Z., EWING, R. E. & ESPEDAL, M. 1994. Multiphase flow simulation with various boundary conditions. *In*: PETERS, A., WITTUM, G., HERRLING, B., MEISSNER, U. BREBBIA, C. A., GRAY, W. G. & PINDER, G. F. (eds) *Computational Methods in Water Resources*. Kluwer Academic Publishers, Dordrecht, 925–932.

DAGAN, G. 1985. Stochastic modeling of groundwater flow by unconditional and conditional probabilities: The inverse problem. *Water Resources Research*, **21** (1), 65–72.

DAGAN, G. 1989. *Flow and Transport in Porous Formations*. Springer-Verlag, Berlin.

DAGAN, G. & RUBIN, Y. 1988. Stochastic identification of recharge, transmissivity and storativity in aquifer unsteady flow: A quasi-steady approach. *Water Resources Research*, **24** (10), 1698–1710.

DAHLE, H. K., ESPEDAL, M. S., EWING, R. E. & SÆVAREID, O. 1990. Characteristic adaptive subdomain methods for reservoir flow problems. *Numerical Methods for Partial Differential Equations*, **6**, 279–309.

DOUGLAS, J. JR. & ARBOGAST, T. 1990. Dual porosity models for flow in naturally fractured reservoirs. *In*: CUSHMAN, J. H. (ed.) *Dynamics of Fluids in Hierarchical Porous Formations*. Academic Press, London, 177–221.

Douglas, J. Jr. & Russell, T. F. 1982. Numerical methods for convection dominated diffusion problems based on combining the method of Characteristics with Finite Element or Finite Difference Procedures. *SIAM Journal of Numerical Analyses*, **19**, 871–88.

DOUGLAS, J. JR., EWING, R. E. & WHEELER, M. F. 1983. A time-discretization procedure for a mixed finite element approximation of miscible displacement in porous media. *RAIRO Analyse* Numerique, 17, 249–265.

DOUGLAS, J. JR., PAES-LEME, P. J., ARBOGAST, T. & SCHMITT, T. 1987. Simulation of flow in naturally-fractured petroleum reservoirs. *In*: *Proceedings of the Ninth SPE Symposium on Reservoir Simulation*. Society of Petroleum Engineers, Dallas, Texas, SPE 16019, 271–279.

DOUGLAS, J. JR., PAES-LEME, P. J. & HENSLEY, J. L. 1991. A limit form of the equations for immiscible displacement in a fractured reservoir. *Transport in Porous Media*, **6**, 549–565.

Douglas, J. Jr., Arbogast, T., Paes-Leme, P. J., Hensley, J. L. & Nunes, N. P. 1993. Immiscible displacement in vertically fractured reservoirs. *Transport in Porous Media*, **12**, 73–1006.

DOUGLAS, J. JR., FURTADO, F. & PEREIRA, F. 1997. On the numerical simulation of waterflooding of heterogeneous petroleum reservoirs. *Computational* Geosciences, 1, 155–190.

DOUGLAS, J. JR., KISCHINHEVSKY, M., PAES-LEME, P. J. & SPAGNUOLO, A. M. 1998. A multiple-porosity model for a single-phase flow through naturally-fractured porous media. *Computing in Applied Mathematics*, **17**, 19–48.

DOUGLAS, J. JR., HUANG, C.-S. & PEREIRA, F. (in press). The modified method of characteristics with adjusted advection. *Numerische Mathematik*.

DU ROUCHET, J. 1981. Stress fluid, a key to oil migration. *AAPG Bulletin*, **65**, 74–85.

ESPEDAL, M. S. & EWING, R. E. 1987. Characteristic Petrov-Galerkin subdomain methods for two-phase immiscible flow. *Computer Methods in Applied Mechanics and Engineering*, **64**, 113–135.

ESPEDAL, M. S., HANSEN, R., LANGLO, P., SÆVAREID, O. & EWING, R. E. 1990. Heterogeneous porous media and domain decomposition methods. *Proceedings 2nd European Conference on the Mathematics of Oil Recovery*. Technip, Paris, 157–163.

ESPEDAL, M. S., LANGLO, P., SÆVAREID, O. & HANSEN, R. 1991. Heterogeneous reservoir models, local refinements, and effective parameters. *Proceedings of Eleventh SPE Symposium on Reservoir* Simulation, Anaheim, CA. SPE 21231, 307–316.

EWING, R. E. 1989*a*. Domain decomposition techniques for efficient adaptive local grid refinement. *In*: CHAN, T. F., GLOWINSKI, R., PERIAUX, J. L. & WIDLUND, O. B. (eds) *Domain Decomposition Methods*. SIAM, Philadelphia, 192–206.

EWING, R. E. 1989*b*. Adaptive grid refinements for transient flow problems. *In*: FLAHERTY, J. E., PASLOW, P. J., SHEPHARD, M. S. & VASILAKIS, J. D. (eds) *Adaptive Methods for Partial Differential Equations*. SIAM, Philadelphia, 194–205.

EWING, R. E. 1991. Operator splitting and Eulerian-Lagrangian localized adjoint methods for multiphase flow. *In*: Whiteman, J. (ed.) *The Mathematics of Finite Elements and Applications VII MAFELAP 1990*. Academic Press, San Diego, 215–237.

EWING, R. E. & GEORGE, J. H. 1984. Viscous fingering in hydrocarbon and recovery processes. *In*: GROSS, K. I. (ed.) *Mathematical Methods in Energy Research*. SIAM, Philadelphia, 194–213.

EWING, R. E. & HEINEMAN, R. F. 1984. Mixed finite element approximation of phase velocities in compositional reservoir simulation. *Computer Methods in Applied Mechanics and* Engineering, **47**, 161–176.

EWING, R. E., RUSSELL, T. F. & WHEELER, M. F. 1983. Simulation of miscible displacement using mixed methods and a modified method of characteristics. *Proceedings of Seventh SPE Symposium on Reservoir Simulation*, San Francisco, CA. SPE 12241, 71–82.

EWING, R. E., RUSSELL, T. F. & WHEELER, M. F. 1989. Convergence analysis of an approximation of miscible displacement in porous media by mixed finite elements and a modified method of characteristics. *Computer Methods in Applied Mechanics and Engineering*, **47**, 73–92.

EWING, R. E., KOEBBE, J. V., GONZALEZ, R. & WHEELER, M. F. 1985. Mixed finite element methods for accurate fluid velocities. *In*: GALLAGHER, R. H., CAREY, G. F., ODEN, J. T. & ZIENKIEWICZ, O. C. (eds) *Finite Elements in Fluids* **6**. Wiley, New York, 233–249.

EWING, R. E., BOYETT, B. A., BABU, D. K. & HEINEMANN, R. F. 1989*a*. Efficient use of locally refined grids for multiphase reservoir simulations. *Proceedings of Tenth SPE Symposium on Reservoir Simulation*. Houston, TX, February 6–8, 1989. SPE 18413, 55–70.

EWING, R. E., RUSSELL, T. F. & YOUNG, L. C. 1989*b*. An anisotropic Coarse-grid dispersion model of heterogeneity and viscous fingering in five-spot miscible displacement that matches experiments and fine-grid simulations. *Proceedings of Tenth SPE Symposium on Reservoir* Simulation, Houston, Texas, February 6–8, 1989, SPE 18441, 447–466.

EWING, R. E., LAZAROV, R. D., RUSSELL, T. F. & VASSILEVSKI, P. S. 1990. Local refinement via domain decomposition techniques for mixed finite element methods with rectangular Raviart-Thomas elements. *In*: CHAN, T. F., GLOWINSKI, R., PERIAUX, J. & WIDLUND, O. B. (eds) *Domain Decomposition Methods for Partial Differential Equations*. SIAM Publications, Philadelphia, 98–114.

EWING, R. E., LAZAROV, R. D. & VASSILEVSKI, P. S. 1991. Local refinement techniques for elliptic problems on cell-centred grids, I: Error analysis. *Mathematics and Computing*, **194**, 437–462.

EWING, R. E., LAZAROV, R. D. & VASSILEVSKI, P. S. 1994. Local refinement techniques for elliptic problems on cell-centered grids, II: Optimal order two-grid iterative methods. *Numerical Linear Algebra and its Applications*, **1** (4), 337–368.

FURTADO, J., GLIMM, J., LINDQUIST, W. B. & PEREIRA, L. F. 1991. Characterization of mixing length growth for flow in heterogeneous porous media. *Proceedings of Eleventh SPE Symposium on Reservoir Simulation*, Anaheim, CA, 317–322.

GELHAR, L. W. & AXNESS, C. L. 1983. Three-dimensional stochastic analysis of macro-dispersion in aquifers. *Water Resources Research*, **19** (1), 161–180.

GINN, T. R. & CUSHMAN, J. H. 1990. Inverse methods for subsurface flow: A critical review of stochastic techniques. *Stochastic Hydrology and Hydraulics*, **4**, 1–26.

GLIMM, J. & LINDQUIST, W. B. 1992. Scaling laws for macrodispersion. *Proceedings of Ninth International Conference on Computational Methods in Water Resources*, **2**, 35–49.

GLIMM, J., LINDQUIST, W. B., PEREIRA, F. & ZHANG, Q. 1993. A theory of macrodispersion for the scale up problem. *Transport in Porous Media*, **13**, 97–122.

HALDORSEN, H. & DAMSETH, E. 1990. Stochastic modeling. *Journal of Petroleum Technology*, **42**, 404–413.

KITANIDIS, P. K. & VOMVORIS, E. G. 1983. A geostatistical approach to the inverse problem in groundwater modeling (steady state) and one-dimensional simulations. *Water Resources Research*, **19** (3), 677–690.

LANGLO, P. & ESPEDAL, M. S. 1994. Macrodispersion for two-phase, immiscible flow in porous media. *Advances in Water Resources*, **17**, 297–316.

LASSETER, T. J., WAGGONER, J. R. & LAKE, L. W. 1986. Reservoir heterogeneities and their influence on ultimate recovery. *In*: LAKE, L. W. & CARROLL, H. B. (eds) *Reservoir Characterization*. Academic Press, New York, 545–560.

MEISSNER, F. 1978. Petroleum geology of the Bakken Formation, Walliston Basin, North Dakota and Montana. *Williston Basin Symposium Proceedings*. Montana Geological Society, 207–227.

NAKAYAMA, K. & LERCHE, I. 1987. Basin analysis by model simulation: effects of geological parameters on 1D and 2D fluid flow systems with applications to an oil field. *Gulf Coast Association of Geological Societies Transaction*, **37**, 175–184.

NEUMAN, S. P. 1980. A statistical approach to the inverse problem of aquifer hydrology, III. Improved solution method and added perspective. *Water Resources Research*, **16** (2), 331–346.

NEUMAN, S. P. & YAKOWITZ, S. 1979. A statistical approach to the inverse problem of aquifer hydrology, I. Theory. *Water Resources Research*, **15** (4), 845–860.

NEUMAN, S. P. & ZHANG, Y. K. 1990. A quasi-linear theory of non-Fickian and Fickian subsurface dispersion, 1. Theoretical analysis with application to isotropic media. *Water Resources Research*, **26** (5), 887–902.

NEUMAN, S. P., FOGG, A. & JACOBSEN, E. 1980. A statistical approach to the inverse problem of aquifer hydrology, II. Case Study. *Water Resources Research*, **16**, 33–58.

NEUMANN, S. P., WINTER, C. L. & NEWMAN, C. M. 1987. Stochastic theory of field-scale dispersion in anisotropic porous media. *Water Resources Research*, **23** (3), 453–466.

PALCRAUSKAS, V. V. & DOMENICO, P. A. 1980. Microfacture development in compacting sediments: relation to hydrocarbon maturation kinetics. *AAPG Bulletin*, **52**, 57–65.

PERKINS, T. K. & JOHNSTON, O. C. 1963. A review of diffusion and dispersion in porous media. *Society of Petroleum Engineers Journal*, **3**, 70–84.

RUBIN, Y. 1990. Stochastic modeling of macrodispersion in heterogeneous porous media. *Water Resources Research*, **26**, 133–141.

RUBIN, Y. & DAGAN, G. 1987*a*. Stochastic identification of transmissivity and effective recharge in steady state groundwater flow, 1. Theory. *Water Resources Research*, **23** (7), 1185–1192.

RUBIN, Y. & DAGAN, G. 1987*b*. Stochastic identification of transmissivity and effective recharge in steady state groundwater flow, 2. Case study. *Water Resources Research*, **23** (7), 1192–1200.

RUBIN, Y., MAVKO, G. & HARRIS, J. 1992. Mapping permeability in heterogeneous aquifers using hydrological and seismic data. *Water Resources Research*, **28** (7), 1809–1816.

RUSSELL, T. F. 1985. The time-stepping along characteristics with incomplete iteration for Galerkin approximation of miscible displacement in porous media. *SIAM Journal of Numerical Analysis*, **22**, 970–1013.

RUSSELL, T. F. & WHEELER, M. F. 1983. Finite element and finite difference methods for continuous flows in porous media. *In*: EWING, R. E. (ed.) *The Mathematics of Reservoir Simulation*. Frontiers in Applied Mathematics, SIAM, Philadelphia, 35–106.

SLATTERY, J. C. 1969. Single-phase flow through porous media. *AIChE Journal*, **15**, 866–872.

SLATTERY, J. C. 1970. Two-phase flow through porous media. *AIChE Journal*, **15**, 345–352.

SMITH, J. E. 1971. The dynamics of shale compaction and evolution of fluid pressure. *Mathematical Geology*, **3** (3), 239–269.

SPAGNUOLO, A. M. (in press) Part I: Convergence analysis of an approximation of contaminant transport in porous media by mixed finite elements and a modified method of characteristics with adjusted advection.

TODD, M. R. & LONGSTAFF, W. J. 1972. The development, testing, and application of a numerical simulator for predicting miscible flood performance. *Journal of Petroleum Technology*, **253**, 874–882.

UNGERER, P., BURRUS, J., DOLIGEZ, B., CHENET, P. Y. & BESSIS, F. 1990. Basin evaluation by integrated two-dimensional modeling of heat transfer, fluid flow, hydrocarbon generation, and migration. *AAPG Bulletin*, **74** (3), 309–335.

WARREN, J. E. & ROOT, P. J. 1963. The behavior of naturally fractured reservoirs. *Society of Petroleum Engineers Journal*, **3**, 245–255.

WARREN, J. E. & SKIBA, F. F. 1964. Macroscopic dispersion. *Society of Petroleum Engineers Journal*, **4**, 215–230.

WELTE, D. H. & YUKLER, M. A. 1981. Petroleum origin and accumulation in basin evolution – a quantitative model. *AAPG Bulletin*, **65**, 1387–1396.

YEH, W.W.-G. 1986. Review of parameter identification procedures in groundwater hydrology: The inverse problem. *Water Resources Research*, **22** (2), 95–108.

YOUNG, L. C. 1984. A study of spatial approximations for simulating fluid displacements in petroleum reservoirs. *Computer Methods in Applied Mechanics and Engineering*, **47**, 3–46.

Simulation of water and gas flow in fractured porous media

ABDELAZIZ L. AL-KHLAIFAT[1] & HAMID ARASTOOPOUR[2]

[1]*Department of Chemical Engineering, Faculty of Engineering, Mutah University, Mutah 61710 Jordan (email: khlaifat@mutah.edu.jo)*

[2]*Department of Chemical and Environmental Engineering, Illinois Institute of Technology, Chicago, IL 60616, USA*

Abstract: The porous media considered in this paper belong to gas reservoirs characterized by a low-permeability matrix, combined with a high-permeability fracture system. The transient, three-dimensional, two-phase numerical model is presented for simulating the simultaneous flow of gas and water through porous and fractured media. The specific problem to which the numerical simulation is applied assumes a single horizontal fracture perpendicular to the flow direction in a quite small geometry. The equations solved treat both fluids as incompressible and immiscible. Accounted for are the effect of capillary pressure and relative permeability. Technically, the finite volume approach is used to discretize the equations. The set of discretized and linearized equations is solved using the IPSA (inter-phase slip algorithm) method. The results of this study indicate that the multi-layer reservoir provides better estimates of post-fracture performance compared to a more conventional, single-layer reservoir description.

The prediction of flow patterns in fractured porous media has important technological applications. Geothermal and petroleum reservoir engineering, the prediction of the impact of nuclear waste repository emplacements and many other applications justify the growing interest in this subject.

Simulation of fractured reservoirs is a challenging task from both a reservoir description and a numerical standpoint. Flow of fluids through the reservoir primarily is through the high-permeability, low-effective-porosity fractures surrounding individual matrix blocks. The matrix blocks contain the majority of the reservoir reserves and act as source or sink terms to the fracture.

Economic production of gas from the rock matrix generally requires that flow to the well bore be aided by natural or induced fracture systems. However, production rate is still set by the rate of flow of gas from the matrix into the fracture system. Therefore knowledge of fluid flow in the matrix is required. There are several experimental studies in the literature for flow properties in tight sand such as (Jones & Owens 1980; Randolph 1983; Walls *et al.* 1982; Freeman & Bush 1983; Soeder & Arastoopour 1986; Al-Khlaifat & Arastoopour 1999; Al-Khlaifat *et al.* 2001); however, there is a need for more experimental results to determine the behaviour of fractured tight sand formations.

In the literature there are few studies regarding the theoretical analysis in single and multiphase flow in tight sand formations. Newberg & Arastoopour (1986) and Chowdiah & Arastoopour (1983) were among the first to develop a mathematical model to describe transient flow of gas through low-permeability tight sand media. Fractured reservoirs have been studied extensively over the last decades. Kazemi (1969) simulated the behaviour of naturally fractured reservoirs. A good review by Lee & Tan (1987) gave a summary of work on multiphase numerical simulation of fractured reservoirs. Al-Khlaifat & Arastoopour (1997) were the first to modify the multifluid model of CFX-F3D program to account for capillary pressure effect. In their modification the fluid velocities were assumed to follow Darcy and non-Darcy equations for the water and gas phase, respectively.

Usually fractured reservoirs are pictured as double porosity and permeability media, namely matrix and fracture. The permeability and porosity of the fracture, particularly in tight sand formations, are significantly higher than the matrix. It is believed that the interconnected fracture network provides the main path for fluid flow through the reservoir (Codreanu *et al.* 1985). Based on this concept Kazemi (1976) and Thomas *et al.* (1983) developed dual-porosity, dual-permeability models which consist of two

From: AMEEN, M. (ed.) *Fracture and In-Situ Stress Characterization of Hydrocarbon Reservoirs*. Geological Society, London, Special Publications, **209**, 201–212. 0305–8719/03/$15.00
© The Geological Society of London 2003.

systems of flow equations, one for matrix and another for fracture, with a source term which accounts for the matrix fracture flow.

This paper presents the development of a model to simulate two-phase flow through fractured porous media. The three-dimensional, isothermal, transient two-phase flow model was solved using the CFX-F3D (1995) computer code. Our model includes an experimentally based expression for the capillary pressure as a function of concentration of each phase, Darcy and non-Darcy expressions for water and gas phases, and empirical expressions of the relative permeability of each phase. This model was solved for the case when the dimensions of matrix blocks are very small.

Mathematical formulation

A three-dimensional, isothermal, two-phase flow through porous media model was developed using the volume averaging technique (Al-Khlaifat 1998) to study the behaviour of homogeneous and rigid tight porous media. In order to simulate this phenomenon, we had to modify the multi-fluid model of CFX-F3D to account for the capillary pressure effect. The experimental values of capillary pressure as a function of concentration of each phase were included in the code. Darcy and non-Darcy expressions for water and gas phases were used as interfacial forces between the phases and porous media.

Continuity equation:

$$\frac{\partial}{\partial t}(\phi S_\alpha \rho_\alpha)+\nabla(\phi S_\alpha \rho_\alpha \mathbf{U}_\alpha)=0 \tag{1}$$

Momentum equation:

$$\frac{\partial}{\partial t}(\phi S_\alpha \rho_\alpha \mathrm{U}_\alpha)+\nabla(\phi S_\alpha(\rho \mathrm{U}_\alpha \otimes \mathrm{U}_\alpha - \tau_\alpha)=\phi S_\alpha(\mathrm{B}_{m\alpha}-\nabla P_\alpha) \tag{2}$$

where α denotes phase (water or gas), ϕ is porosity, ρ_α density, S_α saturation, $\mathbf{U}_\alpha$ velocity vector, τ_α viscous stress tensor, $\mathbf{B}_{m\alpha}$ interfacial momentum transfer between the rock matrix and phase.

The viscous stress tensor may be expressed as:

$$\tau_\alpha=\mu_\alpha(\nabla\mathbf{U}_\alpha+(\nabla\mathbf{U}_\alpha)^T) \tag{3}$$

where T denotes transpose and equation of state as:

$$\rho_\alpha=\rho_\alpha(T_\alpha,P_\alpha) \tag{4}$$

where T denotes temperature. In addition:

$$S_w+S_g=1 \tag{5}$$

The liquid-rock interfacial force was expressed in the form of Darcy's equation:

$$\mathbf{B}_{rw}=\frac{\mu_w\varepsilon_w}{k_w}\mathbf{U}_w \tag{6}$$

and the gas-rock interfacial force was expressed in the following form:

$$\mathbf{B}_{rg}=\frac{\mu_g\varepsilon_g}{k_g}\mathbf{U}_g+\beta_g\rho_g\varepsilon_g(\mathbf{U}_g)^2 \tag{7}$$

where ε_α is the volume fraction ($\varepsilon_\alpha=\phi S_\alpha$), k_α, effective permeability, and β_α is a coefficient accounting for local flow non-uniformities, and usually is a function of pore size, pore size distribution, and pore structures.

Equations 1, 2, 4, and 5 represent 11 equations and 12 unknowns (ρ_g, ρ_w, U_g, V_g, W_g, U_w, V_w, W_w, P_g, P_w, S_g and S_w) where U, V and W are the x, y and z velocity components, respectively. We need one more equation to close the system. This equation is given by:

$$P_c=P_g-P_w \tag{8}$$

where is the capillary pressure. In our study, the capillary pressure was defined experimentally (Holditch *et al.* 1989) as a function of wetting fluid (water) saturation as shown in the following expression:

$$P_c=\frac{97480}{(S_w)^{2.2658}} \tag{9}$$

where the unit of the capillary pressure is Pascal.

The effects of capillary pressure and interfacial momentum transfer were incorporated into the code through the subroutine 'user body force.' Moreover, we took into consideration the fact that the presence of the non-aqueous fluid reduces the effective permeability of the mixture to the aqueous fluid and vice versa. Thus we can no longer consider k_w and k_g to be rock properties alone, but we must allow them to depend on the degree to which each fluid phase blocks the flow of the other. The most obvious phenomenological postulate allowing for this dependence is to stipulate that k_w and k_g depend on the fluid saturations, the effective permeability of each fluid phase increasing with its saturation. Thus we have

$$k_w = k\,k_{rw}(S_w) \tag{10}$$

$$k_g = k\,k_{rg}(S_w) \tag{11}$$

The functions $k_{r\alpha}(S_w)$ are called relative permeabilities; they obey the bounds $0 \le k_{r\alpha} \le 1$. The most widely used empirical equation for relative permeability curves is defined by Brook & Corey (1966) and described as follows:

$$k_{rw} = S_e^{(2+3\eta)/\eta} \tag{12}$$

and

$$k_{rg} = (1-S_e)^2(1-S_e^{(2+\eta)/\eta}) \tag{13}$$

where S_e is effective saturation defined as:

$$S_e = \frac{S_w - S_r}{1 - S_r} \tag{14}$$

and η is a pore size index, S_w is wetting fluid saturation and S_r is the residual wetting fluid saturation. The pore size index and residual wetting fluid saturation are both empirical values and strongly depend on the type of sandstone and pore size distribution.

Normally, the amount of residual wetting fluid saturation depends on the permeability of the porous medium, and the pore size index depends on the pore size distribution.

Discretization and solution algorithm

The discretization technique used in CFX is based on the finite volume approach. The hydrodynamic equations are in a non-staggered Cartesian or body-fitted grid where all variables are defined at the centre of control volumes (cells). Each equation is integrated over each control volume to obtain a discrete equation that connects the variable at the centre of a control volume with its neighbours. The set of discretized and linearized equations is then solved using the IPSA (inter-phase slip algorithm) method.

Discretization of the Transport Equations

The discretization of the transport equations was done using the standard finite volume technique. Diffusion terms are discretized using a backward first-order differencing scheme. For transient situations, the time-stepping procedure with a fully implicit first-order backward differencing scheme was used.

The convective terms in the continuity equations were solved by the first-order upwind scheme. Although less accurate, the first-order scheme ensured that the volume fractions remained bounded which may not be the case with higher-order schemes. First-order schemes are also more robust and converge faster.

Solution Algorithms

A number of different methods are available for the solution of linearized transport equations. The linear equations were derived by integrating transport equations over control volumes (cells). Iteration was used at two levels: an inner iteration to solve for the spatial coupling for each variable, and an outer iteration to solve for the coupling between variables. Thus each variable was taken in sequence, regarding all other variables as fixed, a discrete transport equation for that variable is formed for every cell in the flow domain, and the problem was handed over to a linear equation solver which returns the updated values of the variables. The non-linearity of the original equations was simulated by reforming the coefficients of the discrete equations, using the most recently calculated values of the variables before each outer iteration.

The treatment of pressure was slightly different from the foregoing description, since it does not obey a transport equation. Instead simplified versions of the discrete momentum equations were used to derive a functional relationship between a correction to the pressure and corrections to the velocity components in each cell. Substitution of these expressions into the continuity equation leads to an equation linking the pressure correction with the continuity error in the cell. This set of simultaneous equations was passed, as before, to a linear equation solver. The solution was used both to update pressure and to correct the velocity field through the functional relationship in order to enforce mass conservation. The exact implementation of this pressure-correction step can give rise to several different velocity–pressure coupling algorithms, of which the SIMPLE algorithm is the best known.

The errors in mass continuity for the nodal velocity are referred to as Rhie-Chow errors. As they are proportional to the grid increment cubed multiplied by a fourth derivative of pressure, they are small, provided that the pressure field is sufficiently smooth. They may be large, however, if there are discontinuities in pressure or pressure gradient. The latter situation occurs when the pressure gradient has to balance a strongly varying body force. This may occur for flow through porous media with discontinuities in resistance. Discontinuities in porosity cause discontinuities in velocity, due to mass continuity, and these, in turn, cause discontinuities in the pressure field. This problem is solved by using the improved Rhie-Chow interpolation method that tries to mimic the staggered grid situation by ensuring that body forces $\mathbf{B}_{m\alpha}$ are defined primitively at cell faces, and by ensuring that $-\nabla_{P\alpha}$ in the Rhie-Chow formula is replaced by the smooth $\mathbf{B}_{m\alpha}-\nabla_{P\alpha}$.

Geometry and topology

The computational domain is a collection of blocks that are connected together across inter-block boundaries. The underlying grid structure of each block is contained in a region that is topologically a cuboid. This means that each block may be considered conceptually as a rectangular array of 'bed strings,' which may be distorted to fit the boundary of a desired physical domain. This distortion defines a 'mapping' from 'computational space,' with coordinates given by the integer indices (I, J, K) of cell vertices, to the desired region of 'physical space'. The only restriction to the mapping is that cells do not overlap.

We created the required physical domain for our simulation using Mesh-Build. The created geometry has the same dimensions of the sample used in our experimental studies as shown in Figure 1. We have built the geometry in such away that in general we have two similar low-permeability cylindrical blocks (porous-1 and porous-3) of 3.81 cm in diameter and 4.35 cm in length and a grid of 10×10×20. In between these two blocks we created a very tiny high permeability block (porous-2) of the same diameter and of 0.05 cm in length (fracture) and a grid of 10×10×1. All the simulations conducted for this geometry were three-dimensional, with neglected gravity effect, and we assumed isothermal, laminar, transient, and incompressible flow. Inlet and outlet pressures were specified to be 4.237 MPa and 0.1 MPa, respectively. Coefficient β was found experimentally to be 1.14×10^{14} m^{-1}.

The flow parameters for low-permeability porous media were found based on single phase experiments at an overburden pressure of 20.7 MPa ($R_w=1.547\times10^{14}$ Pa s m^{-2} and $R_g=6.785\times10^{11}$ Pa s m^{-2} for liquid and gas phases, respectively). The characteristic of the high-permeability portion (middle block (fracture), porous-2) was specified to be similar to the characteristic of Berea Sandstone with porosity

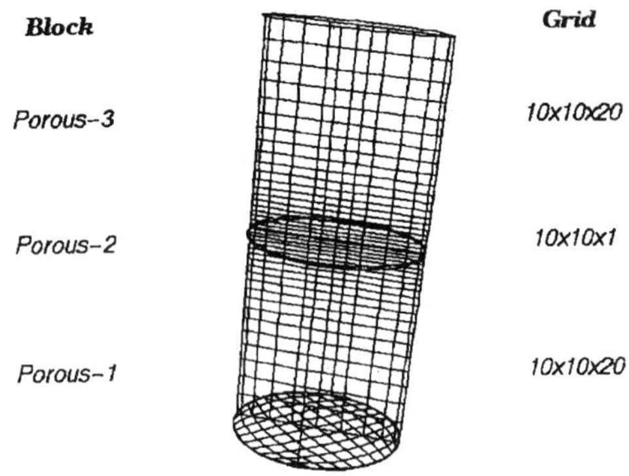

Fig. 1. Schematic representation of fractured porous media with grid definition.

of 0.28 found experimentally at an overburden pressure of 20.7 MPa ($R_w=5.729\times10^{10}$ Pa s m^{-2} and $R_g=5.495\times10^{8}$ Pa s m^{-2} for liquid and gas phases, respectively).

Initial and boundary conditions

In all numerical studies carried out for this geometry, we defined the initial conditions to be the same. Initial water and gas saturation was set to be 0.4 and 0.6 ($S_{w\text{-initially}}=0.4$, $t=0$), respectively, in all three blocks. The inlet saturation was specified to be 0.6 and 0.4 for water and gas phases ($S_w=0.6$, $t=0$), respectively.

At the wall the standard no-slip conditions were used. We used the inflow and outflow pressure boundaries because only pressure was known experimentally. For pressure boundary, pressure is specified and Neumann conditions are applied to the velocities, $\partial\mathbf{U}/\partial n=0$, where n is the normal direction to the inlet or outlet cells. Inlet and outlet pressures were specified to be 4.237 MPa and 0.1 MPa, respectively.

Results and discussion

The mathematical model described above was solved using CFX-F3D for the geometry shown in Figure 1. For the physical domain of three porous blocks in series we have performed three kinds of simulations; the following is a systematic discussion of each simulation.

Two-phase flow simulation through porous media consisting of three porous blocks where the middle block is highly permeable

In this simulation, the first and third porous blocks are a low-permeability porous type and have a porosity of 7% and water and gas resistance of 1.547×10^{14} Pa s m^{-2} and 6.785×10^{11} Pa s m^{-2}, respectively. The second porous block (middle block, porous-2) is a high-permeability type and has a resistance of 5.729×10^{10} Pa s m^{-2} and 5.495×10^{8} Pa s m^{-2} for the water and gas phases, respectively, and a porosity of 28%. The interfacial forces between both phases and the porous medium for the portion of high permeability were expressed by Darcy's equation, while low-permeability portions were expressed by the non-Darcy equation.

The water phase saturation profile versus the length of the sample is shown in Figure 2. From Figure 2 one can see that as time goes on, the water front continuity wave travels along the medium. The inflection in the curve of the water saturation profiles defines the location of the water-gas front throughout the porous blocks as a function of time. The increase in water saturation continued until reaching the specified inlet saturation.

The liquid and capillary pressure profiles are shown in Figure 3. Because of the relationship between capillary pressure and fluid saturation, capillary pressure behaved similar to saturation behaviour. Liquid pressure behaviour was almost linear, except for the change of slope across the high-permeability, high-porosity porous block (porous-2). Lower resistance and pressure drop in the fracture zone cause this change of slope.

Figures 4 and 5 show the calculated liquid and gas velocity profiles, respectively, versus the length of the sample. The velocity profiles for

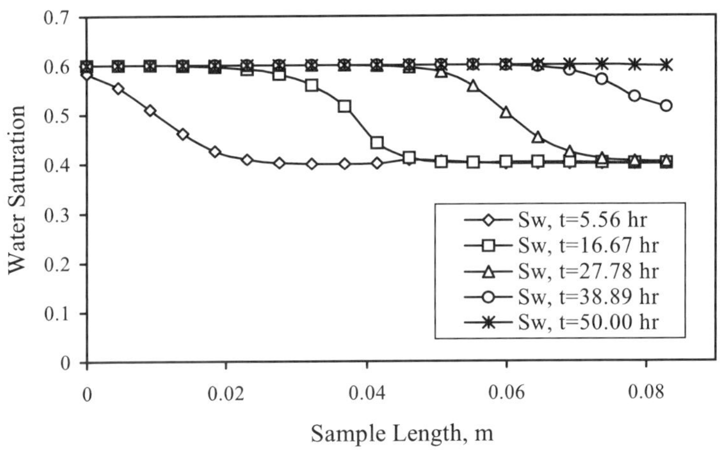

Fig. 2. Water phase saturation distribution of two-phase flow simulation (middle block is highly permeable.

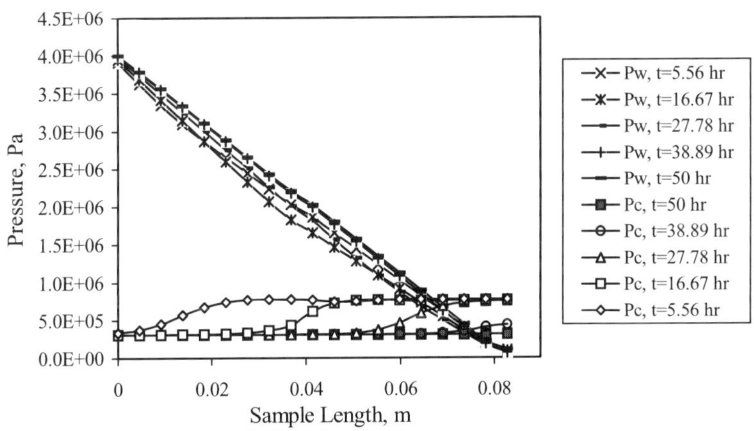

Fig. 3. Pressure distribution of two-phase flow simulation (middle block is highly permeable).

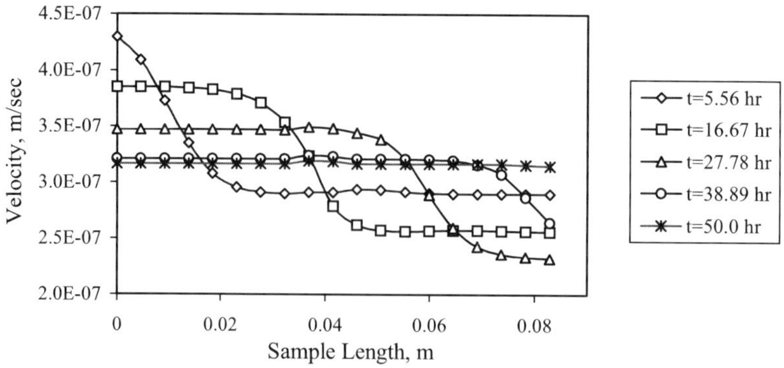

Fig. 4. Water velocity distribution of two-phase flow simulation (middle block is highly permeable).

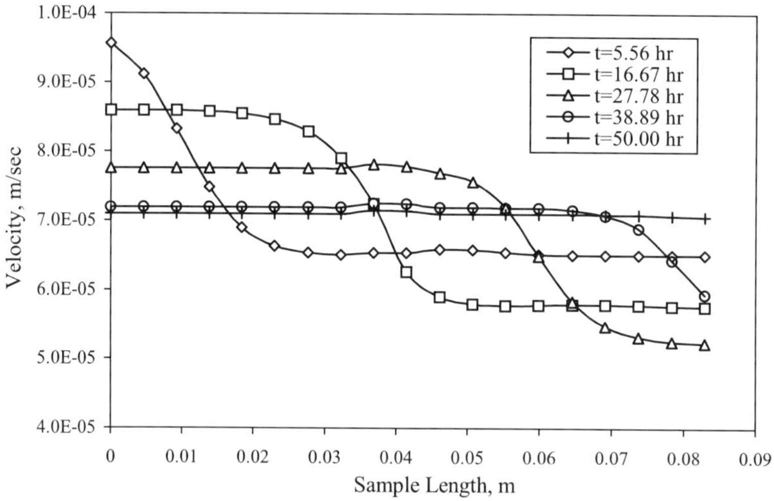

Fig. 5. Gas velocity distribution of two-phase flow simulation (middle block is highly permeable).

both water and gas phases behaved similarly, except that the gas velocity is almost two orders of magnitude higher than the water velocity. This is due to higher gas permeability compared to the liquid phase. The higher permeability zone causes a slight increase in gas and water velocities due to lower resistance to flow. This can be seen clearly at the steady-state condition.

Figure 6 shows the water and gas flow rates at the exit for this simulation that contained initial gas saturation of 0.6. The gas flow rate decreased gradually until it reached steady state after about 50 hours. The low water flow rate during the first 35 hours as shown in Figure 6 was due to very low water saturation, and in turn high pore entry pressure for the water phase. When the water–gas front moved closer to the outlet, the water flow rate began to increase at a reasonably high rate until it reached a steady–state value.

We also studied the effects of different high-permeability media between two low-permeability sections. Figures 7 and 8 show water and gas flow rates, respectively, versus time for the three values of resistance shown in Table 1.

Table 1. *Values of middle block (Porous-2) resistance (Pa s m^{-2})*

	Water Phase	Gas Phase
Resistance-1	5.729×10^{10}	5.495×10^{8}
Resistance-2	5.729×10^{9}	5.495×10^{7}
Resistance-3	5.729×10^{8}	5.495×10^{6}

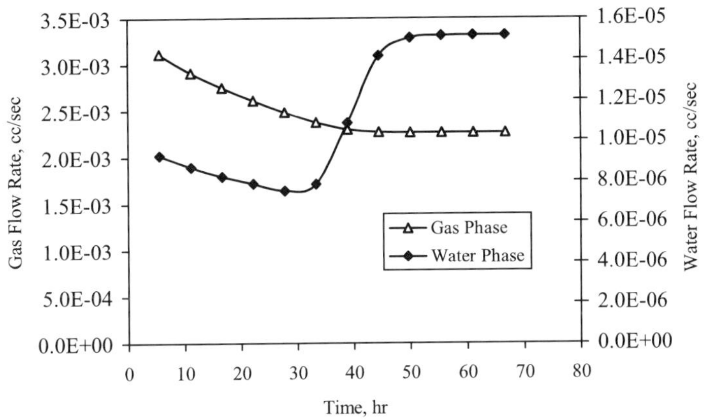

Fig. 6. Water and gas flow rates of two-phase flow simulation (middle block is highly permeable).

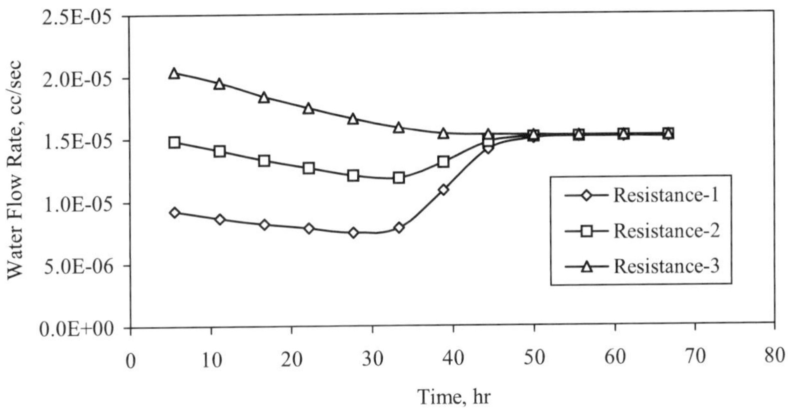

Fig. 7. Effect of resistance of the middle block on the water flow rate of two-phase flow simulation (Resistance-1>Resistance-2>Resistance-3).

From Figure 7 it can be concluded that lower resistance of the middle block (porous-2) results in higher exit water flow rate during the transient condition. However, under steady-state conditions (e.g. after 45 hours), the change in permeability of high-permeability media does not have appreciable influence on the flow. This means that under steady state, water flow rate is controlled only by the lower-permeability zone, and only at the transient condition, when the presence of the higher-permeability zone has a significant effect on water production. The gas flow rate increased slightly under both transient and steady-state conditions, with an increase in the permeability of the high-permeability zone (see Figure 8).

Two-phase flow through three porous blocks where low permeability is followed by high permeability porous media

In this geometry we defined the first block to be of low permeability and the second and third blocks to be of high permeability. Initial gas saturation of all blocks was 0.6 and inlet gas and water saturations were set at 0.4 and 0.6 for the gas and liquid phases, respectively. The first block is of low permeability and has a length of 4.35 cm, while the second and third blocks are of high permeability and have lengths of 0.05 and 4.35 cm, respectively. Figure 9 shows the water phase saturation profile versus the sample length.

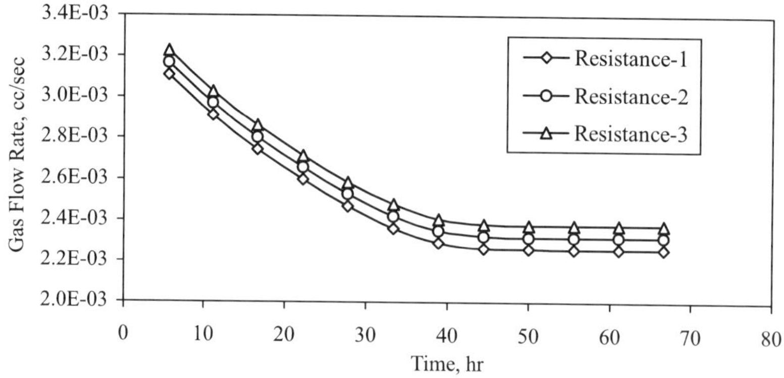

Fig. 8. Effect of resistance of the middle block on the gas flow rate of two-phase flow simulation (Resistance-1>Resistance-2>Resistance-3).

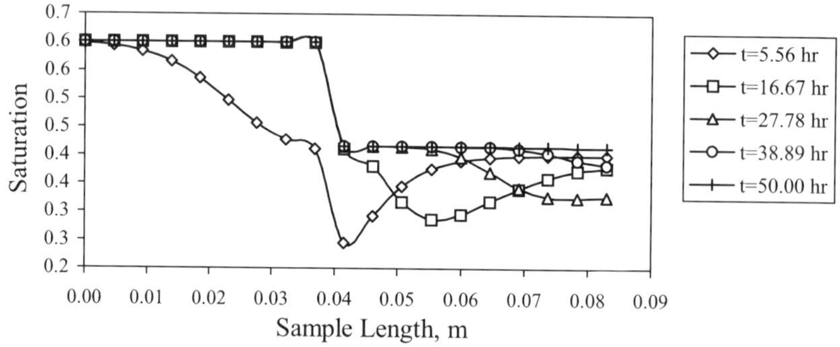

Fig. 9. Water phase saturation distribution of two-phase flow simulation when the first block has low permeability while the second and third blocks have high permeability.

The behaviour observed in Figure 9 shows higher water saturation in the lower permeability zone, which contains smaller pores. This confirms the fact that when two phases (wetting and non-wetting fluid) flow simultaneously through a low-permeability porous medium, the wetting fluid has a tendency to occupy the smaller pores (lower-permeability zone), while the non-wetting fluid occupies the larger pores (higher-permeability zone).

Because of the low permeability of the first porous block, the entire pressure drop was spent in overcoming the flow resistance of this block (see Fig. 10). The pressure drop in the second and third porous blocks is small compared to the first block pressure drop and is about 800 Pa at steady state. Capillary pressure behaves similarly to fluid saturation. As the wetting fluid saturation increases, the capillary pressure decreases (see Equation 9 and Fig. 11). Conversely, when the wetting fluid saturation decreases, the capillary pressure increases. Figure 11 clearly shows lower capillary pressure in the lower-permeability zone due to high water saturation.

Exit gas and water flow rate profiles are shown in Figure 12. From this figure we can see that the gas flow rate reached a steady-state value much faster (about 30 hours) than the water phase flow rate. This is probably due to the fact that most of the porous blocks were initially saturated with gas phase.

Two-phase flow through three porous blocks where high permeability is followed by low permeability porous media

This case is the opposite of the previous one. The first and second porous blocks were of high

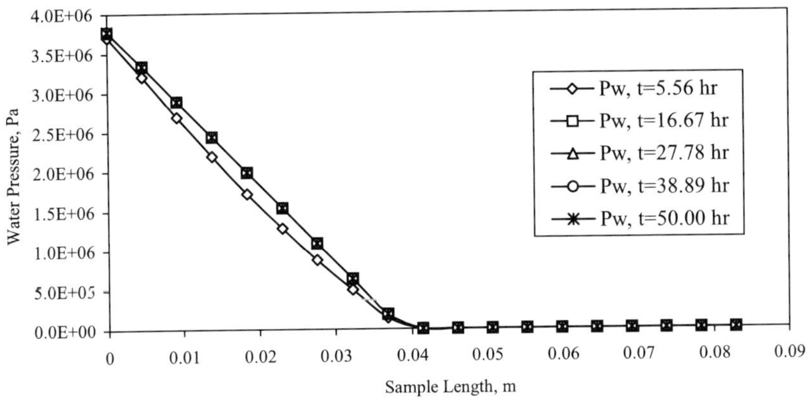

Fig. 10. Water pressure distribution of two-phase flow simulation when first block has low permeability and second and third blocks have high permeablility.

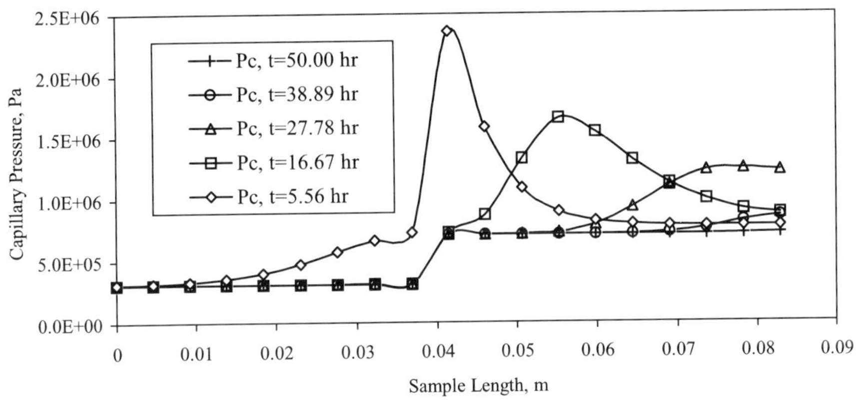

Fig. 11. Capillary pressure distribution of two-phase flow simulation when first block has low permeability and second and third blocks have high permeability.

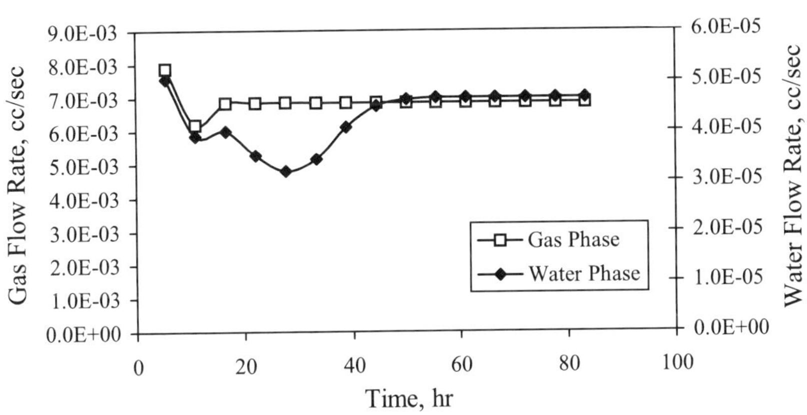

Fig. 12. Water and gas flow rates of two-phase flow simulation when first block has low permeability and second and third blocks have high permeable.

permeability while the third block was of low permeability porous media. Initial gas saturation of all blocks was 0.6, and inlet gas and water saturation were set at 0.4 and 0.6 for the gas and liquid phases, respectively. The first and second blocks were of high permeability and had a length of 4.35 and 0.05 cm, respectively, while the third block was of low permeability and had a length of 4.35 cm. Figure 13 shows the water phase saturation profile versus the sample length.

The inlet water saturation of 0.6 was obtained throughout the high-permeability portion of the media under steady-state condition. Water saturation reached at steady state in the low-permeability portion is higher than that reached in the high-permeability portion. This once again proves the tendency of the wetting fluid to occupy the smaller (tight) pores rather than the bigger ones.

Figure 14 shows the pressure distribution profile. As in the previous case, almost all the pressure drop was obtained in the low-permeability portion of the media. Figure 15 shows capillary pressure profiles. Capillary pressure was lower in the high water saturation zone.

Figure 16 shows water and gas flow rate profiles. The water flow rate was increased until it reached steady state after about 30 hours, while the gas flow rate decreased until it reached steady state after about 20 hours. The 10 hour difference in reaching steady state may be attributed to the fact that the gas phase flows in higher pores and has higher permeability compared to the water phase.

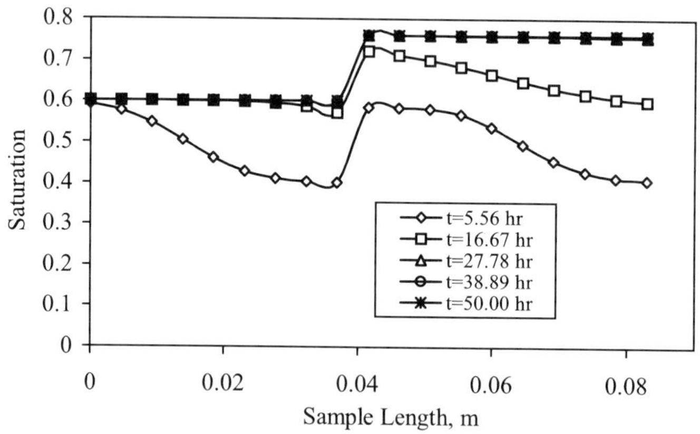

Fig. 13. Water phase saturation distribution of two-phase flow simulation when first and second blocks have high permeability and thid block has low permeability.

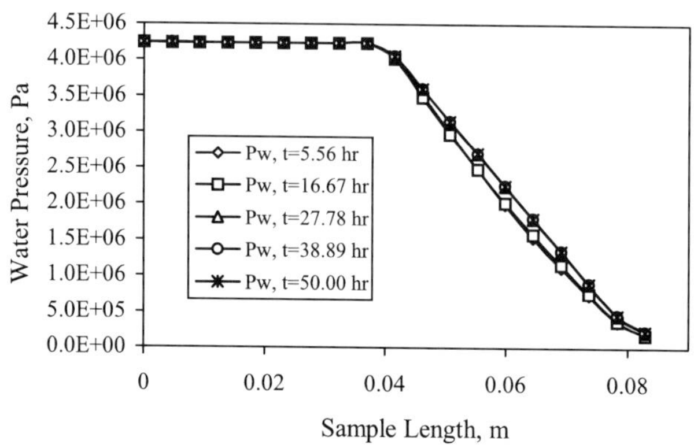

Fig. 14. Water pressure distribution of two-phase flow simulation when first and second blocks have high permeability and third block has low permeability.

Comparison between gas and water flow rate profiles of Figure 16 (a medium consisting of a high permeability zone followed by a low permeability zone) with the gas and water flow rate profiles of Figure 12 (a medium consisting of a low permeability zone followed by a high permeability zone) demonstrates that the steady-state gas and water flow rates are higher in Figure 12 than in Figure 16. The gas flow rate at steady state is 156% higher, while the water flow is 34% higher. This means that when the two phases of gas and water flow toward the bottom hole zone of the well in a gas reservoir, first through low-permeability media then through high-permeability media, the production of gas from this gas reservoir is more feasible than when the opposite occurs. Therefore, this might be a good criterion for deciding the drilling location of the gas well.

Conclusions

A transient three-dimensional, isothermal, two-phase flow model has been developed for the vertical flow of gas and water through low-permeability fractured porous media. The differential equation has been solved under conditions that had been determined experimentally. The solution was accomplished through the modification of the CFX-F3D code. The modified code considered the Darcy equation for the liquid phase and the non-Darcy equation for the gas phase as interfacial momentum transfer terms between the low-permeability solid matrix and each of the fluids. For flow through low permeability porous media, capillary force has significant effects on the flow patterns of both gas and water phases and was accounted for.

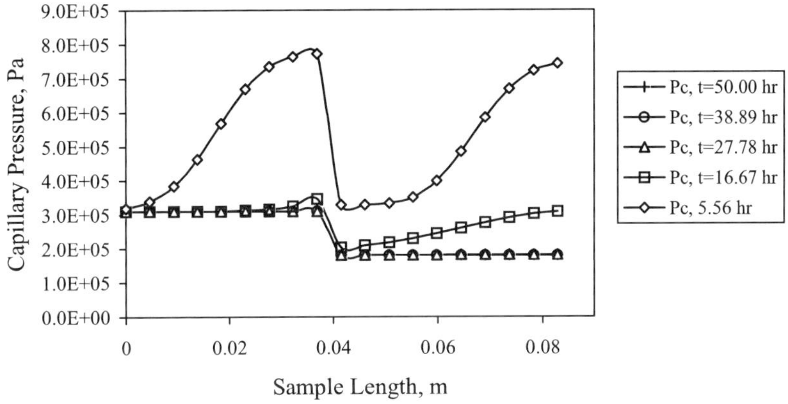

Fig. 15. Capillary pressure distribution of two-phase flow simulation when first and second blocks have high permeability and third block has low permeability.

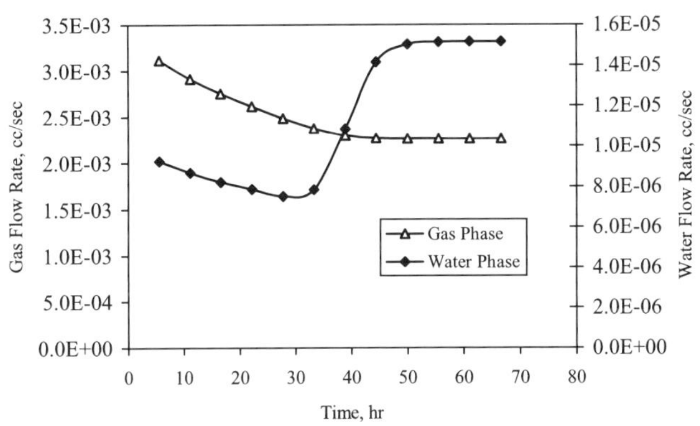

Fig. 16. Water and gas flow rates of two-phase flow simulation when first and second blocks have high permeability and third block has low permeability.

Performing the simulation of two-phase flow through a porous medium consisting of several porous blocks with different permeabilities and porosities showed that the production is higher when the fluid flows first through a low-permeability medium followed by a high-permeability medium. This means higher fracturing around the bottom hole zone of the production well significantly enhances gas production from tight sand formations.

Nomenclature

B interfacial momentum transfer between rock matrix and each phase
k permeability
n normal direction to the inlet or outlet cells
P pressure
R resistance constant
S saturation
t time
U velocity vector

Greek Letters

β coefficient accounting for local flow non-uniformities
τ viscous stress tensor
ε volume fraction
ϕ volume porosity
η pore size index
μ fluid viscosity
ρ fluid density

Subscripts

c capillary
e effective
g gas
m matrix
r rock
w water
α water (w) or gas (g) phase

References

AL-KHLAIFAT, A. L. 1998. *Two Phase Flow through Low Permeability Fractured Tight Sand Porous Media*. PhD thesis, Illinois Institute of Technology, Chicago.

AL-KHLAIFAT, A. L. & ARASTOOPOUR, H. 1997. Simulation of two-phase flow through low permeability porous media. *AEA Technology International Users Conference Proceedings*. AEA, Chicago, 31–42.

AL-KHLAIFAT, A. L. & ARASTOOPOUR, H. 1999. Experimental study of drainage, imbibition, and two phase flow through low permeability tight sand fractured porous media. *Journal of the Chinese Institute of Chemical Engineers*, **30** (4), 297–302.

AL-KHLAIFAT, A. L., PARTO, P. & ARASTOOPOUR, H. 2001. Measurement and analysis of single phase water flow through fractured and non-fractured low permeability porous media. *Journal of the Chinese Institute of Chemical Engineers*, **32** (1), 57–61.

BROOK, R. H. & COREY, A. T. 1966. Irrigation and drainage division. *Proceedings of the American Society of Civil Engineers*, 61–85.

CFX-F3D, 1995. *User Manual, Computational Fluid Dynamics Series, Version 4.1*. AEA, Harwell.

CHOWDIAH, P. & ARASTOOPOUR, H. 1983. Measurement and analysis of transient flow of gas through tight sand core samples. *Proceedings of 18TH Intersociety Energy Conversion Engineering Conference*. 480–485.

CODREANU, D. B., BIA, R. B. & SOHATHIER, J. C. 1985. The check model, an improvement in modeling naturally fractured reservoirs with a tridimensional, triphasic, black oil numerical model. *Society of Petroleum Engineers Journal*, **25**, 743–756.

FREEMAN, D. L. & BUSH, D. C. 1983. Low permeability laboratory measurements by nonsteady state and conventional methods. *Society of Petroleum Engineers Journal*, **23**, 928–936.

HOLDITCH, S. A. *et al.* 1989. *Staged Field Experiment No. 2. Application of the Advanced Geological, Petrophysical and Engineering Technologies to Evaluate and Improve Gas Recovery From Low Permeability Sand Stone Reservoirs*. Gas Research Institute, Chicago, Report No. GRI-89/0140.

JONES, F. O. & OWENS, W. W. 1980. A laboratory study of low permeability gas sands. *Journal of Petroleum Technology*, **32**, 1631–1640.

KAZEMI, H. 1969. Pressure transient analysis of naturally fractured reservoirs with uniform fracture distribution. *Society of Petroleum Engineers Journal*, **9**, 451–462.

KAZEMI, H. 1976. Numerical simulation of water oil flow in naturally fractured reservoir. *Society of Petroleum Engineers Journal*, **16**, 317–326.

LEE, B. Y. Q. & TAN, T.B.S. 1987. Application of multiple porosity/permeability simulators in fractured reservoir simulation. *Society of Petroleum Engineers Journal*, **27**, 181–192.

NEWBERG, M. A. & ARASTOOPOUR, H. 1986. Analysis of the flow of gas through low permeability porous media. *SPE Reservoir Engineering*, **3**, 647–653.

RANDOLPH, P. L. 1983. Porosity and Permeability of Mesaverde Sandstone Core from the U.S. DOE Mulliwell Experiment, Garfield Country, Colorado. *Proceedings of the SPE/DOE Joint Symposium on Low Permeability Gas Reservoirs*, 449–460.

SOEDER, D. J. & ARASTOOPOUR, H. 1986. Laboratory drying procedure and permeability of tight sandstone core. *Society of Petroleum Engineers Formation Evaluation*. **1**, 16–22.

THOMAS, L. K., THOMAS, N. D. & RAY, G. P. 1983. Fractured reservoir simulation. *Society of Petroleum Engineers Journal*, **23**, 42–54.

WALLS, J. D., NUR, A. M. & BOURBIE, T. 1982. Effect of pressure and partial water saturation on gas permeability in tight sands. *Journal of Petroleum Technology*, **34**, 930–936.

Index

Page numbers in *italics* refer to figures.

TREMOUGH
Library
CAMPUS